Hydrological Research in China: *Process Studies, Modelling Approaches and Applications*

Other IAHS Publications

Hydrological Sciences for Managing Water Resources in the Asian Developing World

edited by Xiaohong Chen, Yongqin David Chen, Jun Xia & Hailun Zhang

Many regions in Asia are experiencing unprecedented rapid development resulting in great pressures on environmental quality and sustainable management of natural resources. China has traditionally emphasised water shortages in the Yellow River basin and flooding by the Yangtze River, but water problems in South China, and especially the Pearl River (Zhujiang) basin have attracted attention in recent years. Many of the contributions are from China, especially the south, and provide an insight to the innovative work that is on-going there.

- Novel Techniques for Hydrological Analysis
- Hydrological Modelling
- Hydrological Impacts of Global Change and Human Activities
- Optimal Allocation of Water Resources
- Water Resources and Watershed Management
- Water Environment and Aquatic Ecosystems

Publ. 319 (*2008*) ISBN 978-1-901502-44-2, 422 + x pp. £72.00

Reducing the Vulnerability of Societies to Water Related Risks at the Basin Scale

edited by Andreas Schumann & Markus Pahlow

The International Commission on Water Resources Systems (ICWRS) of IAHS has, for many years, embraced Integrated Water Resources Management (IWRM) as the main topic for its research. The need for integrated approaches to deal with complex water resources management issues in the developed and developing world is well recognised, yet, in many places, IWRM is still only a concept and not established practice. This volume, comprised of peer-reviewed papers selected from the Third International Symposium on IWRM (September 2006, Ruhr-University Bochum, Germany), accounts for the heterogeneity of world water problems by addressing the following important questions: What has to be integrated? How can it be accomplished? What are the options to balance the different views? This symposium strived to not only identify problems, but to provide practical solutions. The subject of how to cope with water-related vulnerability of societies formed the overarching theme. Together, the papers provide an excellent overview of current IWRM research worldwide.

Publ. 317 (*2007*) ISBN 978-1-901502-29-9, 450 + xiv pp. £78.00

A New Focus on Groundwater–Seawater Interactions

edited by Ward Sanford, Christian Langevin, Maurizio Polemio & Pavel Povinec

Water and chemical fluxes across the sea floor provide an important linkage between terrestrial and marine environments. Oceanographers recognize that these fluxes may act as a source of nutrients or harmful contaminants to marine systems. They may also act as a beneficial source of freshwater for coastal estuaries that require relatively low salinities. Hydrologists and hydrogeologists recognize that fluxes across the sea floor comprise an important part of the water balance for coastal aquifers. Most fresh groundwater discharge to the ocean is derived from terrestrial aquifer recharge. However, excessive groundwater withdrawals from coastal aquifers can cause saltwater intrusion by intercepting the seaward flux. The papers in this volume present research by those working from the marine and terrestrial sides of the issues, and cover a variety of investigative approaches applied at locations worldwide. Together, they form an important contribution to the literature.

Publ. 312 (*2007*) ISBN 978-1-901502-04-6; 344 + x pp; £64.00

Methodology in Hydrology

edited by Liliang Ren, Qiongfang Li, Danrong Zhang & Jun Xia

This volume provides a unique overview, in English, of current hydrological science in China, detailing issues, approaches, innovations and achievements. The 97 contributions, primarily from China, were selected from a major conference held in Nanjing, and are grouped as follows:

- Hydrological Modelling and Flood Forecasting
- Stochastic Hydrology
- Water Quality Modelling and Analysis
- Water Resources Management and Water Economy
- Interdisciplinary Hydrology

Publ. 311 *(2007)* ISBN 978-1-901502-93-0 654 + xii pp, £110.00

Predictions in Ungauged Basins: Promises and Progress

edited by Murugesu Sivapalan, Thorsten Wagener, Stefan Uhlenbrook, Erwin Zehe, Venkat Lakshmi, Xu Liang, Yasuto Tachikawa & Praveen Kumar

This book presents peer-reviewed papers selected from a symposium held at Foz do Iguaçu, Brazil, in April 2005, which coincided with the completion of Phase 1 (the preparation phase) of PUB and provided the opportunity for the global PUB community to get together to discuss the progress that had been achieved, to identify areas of weakness or lack of activity, re-iterate the plans and promises of PUB, and to make necessary modifications or refinements to the PUB implementation activities with a view to achieving faster and more demonstrable progress in the future. The papers are organized into five sections that focus on the distinct PUB themes:

- Model Improvements Through Detailed Process Studies;
- Model Evaluation and Comparison: Uncertainty Analysis and Diagnostics;
- New Data Collection Approaches and Model Development;
- New Modelling Approaches and Methods for Testing Models Against Observations;
- Progress in PUB Implementation: PUB Working Groups.

Publ. 303 (*2006*) ISBN 1-901-502-48-1; 520 + viii pp; £84.00

Sustainability of Groundwater Resources and its Indicators

edited by Bruce W. Webb, Ricardo Hirata, Eduardo Kruse & Jaroslav Vrba

Groundwater indicators are based on monitoring and assessment programmes and are helpful for several reasons, e.g. they may be used to identify the state of the groundwater system, both in terms of its quantity and quality, and its evaluation over time and space; they also facilitate communication between policy makers and the general public. Indicators should be selected and developed which provide information on the threat to groundwater systems from the point of view of societal and ecological needs.

Publ. 302 (*2006*) ISBN 978-1-901502 43-5; 220 + viii pp; £45.00

Bringing Groundwater Quality Research to the Watershed Scale

edited by Neil R. Thomson

Sustainable economic growth requires plentiful supplies of high-quality water. Pressure on resources globally has forced the international groundwater science, engineering and regulatory community to recognize current limitations of research and management. Integrated and technically feasible approaches tackling local issues and watershed-scale concerns concurrently are required. This volume, an outcome of GQ2004 (Waterloo, Canada) deals with Global and national perspectives, Contaminant input processes; Site characterization; Management and decision making; Natural attenuation processes and applications; *In situ* remediation; and Flow and transport modelling at various scales.

Publ. 297 (*2005*) ISBN 978-1-901502-18-3; 576 + xiv pp; £85.00

More details at **www.iahs.info**

Please send orders and enquiries about books to:

Mrs Jill Gash, IAHS Press,
Centre for Ecology and Hydrology, Wallingford,
Oxfordshire OX10 8BB, UK
tel: +44 1491 692442
fax: +44 1491 692448/692424
jilly@iahs.demon.co.uk

IAHS members benefit from discounts on book prices: 25% for all members, and 80% discount for members in financially disadvantaged countries; IAHS membership is free of charge.

All prices include delivery by surface mail

Hydrological Research in China: *Process Studies, Modelling Approaches and Applications*

Edited by:

DAWEN YANG
State Key Laboratory of Hydro-Science and Engineering
Department of Hydraulic Engineering
Tsinghua University
Beijing, China

FUQIANG TIAN
State Key Laboratory of Hydro-Science and Engineering
Department of Hydraulic Engineering
Tsinghua University
Beijing, China

LIHUA TANG
State Key Laboratory of Hydro-Science and Engineering
Department of Hydraulic Engineering
Tsinghua University
Beijing, China

ZHIYU LIU
Bureau of Hydrology, Ministry of Water Resources
Beijing, China

IAHS Publication 322
in the IAHS Series of Proceedings and Reports

Published by the International Association of Hydrological Sciences 2008

IAHS Publication 322

ISBN 978-1-901502-64-0

British Library Cataloguing-in-Publication Data.
A catalogue record for this book is available from the British Library.

The papers included in this volume have been peer-reviewed and some were extensively revised by the Editors, in collaboration with the authors, prior to publication.

IAHS is indebted to the employers of the Editors for the invaluable support and services provided that enabled them to carry out their task effectively and efficiently.

Publications in the series of Proceedings and Reports are available from:
IAHS Press, Centre for Ecology and Hydrology, Wallingford, Oxfordshire OX10 8BB, UK
tel.: +44 1491 692442; fax: +44 1491 692448; e-mail: jilly@iahs.demon.co.uk

Printed by Alden Group, Oxford, UK.

Preface

There is a broad range of climate in China, varying from arid to semiarid and semi-humid to humid, which induces a variety of hydrological phenomena. The Chinese population has tripled during the last 50 years, now reaching 1.3 billion, which is currently the largest in the world. Under the pressure of the increasing population, in northern China the scarcity of water resources is becoming a bottleneck for social and economic developments, and has led to the widespread degradation of natural ecosystems and the environment, while in south China flooding is a great threat whose potential risks are elevating steadily with the growth of the population as well as the economy. Accurate predictions of droughts and floods are extremely important to the social and economic safety of all these regions. China is on the way to changing from traditional water resources development to water resources management for sustainable development. The advanced hydrology can play an important role in the water resources management with regard to hydrological predictions. There are increasing demands for the application of PUB (Prediction in Ungauged Basins), which was launched by IAHS from 2003 to 2012, for better management of water resources in China.

The China PUB organization was formed in 2004. As a result of the efforts of the China PUB organization, the China PUB Working Groups were established, which focus on the new methodology of hydrological simulation and prediction under natural and human-induced global changes. To promote exchanges between China and IAHS, and with help from the international IAHS-PUB Scientific Steering Group (SSG), the China PUB National Steering Committee organized the International Symposium on Flood Forecasting and Water Resources Assessment for IAHS-PUB held in September 2006. Taking this opportunity, the symposium organizers would like to express sincere thanks to Prof. Murugesu Sivapalan (University of Illinois at Urbana-Champaign), Prof. Jeffrey McDonnell (Oregon State University) and Prof. Kuniyoshi Takeuchi (Yamanashi University) for their enthusiastic and valuable support.

There were 10 plenary lectures by eminent researchers, 70 research presentations and five poster presentations. All the papers were published in 2006 as the proceedings of the symposium by the China WaterPower Press. From this published proceedings, we selected 43 papers and proceeded with a peer-review process. Finally 37 papers were published as this IAHS publication.

The editors gratefully acknowledge the assistance of the reviewers who made valuable contributions to this volume. We also thank Penny Perrins and Cate Gardner of IAHS Press for their professional approach and help with the processing of the manuscripts. Mrs Huafang Lv (staff in Tsinghua University) and Mr Hongyi Li (PhD student in University of Illinois at Urbana-Champaign) served as secretaries to the editors' board.

Dawen Yang, Fuqiang Tian & Lihua Tang
State Key Laboratory of Hydro-Science and Engineering
Department of Hydraulic Engineering,
Tsinghua University
Beijing, China

Zhiyu Liu
Bureau of Hydrology, Ministry of Water Resources
Beijing, China

Contents

3 Applications

1 Process studies

Evaporation paradox in the Yellow River Basin, China

ZHENTAO CONG, DAWEN YANG, FUBAO SUN & GUANGHENG NI

Department of Hydraulic Engineering, Tsinghua University, Beijing 100084, China

congzht@tsinghua.edu.cn

Abstract It is well known that the surface of the Earth has warmed over the past 50 years. It is generally expected that the air will become drier and that evaporation will increase. However, many observations show that pan evaporation has been steadily decreasing all over the world. The contrast is called the evaporation paradox. Using climate data from 140 weather stations and hydrological data from 71 stations, it was found that potential evaporation, both pan evaporation and reference potential evaporation, show a significant decreasing trend, which indicates that the evaporation paradox does exist in the Yellow River Basin. At the same time, actual evaporation estimated by water balance decreased with the decrease in precipitation in most sub-basins. With the increase of precipitation, potential evaporation and actual evaporation have a complementary relationship; while based on yearly results, potential evaporation and actual evaporation have a proportional relationship.

Key words evaporation paradox; Yellow River; potential evaporation; actual evaporation; evaporation

INTRODUCTION

Evaporation paradox

It is now well established that global surface temperatures have increased between 0.4 and 0.8°C since the late 19th century and the rate of temperature increase since 1976 has been over 0.15°C per decade (IPCC, 2001). An expected consequence of this warming is that the air near the surface should be drier, which should result in an increase in the rate of evaporation from terrestrial open water bodies. However, many observations show that the rate of evaporation from open pans of water has been steadily decreasing all over the world over the past 50 years, and the same is apparent for reference evapotranspiration. This trend was found to be generally evident in the United States and the former Soviet Union (Peterson *et al.*, 1995), in India (Chattopadhyay *et al.*, 1997), in China (Thomas, 2000; Liu *et al.*, 2004), in Italy (Moonen *et al.*, 2002), and in Australia (Roderick *et al.*, 2004), but not universal in Israel (Cohen *et al.*, 2002) and in east Asia (Xu, 2001).

The contradiction between expectation and observation is called the evaporation paradox (Roderick *et al.*, 2002). The decrease in pan evaporation may be caused by the reduction in solar radiation resulting from more clouds and/or aerosols (Peterson *et al.*, 1995; Stanhill *et al.*, 2001; Roderick *et al.*, 2002), by the reduction in vapour pressure deficit due to the addition of air humidity (Chattopadhyay *et al.*, 1997), or by the reduction in wind speed (Cohen *et al.*, 2002). Evaporation and transpiration play an important role in the global water and energy cycle, and their change will make a significant effect on global climate, hydrological cycle, water resources, crop growth, and the ecological environment (Chahine *et al.*, 1992; Stanhill, *et al.* 2001; Sun *et al.*, 2001; Roderick *et al.*, 2002).

It is disputed whether the decrease of pan/reference evaporation predicates the decrease of actual evaporation. Peterson *et al.* (1995) concluded that a decrease in pan evaporation indicates a decrease in actual evaporation, which was proven with the increasing runoff of rivers in the former Soviet Union and the United States in the past 20 years. On the other hand, Brutsaert *et al.* (1998) concluded that a decrease in pan evaporation indicates an increase in actual evaporation from the surrounding non-humid environments, which indicates that actual evaporation and pan evaporation exhibit a complementary relationship rather than proportional behaviour.

This study aims to know the behaviour of the evaporation paradox in the Yellow River Basin through understanding the spatial and temporal variability of potential and actual evaporation over the basin.

Yellow River Basin

The Yellow River, also called Huanghe in Chinese, is the second longest river in China, with a length of 5646 km, and the basin is the second largest, with a drainage area of 753 000 km^2. In the 1990s the drying up of the main river along the lower reach in the Yellow River drew much attention from all over the world.

Annual precipitation showed a decreasing trend of –45.3 mm/50 years; annual mean temperature had an increasing trend of 1.28°C/50 years; and river discharge had significant decreasing trends ranging from –28 mm/50 years to –61.5 mm/50 years (Yang *et al.*, 2004).

METHODOLOGY

Potential evaporation (ET_p)

Potential evaporation can be expressed by pan evaporation (ET_{pan}) or potential evaporation of extensive free water surface (ET_0). ET_0 is estimated using the Penman-Monteith method (Shuttleworth, 1993; Allen *et al.*, 1998) as follows:

$$ET_0 = \frac{\Delta}{\Delta+\gamma}\left(R_n + A_h\right) + \frac{\gamma}{\Delta+\gamma}\frac{6.43\left(1+0.536U_2\right)D}{\lambda} \tag{1}$$

where, R_n = net radiation exchange for the free water surface, mm day^{-1}; A_h = energy advected to the water body, if significant; U_2 = wind speed measured at 2 m, m s^{-1}; D = vapour pressure deficit, kpa; Δ = slope of saturation vapour pressure curve, kPa°C^{-1}; γ= psychrometric constant, kPa°C^{-1}; λ = latent heat of vaporization, MJ kg^{-1}.

Actual Evaporation (ET_a)

Precipitation (P) and potential evaporation can be measured or calculated at each weather station, and runoff (R) can be measured at each hydrological station in the mouth of the basin. With interpolation, the precipitation and the potential evaporation of sub-basins can be evaluated, and then actual evaporation (ET_a) is estimated using water balance:

$$ET_a = P - R \tag{2}$$

Linear regression and Kendall correlation coefficient

The linear regression model is:

$$y_i = ax_i + b \qquad i = 1, 2, \ldots, n \tag{3}$$

where, y_i = ith observation of the response variable; x_i = ith observation of the explanatory variable; a = slope; b = intercept.

In addition, slope a can be gained by the following equation:

$$a = \sum_{i=1}^{n}\left(x_i - \overline{X}\right)\left(y_i - \overline{Y}\right) \Big/ \sum_{i=1}^{n}\left(x_i - \overline{X}\right)^2 \tag{4}$$

The Kendall correlation coefficient (Kendall's τ) between two variables can be computed with the method presented in the *Handbook of Hydrology* (Salas, 1993).

DATA DESCRIPTION

Daily climate data from 1951 to 1998 from 140 meteorological stations, covering most regions of the Yellow River Basin have been collected; of which 64 stations were chosen to be used as they had records of more than 30 years. The data include precipitation, maximum air temperature,

Fig. 1 Weather stations and hydrological stations in the Yellow River Basin.

Table 1 Trend of potential evaporation in the Yellow River Basin in the past 50 years

Period	Pan evaporation (ET_{pan})		Potential evaporation (ET_0)	
	Increase ($a > 0$)	Decrease ($a < 0$)	Increase ($a > 0$)	Decrease ($a < 0$)
1951–1998	9	55	24	40
1951–1984	16	48	31	33
1985–1998	35	29	44	20

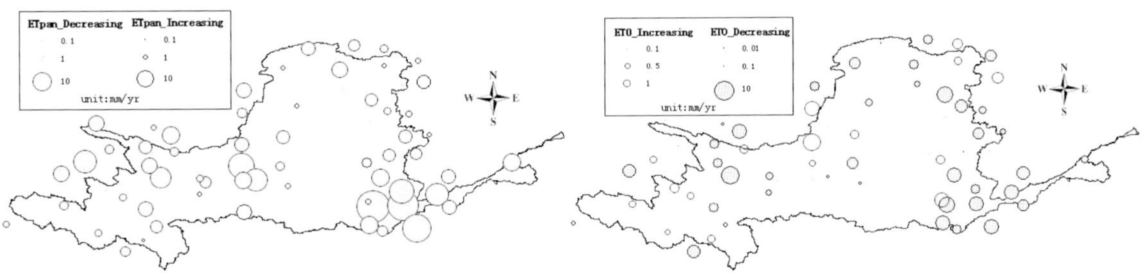

Fig. 2 Trend of potential evaporation in the Yellow River Basin from 1951 to 1998.

minimum air temperature, wind speed, pan evaporation, vapour, sun time, and hemi; from which potential evapotranspiration can be calculated with the Penman-Monteith method. Annual runoff data from 1951 to 1998 from 71 hydrological stations, controlling 71 sub-basins of the Yellow River Basin, have been collected to simulate actual evapotranspiration. The weather stations and the hydrological stations are shown in Fig. 1.

RESULTS AND DISCUSSION

Spatial and temporal variability in potential evaporation

The results of linear regression regarding potential evaporation are listed in Table 1, which indicated that in contradiction with the increase in temperature, both ET_{pan} and ET_0 had decreased at most stations in the Yellow River Basin during the past 50 years, which is also shown in Fig. 2. The trend of pan evaporation is more obvious and decrease occurred in 55 of the total 64 stations with an average Kendall's τ of 0.28. The trend is that ET_{pan} in these 55 stations decreased 28.5 mm/10 years in arithmetic average. Thus, we can conclude that the evaporation paradox occurred in the Yellow River Basin.

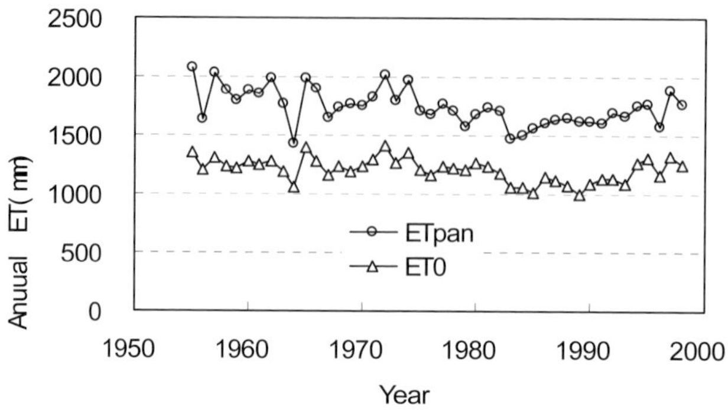

Fig. 3 Trend of potential evaporation at the Taiyuan station in the past 50 years.

Table 2 Trend of actual evaporation in the Yellow River Basin in the past 50 years.

Period	ET_0		Precipitation		Runoff		Actual evaporation	
	$a > 0$	$a < 0$	$a > 0$	$a < 0$	$a > 0$	$a < 0$	$a > 0$	$a < 0$
1951–1998	26	45	10	61	15	56	16	55
1951–1984	36	34	24	41	13	52	33	32
1985–1998	65	6	15	51	38	28	16	50

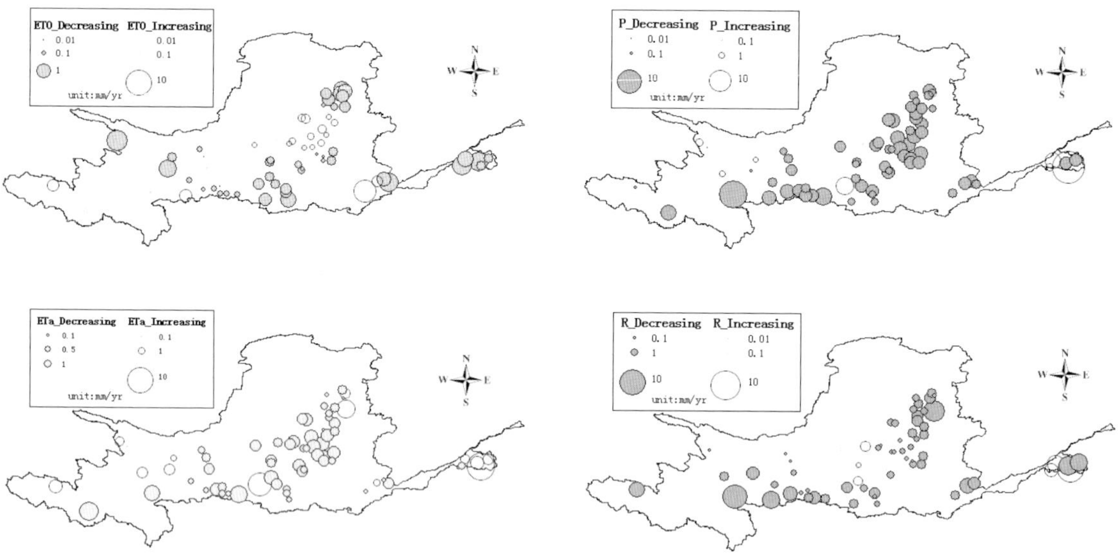

Fig. 4 Trend of actual evaporation in the Yellow River Basin from 1951 to 1998.

In addition we can note that the trend changed after 1985, which was also found by Yang (2004). The change is also due to the increase in solar radiation after 1985, but this is not clearly understood. The typical trend in the past 50 years is shown in Fig. 3, using data from Taiyuan station.

The spatial distribution of the trend is shown in Fig. 2. The decreasing trend is more obvious in the upstream basin and the downstream basin than in the middle stream basin.

Spatial and temporal variability in actual evaporation

The results of linear regression regarding reference evapotranspiration, precipitation, runoff, and actual evaporation of catchments are listed in Table 2. Actual evaporation, estimated with water balance (equation (2)), had reduced continuously in the past 50 years, even with the increase in runoff, due to the decrease in precipitation. The decreasing trend of actual evaporation, about 10.5 mm/10 years in area weighted average, is smaller than that of potential evaporation, about 15.4 mm/10 years. The decreasing trend of precipitation, about 18.4 mm/10 years, caused serious aridity in the Yellow River Basin, but it also reduced the water supply, so actual evaporation decreased in the past 50 years. These trends can also be seen in Fig. 4.

Relation between actual evaporation and potential evaporation

From the above results, in general, both actual evaporation and potential evaporation diminished in the past 50 years. From Table 3, we can find: in most basins, both actual evaporation and potential evaporation reduced from 1951 to 1998; after 1985, actual evaporation still declined even if potential evaporation started to increase.

Table 3 Relation between actual evaporation and potential evaporation.

ET_0	$a > 0$	$a < 0$	$a > 0$	$a < 0$
Actual evaporation	$a > 0$	$a < 0$	$a < 0$	$a > 0$
1951–1998	5	34	21	11
1951–1984	13	9	23	20
1985–1998	12	2	48	4

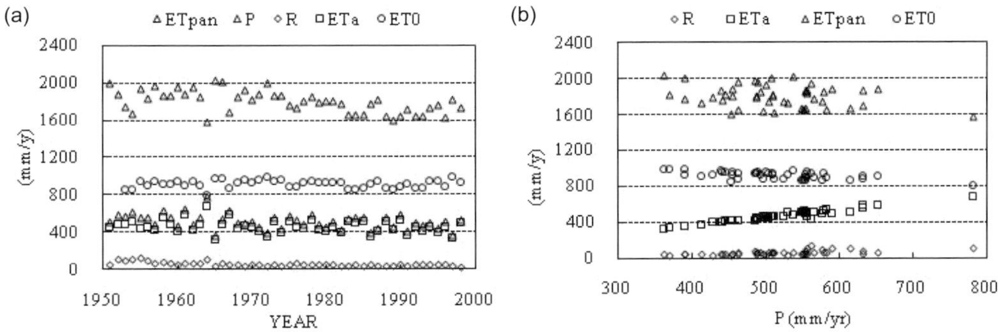

Fig. 5 Relation of actual evaporation and potential evaporation in the Yellow River Basin from 1951 to 1998.

Complementary relationship

The trend of the average pan evaporation (ET_{pan}), potential evaporation (ET_0), actual evaporation (ET_a), precipitation (P), and runoff (R) over the whole basin can be seen in Fig. 5(a). It can be concluded that actual evaporation and potential evaporation have a proportional relationship since they both declined in the past 50 years.

According to the same data, if we make precipitation as the abscissa, it can be found that actual evaporation decreases while pan evaporation and potential evaporation increase, as shown in Fig. 5(b). Therefore, we get a complementary relationship between actual evaporation and potential evaporation.

In general, potential evaporation will increase with decreasing precipitation, but it does not occur in the Yellow River Basin. In this basin, potential evaporation decreased because radiation became smaller; this is also called *global dimming*. Thus, potential evaporation decreased in spite of the reduction of precipitation, as shown in Fig. 5(a). Meanwhile, the influence of radiation reduction is ignored in Fig. 5(b) because the year is confused and the complementary relationship comes back.

CONCLUSION

The major findings can be summarized as follows:

(a) During the past 50 years (1951–1998), both potential evaporation by Penman-Monteith and pan evaporation show a significant downward trend in the Yellow River Basin. At the same time, air temperature shows a significant increasing trend, so the evaporation paradox does occur in the basin.
(b) After 1985, both potential evaporation and pan evaporation show an increasing trend.
(c) During the past 50 years (1951–1998), both precipitation and runoff in most basins show a significant downward trend in the Yellow River Basin.
(d) During the past 50 years (1951–1998), actual evaporation estimated using water balance shows a continuous downward trend in the Yellow River Basin.
(e) In the Yellow River Basin, with the increase of precipitation, potential evaporation and actual evaporation have a complementary relationship; while based on yearly results, potential evaporation and actual evaporation have a proportional relationship.

Acknowledgements This research was supported by the National Natural Science Foundation of China (50509011; 50721140161). The authors would like to express their appreciation for the grants received to aid this research.

REFERENCES

Allen, R. G., Pereira, L. S., Raes, D. & Smith, M. (1998) Crop evapotranspiration. Guidelines for computing crop water requirements. *FAO Irrigation and Drainage Paper 56*, FAO, Rome, Italy.

Brutsaert, W. & Parlange, M. B. (1998) Hydrologic cycle explains the evaporation paradox. *Nature* **396**, 29–30.

Chahine, T. M. (1992) The hydrological cycle and its influence on climate. *Nature* **239**, 373–381.

Chattopadhyay, N. & Hulme, M. (1997) Evaporation and potential evapotranspiration in India under conditions of recent and future climate change. *Agric. For. Met.* **87**, 55–73.

Cohen, S., Ianetz, A. & Stanhill, G. (2002) Evaporative climate changes at Bet Dagan, Israel, 1964–1998. *Agric. For. Met.* **111**, 83–91.

IPCC Third Assessment Report – Climate Change (2001) The Scientific Basic (http://www.ipcc.ch/).

Liu, B., Xu, M. Henderson, M. & Gong, W. (2004) A spatial analysis of pan evaporation trends in China, 1955–2000. *J. Geophys. Res.-Atmospheres* **109**(D15).

Moonen A. C., Ercoli, L., Mariotti, M. & Masoni, A. (2002) Climate change in Italy indicated by agrometeorological indices over 122 years. *Agric. For. Met.* **111**, 13–27.

Peterson, T. C., Golubev, V. S. & Groisman, P. Y. (1995) Evaporation losing its strength. *Nature* **377**, 687–688.

Roderick, M. L. & Farquhar, G. D. (2002) The cause of decreased pan evaporation over the past 50 years. *Science* **298**, 1410–1411.

Roderick, M. L. & Farquhar G. D. (2004) Changes in Australian pan evaporation from 1970 to 2002. *Int. J. Climatol.* **24**, 1077–1090.

Salas, J. D. (1993) Analysis and modeling of hydrologic time series. In: *Handbook of Hydrology* (ed. by D. Maidment), Chapter 19. McGraw-Hill, Inc. New York, USA.

Shuttleworth, W. J. (1993) Evaporation. In: *Handbook of Hydrology* (ed. by R. D. Maidment), Chapter 4. McGraw-Hill, New York, USA.

Stanhill, Gerald & Cohen, Shabtai (2001) Global dimming: a review of the evidence for a widespread and significant reduction in global radiation with discussion of its probable causes and possible agricultural consequences. *Agric. For. Met.* **107**, 255–278.

Sun, Lan & Wu, Guo-Xiong (2001) Influence of land evapotranspiration on climate variations. *Science in China (Series D)* **44**, 838–846.

Thomas, A. (2000) Spatial and temporal characteristics of potential evapotranspiration trends over China. *Int. J. Climatol.* **20**, 381–396.

Xu, J. (2001) An analysis of climatic changes in Eastern Asia using the potential evaporation. *J. Japan Soc. Hydrol. Water Resour.* **14**, 141–170.

Yang, Dawen, Li, Chong, Musiake, Katumi & Kusuda, Tetsuya (2004) Analysis of water resources variability in the Yellow River of China during the last half century using historical data. *Water Resour. Res.* **40**, W065021-W0650212.

Evaluating the advection-aridity model of evaporation using data from field-sized surfaces of HEIFE

SONGJUN HAN, HEPING HU & FUQIANG TIAN

State Key Laboratory of Hydro-science and Engineering, Department of Hydraulic Engineering, Tsinghua University, Beijing 100084, China

huhp@tsinghua.edu.cn

Abstract The advection-aridity model for estimating daily evaporation is evaluated using the data of the HEIFE experiment. In the dimensionless form of the model, the evaporation ratio (the ratio of the actual evaporation (E) to Penman potential evaporation (E_0)) is expressed as a linear function of the proportion of the radiation term (E_{rad}) in E_0. Because the value of the evaporation ratio is between 0 and 1, the advection-aridity model is only applicable under a certain range of E_{rad}/E_0, and the applicability of the model is influenced by the water availability of the surfaces implied by E_{rad}/E_0 from atmospheric conditions. The calculated evaporation ratio is negative for small values of E_{rad}/E_0, and is larger than 1 for large values of E_{rad}/E_0. Significant systemic bias would be presented if the advection-aridity model underestimated the actual evaporation under dry conditions and overestimated under wet conditions. A comparison between the reference values of the mean daily evaporation of the Gobi desert and oasis surfaces obtained by the eddy correlation method and evaporation estimates from the advection-aridity model validated the analysis.

Key words complementary relationship; advection-aridity model; HEIFE

INTRODUCTION

Evaporation is the major component of the hydrological cycle. The accurate estimation of evaporation is necessary for a better understanding of hydrological processes, especially in arid regions that occupy one-third of the entire global land area. The term "evaporation" in this study is used to include both evaporation and transpiration, and thus it is equivalent to the term "evapotranspiration". The actual evaporation is often estimated from the two primary controls: potential evaporation and water availability denoted by soil moisture or precipitation at different time scales (Penman, 1948; Budyko, 1974). On a daily timescale, variables such as soil moisture content or stomatal resistance needed to estimate the actual evaporation, are difficult to obtain. Therefore it is necessary to establish a reliable method to estimate actual evaporation from a few meteorological parameters.

The complementary relationship proposed by Bouchet (1963), in which only the standard meteorological data is needed, is frequently applied in calculating regional evaporation. Different models derived from the concept of complementary relationships exist in the literature, and the advection-aridity model proposed by Brutsaert & Stricker (1979) has been applied to different land surfaces at different timescales. But significant bias in the advection-aridity model, underestimated evaporation under dry conditions and overestimated evaporation for large values under wet conditions was presented, and the bias are more obvious on a daily timescale (Qualls & Gultekin, 1997).

In order to investigate the practicality and reliability of the advection-aridity model, the dimensionless form is used in this study, as the water availability of the surfaces can be detected easily in the dimensionless form (Yang *et al.*, 2006). However, data from surfaces under a wide range of water availability on a daily time scale should be gathered in the analysis to present the significant bias of the advection-aridity model. The requirement can be satisfied with the data from the Gobi desert and oasis surfaces of the HEIFE (Heihe River Basin Field Experiment on Land Surface Processes, 1990–1993, China) experiment used in this study. The data from the Gobi and desert surfaces is mostly under dry conditions, and the data from the oasis surface has a wide range of water availability.

METHODS AND DATA

Method used in the study

When water availability is not limited, evaporation proceeds at the potential evaporation rate, and it is defined as wet environment evaporation. As the surface dries without changing the available energy, the actual and potential evaporation rate depart from the wet environment evaporation with equal but opposite changes in fluxes. Bouchet's complementary relationship is expressed as:

$$E_p + E = 2E_w \tag{1}$$

where E is the actual evaporation, E_p is the potential evaporation, and E_w is the wet environment evaporation. However, many experimental as well as theoretical results suggest that no real complementary relationship between E and E_p exists (McNaughton & Spriggs, 1989; Qualls & Gultekin, 1997; Szilagyi, 2007). And the original formulation (1) was replaced with:

$$(E_p - E_w) = b(E_w - E) \tag{2}$$

where the proportionality b is the constant coefficient which represents the asymmetry.

However, in Bouchet's original formulation, it was not clear how the potential and wet environment evaporation should or could be estimated or measured (Kahler & Brutsaert, 2006). In the advection-aridity model, the potential evaporation is estimated by the Penman equation and E_w is estimated by the Priestley-Taylor equation (Priestley & Taylor, 1972). The Penman potential evaporation contains the radiation term and the drying power term:

$$E_0 = E_{rad} + E_{aero} \tag{3}$$

$$E_{rad} = \frac{\Delta(R_n - G)}{\Delta + \gamma} \tag{4}$$

$$E_{aero} = \frac{\gamma}{\Delta + \gamma} f(u)\left(e_a^* - e_a\right) \tag{5}$$

where E_0 is the Penman potential evaporation, E_{rad} and E_{aero} are the radiation term and the drying power terms, respectively, Δ is the slope of the saturation vapour curve at air temperature, γ is the psychrometric constant, R_n is the net radiation, G is the ground heat flux, e_a^* and e_a are the saturated and actual vapour pressure of the air, respectively, and $f(u)$ is the function of the wind speed at a reference level. In principle, the wind function can be determined by the Monin-Obukhov similarity theory (Crago & Crowley, 2005; Pettijohn & Salvucci, 2006). This function requires further knowledge on displacement and roughness lengths for momentum and water vapour:

$$f(u) = \frac{\rho_a c_p}{\gamma} \frac{\kappa^2 u}{[\ln(\frac{z-d}{z_{0m}}) - \psi_m][\ln(\frac{z-d}{z_{0v}}) - \psi_v]} \tag{6}$$

where ρ_a is the air density, c_p is the specific heat, $\kappa = 0.40$ is the von Karman constant, z is the measurement height, d_0 is the displacement height, z_{0m} and z_{0v} are the roughness lengths for momentum and water vapour, respectively, and ψ_m and ψ_v denote the stability corrections for momentum and water vapour, respectively. At daily or longer time steps, it is often assumed that the atmospheric stability is neutral so that $\psi_m = 0$ and $\psi_v = 0$. The Priestley-Taylor equation is a proportion of the radiation term in the Penman equation:

$$E_{pt} = \alpha E_{rad} \tag{7}$$

where E_{pt} is the Priestley-Taylor evaporation and α is the Priestley-Taylor coefficient.

The advection-aridity model can be expressed in dimensionless form by the substitution of the Penman potential evaporation and Priestley-Taylor evaporation into equation (2) and rearrangement:

$$\frac{E}{E_0} = \alpha(1+\frac{1}{b})\frac{E_{rad}}{E_0} - \frac{1}{b} \tag{8}$$

In this dimensionless form, the complementary relationship was consistent with the Penman and Budyko hypotheses (Yang *et al.*, 2006), and E_{rad}/E_0 reflects the humidity of the evaporating surface from atmospheric conditions. The former form and the dimensionless form of the complementary relationship are both shown in Fig. 1. It then follows that the evaporation ratio (E/E_0) increases linearly with E_{rad}/E_0. The evaporation calculated with (8) is negative when $E_{rad}/E_0 < 1/\alpha$ and is larger than the potential evaporation when $E_{rad}/E_0 > 1/\alpha(b + 1)$. Since the value of the evaporation ratio is between 0 and 1, the advection-aridity model holds only when $E_{rad}/E_0 \in [1/\alpha(b + 1), 1/\alpha]$. However, the value of E_{rad}/E_0 is not limited between $1/\alpha(b + 1)$ and $1/\alpha$. Transitions of the slope of E/E_0 with respect to E_{rad}/E_0 should exist under arid and wet environments.

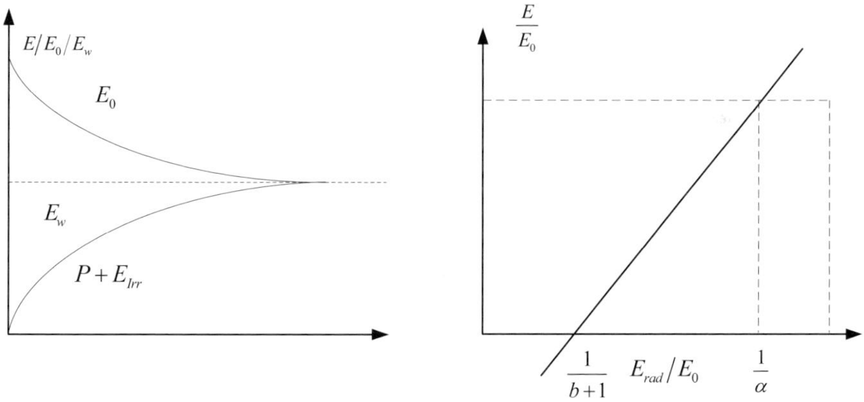

Fig. 1 Schematic representation of the former and the dimensionless form of complementary relationship.

Data used in the study

The HEIFE experiment was carried out in the middle part of the Hexi Corridor area between the Qilian Mountains and the Longshou Mountains, northwest China. A typical annual rainfall is as small as 100 mm, and the land surfaces are covered primarily with sandy and Gobi desert, with some dotted oases and irrigated farm lands (Sugita *et al.*, 2001). The desert station was located at latitude 100°10′E, longitude 39°23′N with an altitude of 1391 m above a surface of fine sand; the Gobi station was located at latitude 100°06′E, longitude 39°09′N with an altitude of about 1400 m above a surface of grit and gravel; and the oasis station was located at latitude 100°26′E, longitude 38°56′N with an altitude of 1483 m above a surface of cropland covered with wheat and maize in summer and bare land in winter.

Formal observations at the three different surfaces consisted of profile measurements on a 20 m tower of air temperature, relative humidity and wind speed, with some other basic meteorological and hydrological variables. Special intensive observations periods (IOPs) of turbulent flux carried out by means of an eddy correlation technique (Tamagawa, 1996) were also conducted for a few weeks in each of the four seasons. Because most of the net radiation and the soil heat flux (G) were not available, it was assumed that $(R_n - G) = LE + H$. Daytime was defined in this study when $(R_n - G) > 0$. The only data that were excluded were those days that recorded missing data for extended periods and whose daytime average values of Q_n or LE were negative. Moreover, air temperature, relative humidity and wind speed measured at $z = 2m$ were used in the analysis. The flux data were archived at 2-hour time steps, while the meteorological data from the tower were attained at 30-min time steps for the desert station and the oasis station. For the Gobi station, all the data were recorded at 1-hour time steps.

In this study, the relationships between the evaporation ratio and the atmospheric wetness index were evaluated at a daily time scale. All the data were first processed into daily means, and then the fluxes were calculated by the wind function derived through the Monin-Obukhov similarity theory, assuming neutral conditions (Crago & Crowley, 2005) with the daily mean variables. As bluff rough surfaces, the surface roughness in the desert and Gobi station is of the order of 10^{-3} m. However, a larger value z_{0m} = 0.0216 m together with z_{0v} = 0.000174 m were used because of the sand dune presented in the study conducted at the same desert station (Tamagawa, 1996; Sugita *et al.*, 2001). A value of z_0 = 0.004 m (Wang *et al.*, 1998) and an assumed value of $z_{0v} \approx 0.1z_{0m}$ were used for the Gobi station. The displacement height d_0 = 0 was used by Tamagawa (1996) and Sugita *et al.* (2001) for the desert station, and this was also used for the Gobi station in this study. For the oasis station, the zero plane displacement height and the roughness length for momentum transfer can be estimated from the crop height, $d = 0.67z_{veg}$, $z_{0n} = 0.123z_{veg}$, and the roughness length governing transfer of heat and vapour can be approximated by $z_{0v} = 0.1z_{0m}$. It was covered with legumes with a height of around 0.40 m in August 1991, and it was covered with wheat with a height of about 0.90 m in May and June 1992.

RESULTS AND DISCUSSIONS

The evaporation ratio of the three different land surfaces plotted with respect to E_{rad}/E_0 are shown in Fig. 2. As the evaporation ratio of the Gobi and desert surfaces are both very small, they are plotted together. For the Gobi and desert sites, there is a critical value of E_{rad}/E_0 around 0.35, before which the evaporation ratio is tiny and increases slowly with E_{rad}/E_0, and after which, evaporation ratio increases rapidly and approximately linearly with respect to E_{rad}/E_0. For the oasis sites, the evaporation ratio increases approximately linearly with E_{rad}/E_0 when the value is between 0.4 and 0.7, which implies that the evaporating surface is neither too dry nor too wet.

The advection-aridity model was only fitted for the observed data during the middle stage. The parameters α and b were optimized by minimizing the mean square error of the estimated evaporation. The optimized value of α is just around the original value 1.26. For the Gobi and desert surfaces, the model yields obvious underestimation, and the actual evaporation calculated with the model is negative even under extremely dry conditions. For the oasis surface, the model performs well as the value of E_{rad}/E_0 is well situated, but the actual evaporation is underestimated under dry conditions and overestimated under wet conditions. However, high correlations and relatively small systematic errors (Table 1) imply that there is a good agreement between the relationships predicted by the advection-aridity model and the observed relationships when the evaporating surface is neither too dry nor too wet.

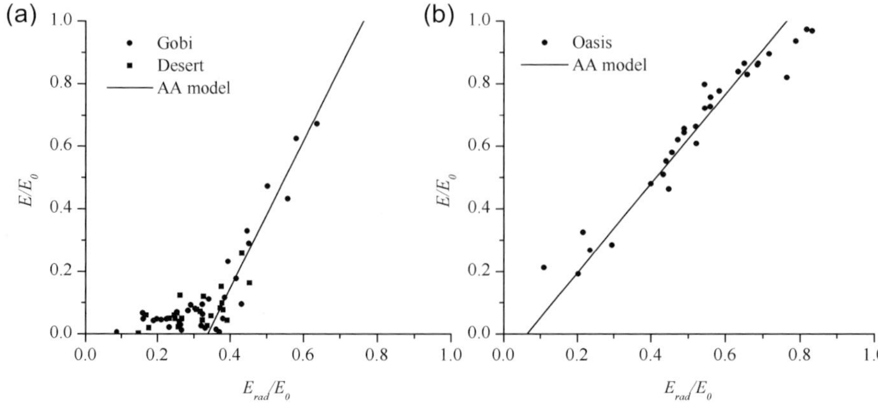

Fig. 2 Plots of evaporation ratio with aspect to the wetness index, compared with the advection-aridity model (AA). Separate graphs are given for the Gobi, desert (a) and oasis (b) surfaces of HEIFE.

Table 1 Performances of the advection-aridity model under conditions neither too dry nor too wet.

Surfaces	α	b	$^*R^2$	$^*MAE$ (W/m^2)	$^*RSME$ (W/m^2)
Gobi and desert	1.31	1.31	0.96	8.29	10.28
Oasis	11.0	1.26	0.98	4.08	5.27

* R^2 is the correlation coefficient; * *MAE* is the Mean Square Error; * *RSME* is the square-root of the Mean Square Error.

Because of the significant systemic bias of the advection-aridity model, it should be re-evaluated with different water availability. It can be supposed that the evaporation ratio is negligible under arid conditions when $x < 1/\alpha(b + 1)$, and evaporation is equal to potential evaporation under wet conditions when $x < 1/\alpha$. The evaporation ratio (E/E_0) only increases linearly with the atmospheric wetness index when $x \in [1/\alpha(b + 1), 1/\alpha]$. This statement leads to the following equation:

$$\frac{E}{E_0} = \begin{cases} 0, & 0 < x \leq 1/[\alpha(1+b)] \\ \alpha(1+\frac{1}{b})x - \frac{1}{b}, & \frac{1}{\alpha(1+b)} < x < \frac{1}{\alpha} \\ 1, & 1/\alpha \leq x < 1 \end{cases} \tag{9}$$

Equation (9) is similar to equations for the evaporation ratio against soil moisture.

CONCLUSIONS

In the dimensionless form of the advection-aridity model, the evaporation ratio is expressed as a linear function of E_{rad}/E_0, and is negative when $E_{rad}/E_0 > 1/\alpha$ and is larger than the potential evaporation when $E_{rad}/E_0 > 1/\alpha(b + 1)$. Since the value of the evaporation ratio is between 0 and 1, the advection-aridity model holds only when $E_{rad}/E_0 \in [1/\alpha(b + 1), 1/\alpha]$. In the advection-aridity model, the water availability of the surfaces are implied by E_{rad}/E_0 from atmospheric conditions. Therefore, the applicability of the model is influenced by water availability.

Using the mean daily evaporation of the Gobi desert and oasis surfaces obtained by the eddy correlation method, it is found that the evaporation is negligible when the evaporating surface is dry, and the evaporation approximates potential evaporation when it is wet, while the evaporation ratio increases with E_{rad}/E_0 approximately linearly. A three-stage pattern similar to the relationship between the evaporation ratio with respect to the soil moisture is found.

The linear relationship suggested by the advection-aridity model is only applicable when the evaporating surface is neither too dry nor too wet. Significant systemic bias would be presented in that the advection-aridity model underestimated the actual evaporation under dry conditions and overestimated under wet conditions. Considering the pattern of the relationship between E/E_0 and E_{rad}/E_0, the advection-aridity model should be re-evaluated or some other new functions should be derived to get a better estimation of the actual evaporation, especially under a wide range of water availability.

Acknowledgements This work was supported by Project 50579031 and 50721140161 sponsored by the National Natural Science Foundation of China. In this study, the data for HEIFE were provided by the "Digital Heihe River Basin" Project (http://heihe.westgis.ac.cn), which was granted by the Cold and Arid Regions Environmental and Engineering Research Institute, Chinese Academy of Sciences.

REFERENCES

Bouchet, R. (1963) Evapotranspiration reelle at potentielle, signification climatique. *Int. Assoc. Sci. Hydro. Pub.* **62**, 134–142.

Brutsaert, W. & Stricker, H. (1979) An advection-aridity approach to estimate actual regional evapotranspiration. *Water Resour. Res.* **15**, 443–450.

Budyko, M. I. (ed.) (1974) *Climate and Life* (translated from Russian by D. H. Miller). Academic Press, San Diego, California, USA.

Crago, R. & Crowley, R. (2005) Complementary relationships for near-instantaneous evaporation. *J. Hydrol.* **300**, 199–211.

Kahler, D. M. & Brutsaert, W. (2006) Complementary relationship between daily evaporation in the environment and pan evaporation. *Water Resour. Res.* **42**, W05413, doi:10.1029/2005WR004541.

McNaughton, K. G. & Spriggs, T. W. (1989) An evaluation of the Priestley and Taylor equation and the complementary relationship using results from a mixed layer model of the convective boundary layer. In: *Estimation of Areal Evapotranspiration* (ed. by T. A. Black), 89–104. IAHS Publ. 177. IAHS Press, Wallingford, UK.

Penman, H. L. (1948) Natural evaporation from open water, bare soil and grass. *Proc. R. Soc. London*, Series A **193**, 120–145.

Pettijohn, J. C. & Salvucci, G. D. (2006) Impact of an unstressed canopy conductance on the Bouchet-Morton complementary relationship. *Water Resour. Res.* **42**, W09418, doi:10.1029/2005WR004385.

Priestley, C. H. B., & R. J. Taylor (1979) On the assessment of surface heat flux and evaporation using large-scale parameters. *Mon. Weather Rev.* **100**, 81–92.

Qualls, R. J. & Gultekin, H. (1997) Influence of components of the advection-aridity approach on evapotranspiration estimation. *J. Hydrol.* **199**, 3–12.

Sugita, M., Usui, J., Tamagawa, I. & Kaihotsu, I. (2001) Complementary relationship with a convective boundary layer to estimate regional evapotranspiration. *Water Resour. Res.* **37**(2), 353–365.

Szilagyi, J. (2007) On the inherent asymmetric nature of the complementary relationship of evaporation. *Geophys. Res. Lett.* **34**, L02405, doi:10.1029/2006GL028708.

Tamagawa, I. (1996) Turbulent characteristics and bulk transfer coefficients at desert HEIFE area. *Boundary Layer Met.* **77**, 1–20.

Wang, J., Bastiaanssen, W. G. M., Ma, Y. & Pelgrum, H. (1998) Aggregation of land surface parameters in the oasis-desert systems of north-west China. *Hydrol. Processes* **12**, 2133–2147.

Yang, D., Sun, F., Liu, Z., Cong, Z. & Lei, Z. (2006), Interpreting the complementary relationship in non-humid environments based on the Budyko and Penman hypotheses, *Geophys. Res. Lett.* **33**, L18402, doi:10.1029/2006GL027657.

An improved analytical approach to estimate *in situ* biodegradation rates

YUAN LIU & XIAOFENG WU

Institute of Hydrology & Water Resources, State Key Laboratory of Hydroscience and Engineering, Tsinghua University, Beijing 100084, China

liuyuan1984@gmail.com

Abstract Biodegradation is a major process in reducing most organic contaminants in aquifers. Estimation of *in situ* biodegradation rates is significant for studying the natural attenuation of contaminants in the field. Buscheck & Alcantar (1995) have developed a method based on field data to estimate biodegradation rates for 1-dimensional (1-D) steady-state plumes. However, in simplifying a field plume, which is usually 2- or 3-D, into a 1-D case, a certain degree of error will occur. In this paper, case studies are performed to assess the errors caused by such simplification. Moreover, an improved analytical method is suggested in steps. Application of the newly developed analytical approach to a field site gives quite good agreement.

Key words biodegradation rate; groundwater contaminant

INTRODUCTION

It is well known that a major factor responsible for natural attenuation and mass reduction of fuel hydrocarbons and other common organic contaminants in groundwater plumes is biodegradation by microorganisms in aquifer materials. First-order decay is usually used to describe the biodegradation process with the parameter–biodegradation rate. Moreover, many theoretical solutions have already been discussed in past decades (Bear, 1979; Domenico, 1987), making the estimation of biodegradation rate a key step to assessing the contribution of biodegradation to attenuation in different sites.

The general range of attenuation rates of some familiar organic chemicals were collected and published by Howard *et al.* (1991). However, the attenuation rates are quite different, not only for different chemicals, but also for the specified situations of the aquifers.

Lots of research has been carried out to estimate the biodegradation rate for a specified site over the last few decades, including laboratory approaches (Rubin *et al.*, 1982; Klecka *et al.*, 1990), field tests (MacIntyre *et al.*, 1993), and a combination of laboratory and field approaches (Chapelle *et al.*, 1996). Such methods have their own pros and cons: laboratory approaches could easily be controlled, but have difficulties in reproducing the *in situ* conditions; results of field tests are always unsatisfactory because of the intricate situations of sites, and it is also hard to distinguish the contribution of biodegradation to attenuation of concentrations from that of dispersion, adsorption, and volatilization. A combination of laboratory and field approaches sounds more reasonable at first glance, however, considering the inherent uncertainties of both approaches, such a combination would hardly represent a unique value of biodegradation rate.

Buscheck & Alcantar (1995) introduced a field-data-based method to estimate biodegradation rates for steady-state plumes, via regression techniques and analytical solutions. This technique is routinely used to estimate biodegradation rates of contaminants in groundwater (McNab & Dooher, 1998). Assuming a first-order decay coefficient as an approximation for the *in situ* biodegradation of the contaminant, Buscheck & Alcantar (1995) showed, for a 1-D idealization, that the degradation rate may be given by:

$$\lambda = (\frac{v_c}{4\alpha_x})[(1-2\alpha_x m)^2 - 1] \quad (1)$$

where m reflects the slope of a regression line fit to the log of contaminant concentration data *versus* the distance along the plume centreline. In fact, $m = k/v_x$, where k is the overall attenuation rate (T^{-1}), v_x is the groundwater linear velocity; v_c is the contaminant velocity along the longitudinal direction (adjusted for retardation, $v_c = v_x/r_f$, r_f is the retardation coefficient); α_x is the longitudinal dispersivity (L), which represents the contribution of dispersion; λ is the biodegradation rate (T^{-1}), which is the main concern in this paper). When biodegradation only occurs in the aqueous phase, v_c is replaced by v_x in equation (1).

Since the information on v_c and α_x for a special site could be achieved through field or laboratory experiments, and slope of the fit line m could be gained through concentration data fitting, then the biodegradation rate λ can be calculated using equation (1).

Though this technique is routinely used to estimate *in situ* degradation rates, its application to field sites is limited by the assumption of the 1-D condition. Since there are usually two or even three dimensions for a real field plume caused by the fuel hydrocarbons or other organic compounds, there is a potential error in estimating *in situ* biodegradation rate when misusing this technique. The objective of this paper is to assess the degree of the errors caused by using the method of Buscheck & Alcantar (1995) in 2-D cases, and to develop an improved analytical approach to estimate biodegradation rates for 2-D plumes.

ASSESSMENT OF THE ERRORS

Concentration distribution along the centreline in a steady-state 3-D site was given by Domenico (1987) as:

$$C(x,0,0) = C_0 \exp\{\frac{x}{2\alpha_x}[1-\sqrt{1+\frac{4\lambda\alpha_x}{v_c}}]\} \bullet erf[\frac{Y}{4\sqrt{\alpha_y x}}] \bullet erf[\frac{Z}{4\sqrt{\alpha_z x}}] \qquad (2)$$

where Y, Z represent the source dimensions; and α_y, α_z are the dispersivities in the transverse and the vertical directions, respectively.

Compared to equation (1), the difference between the 3-D and 1-D cases is the factor $erf[\frac{Y}{4\sqrt{\alpha_y x}}] \bullet erf[\frac{Z}{4\sqrt{\alpha_z x}}]$ at the right side of equation (2), which indicates that the error from simplification is related to the four parameters – the source width, the source thickness, the transverse dispersivity and the vertical dispersivity.

In order to get some quantitative understanding of the error resulting from misusing Buscheck and Alcantar's analytical solutions to estimate biodegradation rates for 2-D, case studies were performed to study the concentration distribution along the centreline for a 2-D plume. The same value of the related parameters provided in the case of Buscheck & Alcantar's paper (1995) were used: $v_x = 0.06$ m day^{-1}, $r_f = 2$, $v_c = 0.03$ m day^{-1} $\alpha_x = 7.5$ m, $C_0 = 3660$ µg L^{-1}, assuming the actual biodegradation rate to be $\lambda = 0.003$ day^{-1}.

For 2-D situations, equation (2) can be simplified as:

$$C(x,0) = C_0 exp\{\frac{x}{2\alpha_x}[1-\sqrt{1+\frac{4\lambda\alpha_x}{v_c}}]\} \bullet erf[\frac{Y}{4\sqrt{\alpha_y x}}] \qquad (3)$$

So the inaccurate estimation of biodegradation rate is related to the two new parameters – the source width Y and transverse dispersivity α_y, in equation (3).

Effect of the source width on the error

To research the effect of the source width on the estimation error, several typical cases were studied.

Although it is practically very difficult to accurately estimate the width of the contaminant source on the field, the width of the contaminant source ranges from several metres to tens of metres in the field in general. For this research, Y is assumed to be 1 m, 5 m and 10 m. Concentration distributions along the centreline of the three cases were calculated separately. In addition, the transverse dispersivity is calculated as a factor of the longitudinal dispersivity, which is known to be 7.5 m.

$$\alpha_y = 0.1 \bullet \alpha_x \text{ (Gelhar } et\ al.\text{, 1992)} \quad (4)$$

Then the contaminant concentration distribution along the centreline in a 2-D plume can be calculated via equation (3), whose results are shown in Fig. 1. It is obvious that for a 2-D plume, the relationship between the centreline concentration and distance from the source is nonlinear in the semi-logarithmic coordinate. However, it is interesting to notice that when the distance from the source is far enough, the curve becomes similar to a linear line, and visually it is almost parallel to the line of the 1-D situation. This phenomenon suggests that using the monitoring data, which are relatively far from the source, to estimate the slope m may give a result close to the 1-D case. Regression techniques were applied to estimate the slope m of the almost linear part in Fig. 1, and equation (1) was used to calculate the biodegradation rate, which includes a certain degree of error. The errors for different values of Y are given in Table 1.

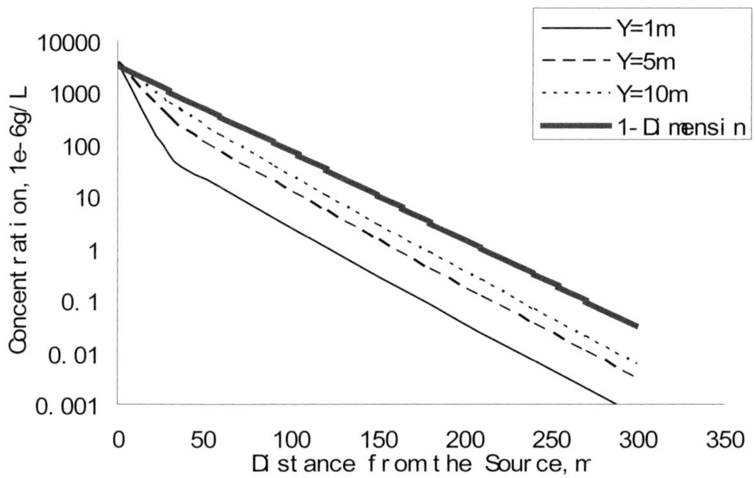

Fig.1 Effect of source width on concentration distribution along centreline.

Table 1 Assessment on the errors for different source widths.

Y (m)	m (m^{-1})	Calculated λ (year^{-1})	Actual λ (year^{-1})	Error (%)
1	–0.04104	1.17527	1.095	7.3%
5	–0.04102	1.17476	1.095	7.3%
10	–0.04098	1.17318	1.095	7.1%

Y: Source width; m: the slope of regression line; λ: biodegradation rate.

The results show that estimating the slope m using the almost linear part of the curves gives a very similar estimation of the biodegradation rates for three different source widths. In other words, estimation of m is not sensitive to the source width, and neither is the further estimation of the biodegradation rate for a 2-D case. However, for a 2-D problem, the biodegradation rate estimated using the method of Buscheck & Alcantar includes an error of about 7%.

Effect of α_y on the error

Following the same considerations, the impact of transverse dispervity α_y on the estimation error is discussed below.

For this case, assume the source width Y to be a constant at 5 m. Equations (4) and (5) are used to express the ratio of α_y to α_x:

$$\alpha_y = 0.33 \bullet \alpha_x \quad \text{(EPA, 2000)} \tag{5}$$

Doing the same calculation and drawing the concentration distribution along the centreline in a semi-logarithmic coordinate, gives the results in Fig. 2 and Table 2. In Fig. 2, very similar phenomena can be observed to Fig. 1. Close to the contamination source, there is a quick decline of the concentration along the centreline. However, when the distance of the monitoring well is far enough from the source, the curves turn almost parallel to the line in the 1-D case. The slopes of the latter part of these curves were calculated based on the data far from the source, and biodegradation rates were calculated using equation (1). The results illustrating the transverse dispersivity's influence on the biodegradation rates are almost the same as the influence caused by the source width. The results show that estimating the slope m using the almost linear part of the curves gives very similar estimation of the biodegradation rates for two different transverse dispesivities. Estimation of m is not sensitive to the transverse dispersivity. However, for a 2-D problem, the Buscheck & Alcantar method (1995) will cause an error of about 7%.

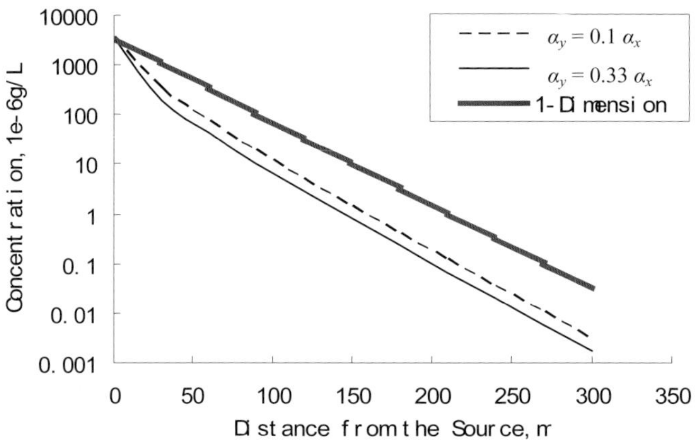

Fig. 2 Effect of transverse dispersity on concentration distribution along the centreline.

Table 2 Assessment on the errors for different ratio of α_y to α_x.

α_y/α_x	m (m^{-1})	Calculated λ (year^{-1})	Actual λ (year^{-1})	Error (%)
0.1	–0.04102	1.17476	1.095	7.3%
0.33	–0.04103	1.17513	1.095	7.3%

IMPROVED ANALYTICAL METHOD

As discussed above, for a 2-D plume, using the almost linear part of the monitoring data to estimate the slope m, the results will not be very sensitive to the source width and the transverse dispersivity, so an assumption of constant source width and transverse dispersivity can be made when field data are not available, such as Y = 10 m and $\alpha_y = 0.1 \bullet \alpha_x$. Under this assumption, an improved analytical approach to estimate the biodegradation rate is developed.

Equation (3) can be changed into the following equation:

$$C(x,0)/erf[\frac{Y}{4\sqrt{\alpha_y x}}] = C_0 \bullet \exp\{\frac{x}{2\alpha_x}[1-\sqrt{1+\frac{4\lambda\alpha_x}{v_c}}]\} \tag{6}$$

The left side of equation (6) can be expressed as a modified concentration $\widetilde{C}(x,0)$:

$$\tilde{C}(x,0) = C(x,0)/erf[\frac{Y}{4\sqrt{\alpha_y x}}] \quad (7)$$

Then equation (6) is given as:

$$\tilde{C}(x,0) = C_0 \bullet \exp\{\frac{x}{2\alpha_x}[1-\sqrt{1+\frac{4\lambda\alpha_x}{v_c}}]\} \quad (8)$$

When biodegradation only occurs in the aqueous phase, v_c should be replaced by v_x.

Obviously, the modified concentration $\tilde{C}(x,0)$ is an exponential function of the distance from the source, and equation (8) can be expressed as:

$$\tilde{C}(x,0) = C_0 \bullet \exp(mx) \quad (9)$$

This is very similar to the method of Buscheck and Alcantar. Substituting the modified concentration $\tilde{C}(x,0)$ for the contaminant concentration $C(x,0)$ in the method of Buscheck and Alcantar, an almost linear line can be plotted with the modified concentration *versus* distance in a semi-logarithmic coordinate, and the slope of the line in such a figure can be obtained. Then the slope m can be used to calculate the biodegradation rate λ.

Comparing equations (8) and (9), λ can be obtained as the same as equation (1):

$$\lambda = (\frac{v_c}{4\alpha_x})[(1-2\alpha_x m)^2 - 1] \quad (10)$$

So with the modified concentration, using the concept developed by Buscheck & Alcantar (1995), it is possible to extend the technique from 1-D situations to 2-D situations. This new analytical approach gives more accurate results for *in situ* biodegradation rates than the method of Buscheck and Alcantar.

APPLYING THE IMPROVED ANALYTICAL APPROACH TO A 2-D FIELD SITE

The improved analytical approach is used to estimate the biodegradation rate of a field site. The site was an industrial plant, which used perchloroethylene (PCE) as the organic solvent and PCE was released to the aquifer about 14 years before the field investigation. At the groundwater flow downgradient of the plant, PCE concentration decrease with the distance from the PCE source was monitored from seven monitoring wells along the centreline (as shown in Fig. 3). The aquifer of the site is mainly composed of sand and gravels, and the thickness of the contaminated aquifer in the site is about 20 m.

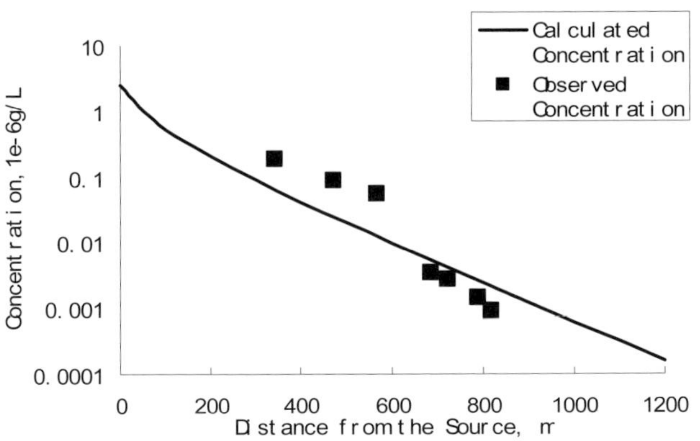

Fig. 3 Comparison of calculated concentration with observed PCE concentration.

Groundwater levels in the site were measured several times in different seasons over two investigation years. Despite a 1-m groundwater level fluctuation, the groundwater gradients were quite stable at 1×10^{-3}. The hydraulic conductivity of the aquifer was obtained from the pumping test result and the data was 3.13×10^{-4} m s^{-1}.

According to the current PCE plume in the aquifer and historical use of PCE in the plant, it is reasonable to simplify the PCE plume as a 2-D plume created by the steady-state groundwater flow from a constant concentration source in the aquifer. Several significant parameters, including effective porosity of the aquifer formation, size of the constant concentration source, and concentration of the source, were obtained through investigation at the field site, as shown in Table 3. Investigation of the field site also illustrated that biodegradation in this site only occurred in the aqueous phase.

Table 3 Parameters of the field site.

Source width, S_w (m)	30	Hydraulic conductivity, K (m s^{-1})	3.13×10^{-4}
Source length, S_L (m)	30	Hydraulic gradient, i (-)	1×10^{-3}
Source thickness, S_T (m)	20	Effective porosity, θe (-)	0.2
Source concentration, C_0 (mg L^{-1})	2.5	Retardation coefficient, R (-)	1.737
Longitudinal dispersivity, α_x (m)	40	Transverse dispersivity, α_y (m)	$0.1\times\alpha_x$

Application of the improved approach could be separated into three steps:

- Calculating the modified concentration $\widetilde{C}(x,0)$ using equation (7).
- Plotting the modified concentration $\widetilde{C}(x,0)$ *versus* the distance in a semi-logarithmic coordinate, and using regression to estimate the slope m of the fitting line.
- Calculating the biodegradation rate λ using equation (1). (Substitute v_x for v_c in equation (1), because investigation showed that biodegradation only occurred in the aqueous phase.)

For the field site, λ is estimated to be 0.389 year^{-1}, namely a half-life of 1.8 years. This result is in accordance with the typical PCE biodegradation half-life range in groundwater – 12 months to 2 years (Howard, 1991). It can be seen from Fig. 3 that the calculated results fit the field data quite well.

CONCLUSIONS

Although the method of Buscheck & Alcantar (1995) is routinely used to estimate *in situ* biodegradation rates, it would lead to some degree of error for any multi-dimensional contamination site, because of the 1-D assumption. For a 2-D plume, two new parameters – the source width and the transverse dispersivity – need to be introduced into the analytical solution. Calculations for several typical cases show that close to the source, there is a quick decline of the contaminant concentration along the centreline. However, when the monitoring data is far from the source, the concentration decline curve almost turns into a straight line, then the line can be used to estimate the biodegradation rate following the method of Buscheck & Alcantar (1995). Research has shown that the estimation of the biodegradation rate is not sensitive to the source width or the transverse dispersivity. For the 2-D situations discussed in this paper, the method of Buscheck & Alcantar (1995) gave an error of about 7%.

An improved analytical approach was developed to estimate the *in situ* biodegradation rate based on the field monitoring data. When the main parameters of the plume have been investigated from field or laboratory experiments, the calculations and regression could be easily performed to get the biodegradation rate. However, there are still some factors needing to be considered: (1) quantities and locations of monitoring wells to represent the concentration decline along the centreline; (2) the rationality of assuming steady-state conditions; (3) reasonable investigation of dispersity and other main parameters for a field site; (4) if the monitoring wells are off the centreline or the field problem is a 3-D one, more intricate equations should be used instead of equation (3) (more details of such equations can be found in Domenico (1987)); (5) for a heterogeneous problem, the simplified analytical equations are not likely to function, and complex numerical simulations would be needed; and (6) for sites where the source location is difficult to identify, the improved analytical solution could not be performed.

REFERENCES

Bear, J. (1979) *Hydraulics of Groundwater*. McGraw-Hill, New York, USA.

Busheck, T. E. & Alcantar, C. M. (1995) Regression techniques and analytical solution to demonstrate intrinsic bioremediation. In: *Intrinsic Bioremediation* (ed. by R. E. Hinchee, J. T. Wilson & D. Downey), 109–116. Battelle Press, Columbus, Ohio, USA.

Chapelle, F. H., Bradley, P. M. Lovley, D. R. & Vroblesdy, D. A. (1996) Measuring rates of biodegradation in a contaminated aquifer using field and laboratory methods. *Ground Water* **34**(4), 691–698.

Todd, D. K. (1995) *Groundwater Hydrology*. John Wiley & Sons, New York, USA.

Domenico, P. A. (1987) An analytical model for multidimensional transport of a decaying contaminant species. *J. Hydrol.* **91**, 49–58.

Fetter, C. W. (1999) *Contaminant Hydrology*. Prentice-Hall, New Jersey, USA.

Gelhar, L. W., Welty, C. & Rehfeldt, K. R. (1992) A critical review of data on field-scale dispersion in aquifers. *Water Resour. Res.* **28**(7), 1955–1974.

Howard, P. H., Boethling, R. S., Jarvis, W. F., Meylan, W. M. & Michalenki, E. M. (1991) *Handbook of Environmental Degradation Rates*. Lewis Publishers, Chalsea, USA.

Klecka, G. M., Davis, J. W., Gray, D. R. & Madsen, S. S. (1990) Natural bioremediation of organic contaminants in ground water: Cliffs-Dow superfund site. *Ground Water* **28**, 534–543.

MacIntyre, W. G., Boggs, M., Antworth, C. P. & Stauffer, T. B. (1993) Degradation kinetics of aromatic organic solutes introduced into a heterogeneous aquifer. *Water Resour. Res.* **20**, 4045–4051.

McNab, W. W. & Dooher, B. P. (1998) A Critique of a steady-state analytical method for estimation contaminant degradation rates. *Ground Water* **36**(6), 983–987.

Pickens, J. F. & Grisak. G. E. (1981) Scale-dependent dispersion in a stratified granular aquifer. *Water Resour. Res.* **17**(4), 1191–1211.

Rubin, H. E., Subba-Rao, R. V. & Alexander, M. (1992), Rates of mineralization of trace concentrations of aromatic compounds in lake water and sewage samples. *Appl. Environ. Microb.* **43**(5), 1131–1138.

United States Environmental Protection Agency (2000). *Biochlor 2.2 Manual*, 1–42. United States Environmental Protection Agency Washington DC, USA.

Estimating unsaturated soil hydraulic parameters using a generalized Richards equation

HAISHEN LÜ, YONGHUA ZHU & ZHONGBO YU

State Key Laboratory of Hydrology-Water Resources and Hydraulic Engineering, Department of Applied Mathematics, Hohai University, Naijing 210098, China

haishen2001@yahoo.com.cn

Abstract Richards equation is widely used as the basis for simulations of water transport in soils. For infiltration into horizontal soil columns, Richards equation predicts that the water content profile is a unique function of the Boltzmann variable (distance)/(time)^q, where $q = 0.5$. However, a number of experiments have found that q is significantly less than 0.5. Scaling with $q < 0.5$ is consistent with a generalized Richards equation that uses a fractional time derivative of the water content. In this paper we consider a generalized Richards equation that incorporates the unsaturated hydraulic conductivity function $k(h)$ of van Genuchten (1980). A new method is proposed for estimating the van Genuchten parameters α and n. Estimates of α and n are expressed as closed form equations that are functions of other parameters such as the length of wetted zone, the sorptivity, and the saturated hydraulic conductivity.

Key words generalized Richards equation; unsaturated hydraulic conductivity function

INTRODUCTION

The hydraulic conductivity of a soil determines the rate at which water can flow in the soil and consequently the rate at which suspended or dissolved materials can be transported. Predicting flow and transport requires accurate knowledge of the hydraulic conductivity. However, determining the unsaturated hydraulic conductivity is difficult and complex. Hydraulic conductivity is a highly nonlinear function of pressure head or water content, and field and laboratory measurements are difficult and time-consuming. Alternative methods for estimating or calculating the conductivity are difficult to validate.

Methods for determining the conductivity generally use some mathematical form for the conductivity function. However, no universal equation exists for the unsaturated hydraulic conductivity. In 1980, van Genuchten put forward an equation that specifies the conductivity as a function of soil water pressure head:

$$k(h) = \frac{k_s\left[1 - \dfrac{(\alpha|h|)^{n-1}}{\left(1+(\alpha|h|)^n\right)^{1-1/n}}\right]^2}{\sqrt{\left(1+(\alpha|h|)^n\right)^{1-1/n}}} \tag{1}$$

where $k(h)$ is the hydraulic conductivity (LT^{-1}), k_s is the saturated conductivity (LT^{-1}), and α (L^{-1}) and n (-) are adjustable parameters. It is difficult to estimate α and n. In this paper, a new method for estimating α and n is presented. The method is based on a generalized Richards equation. Estimates of α and n are expressed as closed form equations that are functions of the characteristic length of the wetted zone, the sorptivity, and the saturated hydraulic conductivity.

GENERALIZED RICHARDS EQUATION

The Richards equation for one dimensional (1-D) horizontal flow may be written as:

$$\frac{\partial \theta}{\partial t} = \frac{\partial}{\partial x}\left[k(h)\frac{\partial h}{\partial x}\right]$$

where θ is the volumetric water content ($L^3 L^{-3}$) and h is the pressure head (L). The Richards equation was obtained originally in 1931 by combining the equation of continuity with the empirically based "Buckingham-Darcy" flux law (Swartzendruber, 1968). Later, Bhattacharya *et al.* (1976) showed that the Richards equation can be derived from physically-based molecular principles. The derivation is based on the assumption that water moves in a Brownian motion in the form of quasi-molecules, an assumption that has also been instrumental in deriving solute diffusion and convective–dispersive transport equations (Bhatacharya & Gupta, 1990).

Both the horizontal water transport equation and the solute diffusion equation conform to Boltzmann scaling, wherein water or solute profiles are unique functions of the Boltzmann variable (distance)/(time)^q, with $q = 0.5$ (Hillel, 1980). However, both water flow and solute transport in porous media have been shown to deviate from Boltzmann scaling and follow a more general scaling law with $q \neq 0.5$ (Neuman, 1990; Hatano & Hatano, 1998; Haggerty *et al.*, 2000; Pachepsky *et al.*, 2001, 2003). A physical model that is consistent with this more general scaling consists of particles undergoing non-Brownian motion due to the structure of, or interactions with, the porous media (Metzler & Klafter, 2000; Pachepsky *et al.*, 2003). Scaling with $q > 0.5$ occurs when particles undergo Levy motion, which is similar to Brownian motion, except that relatively large transitions occur more often, such as may be caused by macropores or fractures. Scaling with $q < 0.5$ occurs when particles remain motionless for extended periods of time. Pachepsky *et al.* (2003) noted that this type of behaviour has long been known in soil wetting, with Nielsen *et al.* (1962) suggesting in 1962 that $q < 0.5$ may occur when the infiltration front experiences sporadic periods of immobility and thus progresses in "jerky movements". A physical model consistent with these dynamics is that of water particles being randomly trapped and having a power law distribution of waiting times. Pachepsky *et al.* (2003) proposed a generalized Richards equation based on this concept.

Particles with a power law distribution of waiting times have an infinite mean waiting time, and it has been suggested that the transport of an ensemble of such particles can be simulated using a fractional time derivative (Meerschaert *et al.*, 2002; Pachepsky *et al.*, 2003). Thus the generalized Richards equation is similar to the original except that the fractional time derivative of water content is used:

$$\frac{\partial^\gamma \theta}{\partial t^\gamma} = \frac{\partial}{\partial x}\left[k(h)\frac{\partial h}{\partial x}\right] \tag{2}$$

where $\gamma(\leq 1)$ is the order of the fractional derivative. Equation (2) reduces to the classical Richards equation when $\gamma = 1$. The fractional derivative is defined by:

$$\frac{\partial^\gamma \theta}{\partial t^\gamma} = \frac{1}{\Gamma(1-\gamma)}\frac{d}{dt}\int_0^t \frac{\theta(\tau,x)}{(t-\tau)^\gamma}d\tau$$

The initial and boundary conditions for horizontal infiltration into a semi-infinite column are:

$$h(x,0) = h_0,\ h(0,t) = 0,\ h(\infty,t) = h_0$$

where h_0 is initial pressure head.

Application of the Boltzmann transform

$$\xi = \frac{x}{t^{\gamma/2}} \tag{3}$$

yields

$$\frac{\partial^\gamma \theta}{\partial t^\gamma} = \frac{d\theta}{d\xi}\frac{\partial^\gamma \xi}{\partial t^\gamma} = \frac{d\theta}{d\xi}x\frac{d^\gamma\left(t^{-\gamma/2}\right)}{dt^\gamma}$$

$$= \frac{d\theta}{d\xi}x\frac{\Gamma(1-\gamma/2)}{\Gamma(1-3\gamma/2)}t^{-\gamma/2-\gamma}$$

$$= \frac{\Gamma(1-\gamma/2)}{\Gamma(1-3\gamma/2)} \frac{x}{t^{\gamma/2+\gamma}} \frac{d\theta}{d\xi}$$

and thus:

$$\frac{\partial}{\partial x}\left[k(h)\frac{\partial h}{\partial x}\right] = \frac{1}{t^{\gamma}} \frac{d}{d\xi}\left[k(h)\frac{dh}{d\xi}\right].$$

Equation (2) becomes:

$$\frac{\Gamma(1-\gamma/2)}{\Gamma(1-3\gamma/2)} \xi \frac{d\theta}{d\xi} = \frac{d}{d\xi}\left[k(h)\frac{dh}{d\xi}\right] \tag{4}$$

with boundary conditions:

$$h(0) = 0,\ h(\infty) = h_0 \tag{5}$$

HYDRAULIC CONDUCTIVITY

In this section, we incorporate the van Genuchten (1980) soil hydraulic conductivity function into the generalized Richards equation and derive closed form expressions useful for estimating van Genuchen parameters. Mualem (1976) presented a statistical pore-size distribution model that related the soil water retention curve to the hydraulic conductivity function. Van Genuchten (1980) proposed a new functional form for the water retention curve and used the model of Mualem (1976) to derive a new expression for the conductivity, equation (1). The water retention function from Van Genuchten (1980) is:

$$\theta(h) = \begin{cases} \theta_r + \dfrac{\theta_s - \theta_r}{\left[1 + |\alpha h|^n\right]^{1-1/n}}, & h < 0 \\ \theta_s, & h \ge 0 \end{cases} \tag{6}$$

where θ_s is the saturated soil water content (L^3 L^{-3}), θ_r is the residual soil water content (L^3 L^{-3}), and α and n are the adjustable parameters discussed above in relation to equation (1). Considerable effort has been directed at developing procedures for estimating values for α and n (e.g. Van Genuchten *et al.*, 1992). In the following, we propose a new method for estimating α and n using the generalized Richards equation and present closed form expressions for the soil hydraulic parameters.

We start by expressing the pressure head as a Maclaurin series:

$$h(\xi) = a_0 + a_1\xi + a_2\xi^2 + \cdots + \cdots$$

Because $h(0) = 0$, we have $a_0 = 0$ and

$$h(\xi) = a_1\xi + a_2\xi^2 + \cdots + \cdots$$

where a_1 is a negative constant. For convenience, we denote $b_1 = a_1$. If ξ is small enough, then $h(\xi) \approx -b_1\xi$. However, for small $\alpha|h|^n$, we have:

$$\left[1 + (\alpha\,|\,h\,|)^n\right]^{1-1/n} \approx 1 + m(\alpha\,|\,h\,|)^n$$

So,

$$\frac{\theta - \theta_r}{\theta_s - \theta_r} = 1 - m(\alpha b_1 \xi)^n \tag{7}$$

where $m = 1 - \frac{1}{n}$. At a characteristic length $\xi = d$ (the depth of the leading edge of the wetting front), the water content is defined $\theta(d) = \theta_0$. Equation (7) then becomes:

$$mb_1^n = \frac{\theta_s - \theta_0}{(\theta_s - \theta_r)(\alpha d)^n} \tag{8}$$

Combining equations (7) and (8) gives:

$$\theta(\xi) = \begin{cases} \theta_s - (\theta_s - \theta_0)\left(\frac{\xi}{d}\right)^n, & 0 < \xi < d \\ \theta_0, & d \le \xi < \infty \end{cases}$$

This water content profile is consistent with the function proposed by Philip (1955) and the experimental data of Shao (2000).

We now consider integrating equation (4). Integration of the right-hand side of equation (4) gives:

$$\int_0^\infty \frac{d}{d\xi}\left[k(h)\frac{dh}{d\xi}\right] = k(h)\frac{dh}{d\xi}\Big|_0^\infty = b_1 k_s$$

This result follows from $\left.\frac{dh}{d\xi}\right|_{\xi=\infty} = 0$, $\left.\frac{dh}{d\xi}\right|_{\xi=0} = -b_1$, $h(0) = 0$, and $k(0) = k_s$. For the left-hand side, we recall the expression given above for $\theta(\xi)$ and obtain

$$\int_0^\infty \frac{\Gamma(1-\gamma/2)}{\Gamma(1-3\gamma/2)}\xi\frac{d\theta}{d\xi}d\xi = \frac{\Gamma(1-\gamma/2)}{\Gamma(1-3\gamma/2)}\int_0^d \xi\frac{d\theta}{d\xi}d\xi$$

$$= -\frac{\Gamma(1-\gamma/2)}{\Gamma(1-3\gamma/2)}\frac{n(\theta_s-\theta_0)d}{(n+1)}$$

Thus if we integrate equation (4) and make the substitution (see equation (8)):

$$b_1 = \frac{1}{\alpha d}\left[\frac{\theta_s - \theta_i}{m(\theta_s - \theta_r)}\right]^{\frac{1}{n}} \tag{9}$$

we obtain an expression for α

$$\alpha = -\frac{\Gamma(1-3\gamma/2)}{\Gamma(1-\gamma/2)}\frac{(n+1)k_s}{n(\theta_s-\theta_0)d^2}\left[\frac{\theta_s-\theta_0}{m(\theta_s-\theta_r)}\right]^{\frac{1}{n}} \tag{10}$$

Next, we seek an expression for the parameter n. At $x = 0$, the soil water flux is:

$$q = -\left(k(h)\frac{\partial h}{\partial x}\right)\Big|_{x=0}.$$

Since $\left.\frac{\partial h}{\partial x}\right|_{x=0} = \left(\frac{\partial h}{\partial \xi}\right)_{x=0}\left(\frac{\partial \xi}{\partial x}\right)_{x=0} = -b_1 t^{-\gamma/2}$ and $k(h)|_{x=0} = k_s$, the flux may also be expressed:

$$q|_{x=0} = k_s b_1 t^{-\frac{\gamma}{2}} \tag{11}$$

For horizontal infiltration, the Phillip infiltration formula has the form:

$$q = \frac{S}{2} t^{-\frac{\gamma}{2}} \tag{12}$$

where $S = \int_{\theta_{dry}}^{\theta_{wet}} \xi d\theta$ is the sorptivity. Combining equations (9), (11), and (12) results in an expression for α:

$$\alpha = \frac{2k_s}{dS} \left[\frac{\theta_s - \theta_0}{m(\theta_s - \theta_r)} \right]^{\frac{1}{n}} \tag{13}$$

Substituting equation (13) into equation (10) yields an expression for n that does not contain α:

$$n = \frac{S\Gamma(1-3\gamma/2)}{-S\Gamma(1-3\gamma/2) - 2\Gamma(1-\gamma/2)(\theta_s - \theta_0)d}. \tag{14}$$

DISCUSSION AND CONCLUSION

Equations (13) and (14) are closed-form expressions for α and n in terms of the parameters k_s, S, d, θ_s, θ_0, θ_r and γ. When these parameter values are available, equation (14) can be used to calculate n and then equation (13) can be used to calculate α. Measurements of horizontal water adsorption can be used to determine γ and S (equation (12)). The parameters k_s, θ_s, θ_0, θ_r, are relatively easy to estimate or measure, whereas the characteristic length d can be estimated as a distance between the surface and the wetting front. The method developed in this paper is a quick and inexpensive approach to estimating hydraulic parameters.

Acknowledgement This research is supported by National Key Basic Research Program of China (2006CB400502) and Jiangsu Planned Projects for Postdoctoral Research Funds. The project (Study on Coupled Processes of Atmosphere-Land-Hydrology) (IRT0717) has been supported by the Program for Changjiang Scholars and Innovative Research Team in University (PCSIRT).

REFERENCES

Bhatacharya, R. & Gupta, V. K. (1990) Application of the central limit theorem to solute transport in saturated porous media: from kinetic to field scales. In: *Dynamics of Fluids in Hierarchical Porous Media* (ed. by J. H. Cushman), 97–124. Academic Press, New York, USA.

Bhattacharya, R. N., Gupta, V. K. & Sposito, G. (1976) A Markovian stochastic basis for the transport of water through unsaturated soil. *Soil Sci. Soc. Am. J.* **40**(2), 465–467.

Haggerty, R., McKenna, S. A. & Meigs, L. C. (2000) On the late-time behavior of tracer test breakthrough curves. *Water Resour. Res.* **36**(12), 3467–3480.

Hatano, Y. & Hatano, N. (1998) Dispersive transport of ions in column experiments: an explanation of long-tailed profiles. *Water Resour. Res.* **34**(3), 1027–1041.

Hillel, D. (1980) *Fundamentals of Soil Physics*. Academic Press, San Diego, USA.

Meerschaert, M. M., Benson, D., Scheffler, H.-P. & Baeumer, B. (2002) Stochastic solution of space-time fractional diffusion equations. *Phys. Rev. E.* **65**, 041103.

Metzler, R. & Klafter, J. (2000) The random walk's guide to anomalous diffusion: a fractional dynamics approach. *Phys. Rep.* **339**(ERI), 1–77.

Mualem, Y. (1976) A new model for predicting the hydraulic conductivity of unsaturated porous media. *Water Resour. Res.* **12**(2), 593–622.

Neuman, S.P. (1990) Universal scaling of hydraulic conductivities and dispersivities in geologic media. *Water Resour. Res.* **26**, 1749–1758.

Nielsen, D. R., Biggar, J. W. & Davidson, J. M. (1962) Experimental considerations of diffusion analysis in unsaturated flow problem. *Soil Sci. Soc. Am. Proc.* **26**, 107–111.

Pachepsky, Ya., Benson, D. & Timlin, D. (2001) Transport of water and solutes in soil as in fractal media. SSSA Special Publication, Soil Science Society of America, Madison, Wisconsin, USA, **56**, 51–75.

Pachepsky, Y., Timlin, D. & Rawls, W. (2003) Generalized Richards' equation to simulate water transport in unsaturated soils. *J. Hydrol.* **272**, 3–13.

Philip, J. R. (1955) Numerical solution of equations of the diffusion type with diffusivity concentration-dependent. *Transactions of the Faraday Society* **51**, 885–892.

Shao, M., Wang, Q. & Robert, H. (2000) A simple infiltration method for estimating soil hydraulic properties of unsaturated soils, theoretical analysis. *Acta Pedologica Sinica* **37**(1), 1–8.

Swartzendruber, D. (1968) The applicability of Darcy's law. *Soil Sci. So. of Am. Proc.* **32**, 11–18.

van Genuchten M. Th. (1980) A closed form equation for predicting the hydraulic conductivity of unsaturated soils. *Soil Sci. Soc. Am. J.* **44**, 892–898.

van Genuchten, M.Th., Leij, F.J., & Lund, L. J. (eds) (1992) *Proceedings of the International Workshop on Indirect Methods for Estimating the Hydraulic Properties of Unsaturated Soils* (Riverside, California, October 11–13, 1989. University of California).

Computation of grid storage capacity from topographic index in hydrological modelling

PENG SHI[1], XIAOFANG RUI[2], SIMIN QU[2] & XI CHEN[1]

1 *State Key Laboratory of Hydrology-Water Resources and Hydraulic Engineering, Nanjing 210098, China*
ship@hhu.edu.cn; ship_sohu@sohu.com

2 *College of Water Resources and Environment, Hohai University, Nanjing 210098, China*

Abstract Recent advances in the field of GIS have made possible the development of distributed hydrological models. However, there are some unsolved problems in full real-time distributed hydrological models. In view of these problems, an alternative conceptual distributed hydrological model is proposed in this paper. Determination of water storage capacity for every grid cell has become an important problem in conceptual distributed hydrological models; taking the similarity between the maximum grid soil moisture deficiency and the grid topographic index into account, it was found that a logarithmic Weibull function relationship exists between them. Based on this, a method to calculate the grid storage capacity from the grid topographic index is proposed. The method was applied in a grid distributed hydrological model with non-compact structure. Results obtained show that in an ungauged basin it is possible to get some of the hydrological parameters from the geographic parameter set. The main conclusion of this paper is that the findings here suggest a promising alternative way for hydrological modelling.

Key words distributed hydrological model; topographic index; grid storage capacity; logarithmic Weibull function

INTRODUCTION

Recent developments in GIS have made it possible to develop a distributed hydrological model since the "blueprint for a physically-based, digitally simulated hydrological response model" was put forward by Freeze & Harlan (1969). As pointed out by Beven (2001), this blueprint was based on some descriptions for hydrological processes. Even though these descriptions about soil water flow, surface flow and river flow may be appropriate on a laboratory scale, when applied (for example, Darcy's law and Richards equation in unsaturated zone) in a distributed hydrological model, we have to determine whether their parameters and variables are consistent in scale, ranging from a few metres to a few kilometres (Bloschl, 2001). Problems will arise if these basic equations, which are used to build physically-based distributed hydrological models, are applied in real basins; there is a fundamental conflict (Beven, 1989, 1996, 2000). Considering the problems of localization, if the equations' localizations are permitted, complicated parameters should be introduced as a consequence. However, these parameters cannot usually be directly observed and some uncertainty exists. The simple implication is that different observed data will be needed to calibrate these parameters in order to study their sensitivity and spatial variability. The way to resolving this conflict does not depend on using more detailed process descriptions or collecting more observed data to determine some parameters and variables, but rather the only viable way is to look for a simple model based on existing information.

An approach, which entails the utilization of existing information, is proposed in this paper. This alternative method calculates some important parameters based on the grid of the DEM. Unlike the traditional conceptual lumped model, which applies on a basin scale, a classical conceptual model on a grid scale is used to form a conceptual distributed hydrological model. The parameters of this kind of model are also spatially variable and cannot be evaluated by just using traditional methods based on model parameter calibrations. So, the problem of rainfall–runoff and flow confluence in a unit grid is changed to a problem of PUB (Predictions in Ungauged Basins). Based on recent studies and experiments, hydrologists and geomorphologists have come to the conclusion that the rainfall–runoff process is controlled, not only by characteristics of precipitation, but also by some distinct relations between the runoff process and the characteristics of the underlying surface (e.g. topography, geomorphy, soil, vegetables, geology, etc.). The appearance of the concept of topographic index (Beven, 1979) made it possible to combine geomorphologic parameters and hydrological parameters well.

METHODS

It is difficult to determine the storage capacity for every grid cell if the conceptual hydrological model (e.g. Xinanjiang model) is applied on a grid scale. The curve of basin storage capacity in the Xin'anjiang model (XAJ) is consistent with the curve of topographic index cumulative frequency in TopModel. Though there are some descriptions about the spatial variability of saturation capacity, it should be noted that areas with large topographic index usually correspond to valleys in a basin prone to producing runoff because of a small water deficit. However, areas with a small topographic index usually correspond to the ridge and invariably cannot produce runoff easily because of large water deficit. Consequently, the topographic index is an indicator of the response of basin hydrological processes (Ren Li-liang, 2000).

This paper addresses the following questions:

- Are there some kind of relations between topographic index and storage capacity?
- How do we determine the relation if it does exist?

Through the analysis of data from nine basins around the Three Gorge of Yangtse River, we found that the shape of the distribution curve of topographic index is similar to that of the basin storage capacity, which is the parabolic curve shape (see Fig. 1). In Fig. 1, the *y*-axis is the difference of grid topographic index and the minimum of basin topographic index; the *x*-axis is a proportion between the numbers of the grid with topographic index less than some level and the total grid number. For purposes of analysis the storage capacity curve, which is calibrated using the lumped model, shall be taken as a known quantity (or statistical real value). In addition, the method of combine or total frequency is adopted where some points from two curves with the same proportion are collected and then fitted in a figure (see Fig. 2).

From Fig. 2 it can be seen that the curve fits the distribution curve of the logarithmic Weibull function well; here, the offset value zero is set to zero. The formula of this curve is given as:

$$\frac{W_i}{WMM} = \exp\left\{-\left[\frac{\ln(TI_i - TI_{\min} + 1)}{\alpha}\right]^{\beta}\right\} \quad (1)$$

where W_i is the grid storage capacity, WMM is the maximum basin storage capacity, TI_i the grid topographic index, $TI_{\min}$ is the minimum basin topographic index, α is a scale parameter, which reflect the grid scale, and β is a shape parameter.

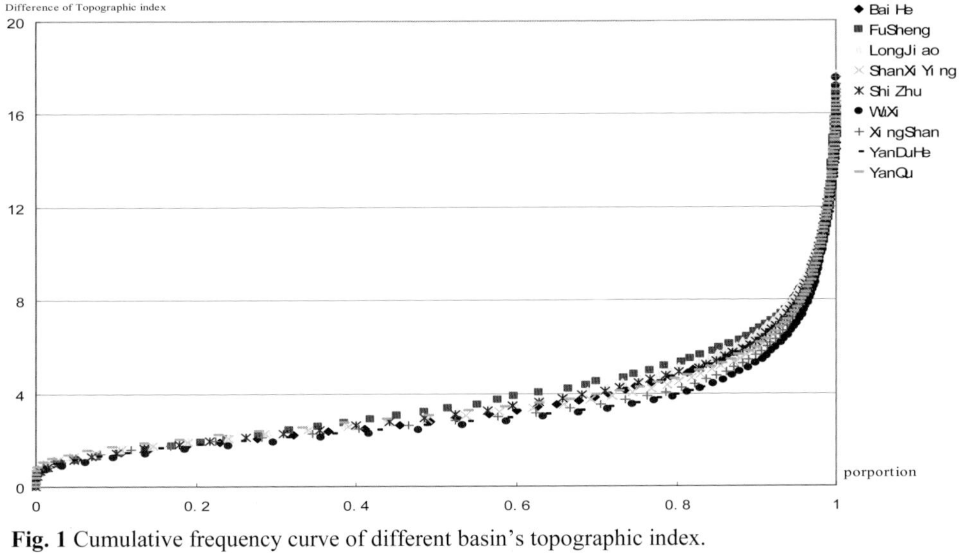

Fig. 1 Cumulative frequency curve of different basin's topographic index.

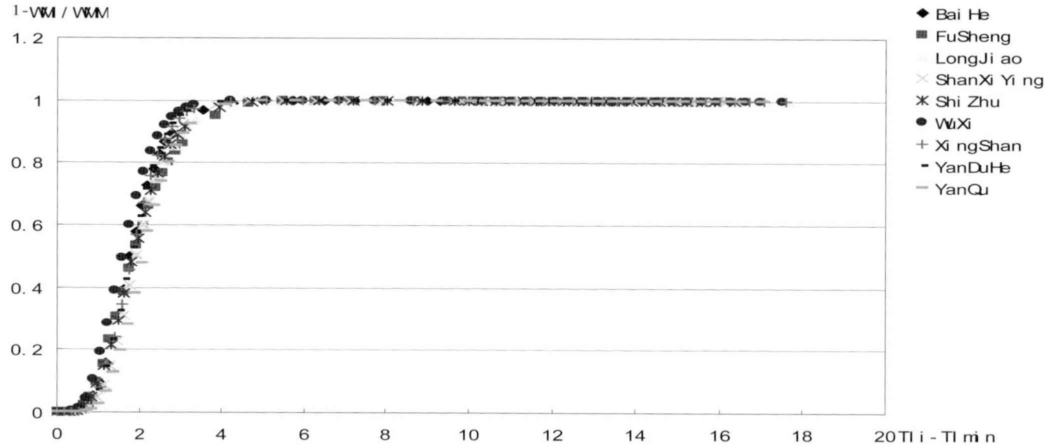

Fig. 2 Relation of topographic index and storage capacity.

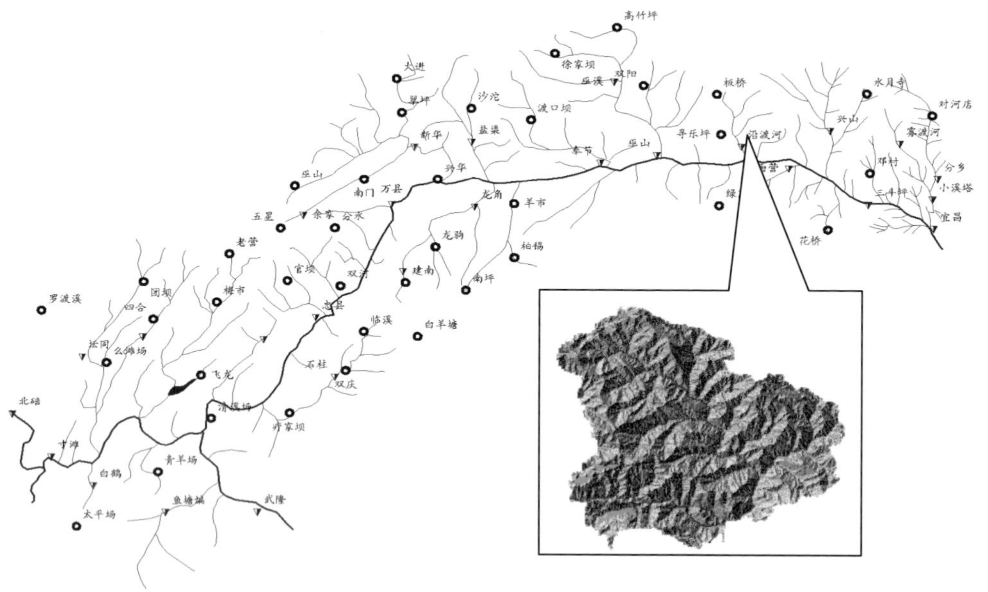

Fig. 3 Yanduhe River Basin.

STUDY SITE

To test the logarithmic Weibull function assumption, the Yanduhe Basin was chosen as the study site. The Yanduhe River Basin, situated in the south of the ShenNongjia virgin forest (Fig. 3), is the first order tributary of Yangtze River in the north bank and covers 601 km^2 of forested land at elevations between 200 m in the southern portion of the catchment and 3052.7 m in the northern parts, which is the highest peak of ShenNongjia virgin Forest. The basin has 26 tributaries with river network density of 0.2 km/km^2, river gradient of 9.5‰ while the sinuosity is 1.4 and vegetation cover stands at about 70%.

RESULTS AND DISCUSSIONS

The maximum basin storage capacity in the Yanduhe Basin, which can be determined by the XAJ model, is 156 while two types of grid resolution for DEMs, 3s and 30s were applied in this study area to obtain the spatial distribution of the basin storage capacity (Fig. 4). For every grid cell, the storage capacity was calculated and the cumulative curve was drawn to obtain the basin storage capacity curve (Fig. 5). The grid storage capacity was then applied to the distributed hydrological model (Shi Peng, 2006) to obtain the results for flood simulation (see Table 1).

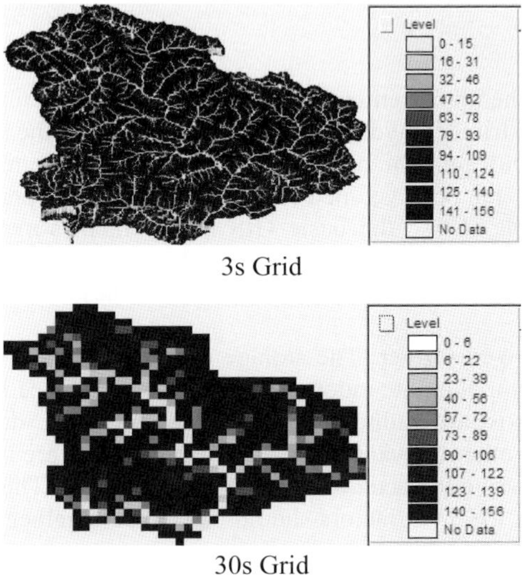

Fig. 4 Calculated distribution of basin storage capacity.

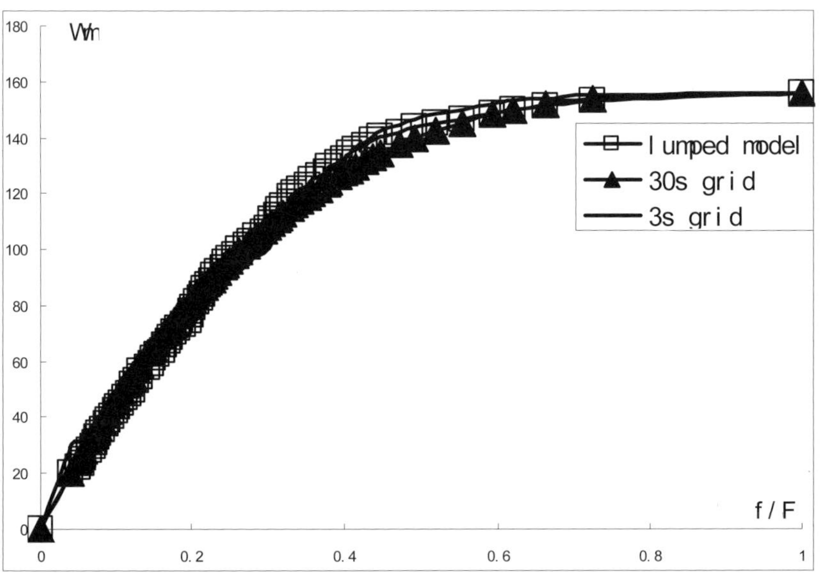

Fig. 5 Calculated cumulative curve of Basin storage capacity.

Table 1 Comparison of simulated and observed flood characteristic for Yanduhe watershed.

Flood Code	Precipi tation	Observed run-off depth (mm)	Calculated run-off depth (mm)	Error of run-off depth (%)	Observed peak flow (m^3/s)	Calculated peak flow (m^3/s)	Error of peak flow (%)	Deterministic coefficients
831004	181.4	176.7	172.3	−1.3	575	576	0.03	0.939
840723	176.1	144.4	148.7	4.3	1060	1062	0.2	0.969
850621	123.5	100.7	97.5	−2.0	476	480	0.9	0.932
860909	194.2	181.9	173	−4.0	844	841	−0.37	0.959
870821	104.2	94.1	91.3	−1.7	556	556	0.1	0.876

CONCLUSIONS

The development of computer technology and the improvement of GIS technology have enhanced the further development and improvement of distributed hydrological models. Based on the results indicated in Figs 3 and 4, the logarithmic Weibull function assumption is correct and the relation between topographic index and storage capacity can be developed via this assumption, as provided by equation (1). The formula makes it possible to use the basic GIS data, which is available everywhere now, and by so doing, makes the model reliable and physically based. The comparisons between the simulated and observed data are listed in Table 1; these results are realistic.

The approach described here can be improved further. The storage capacity was taken as a function of vegetation type or soil type and the topographic index as an exclusive influence factor; this assumption basically differs from the existing traditional knowledge. As such, detailed information about soil, vegetation, geology and other relevant landscape characteristics are needed to determine grid storage capacity that will improve the accuracy of runoff predictions. The main conclusion here is that it is very clear that these essential factors should be studied together in the study of grid storage capacity in the future.

REFERENCES

Beven, K. J. & Kirkby, M. J. (1979) A physically based variable contributing area model of basin hydrology. *Hydrol. Sci. Bull.* **24**(1), 43–69.

Beven, K. J. (1989) Changing ideas in hydrology: the case of physically based models. *J. Hydrol.* **105**, 157–172.

Beven, K. J. (1996) A discussion of distributed modeling, chapter 13A. In: *Distributed Hydrological Modelling* (ed. by J.-C. Refsgaard & M. B. Abbott), 255–278. Kluwer, Dordrecht, The Netherlands.

Beven, K. J. (2000) Uniqueness of place and the representation of hydrological processes. *Hydrol. Earth Sys. Sci.* **4**, 203–213.

Beven, K. J. (2001) How far can we go in distributed hydrological modeling? *Hydrol. Earth Sys. Sci.* **5**(1), 1–12.

Bloschl, G. (2001) Scaling in hydrology. *Hydrol. Processes* **15**, 709–711.

Freeze, R. A. & Harlan, R. L. (1969) Blueprint for a physically based, digitally-simulated hydrologic response model. *J. Hydrol.* **9**, 237–258.

Ren, L. L. & Liu, X. R. (2000) Hydrological processes modeling based on digital elevation model. *Geogr. Res.* **19**(4), 369–376.

Shi, P. (2006) Research of grid distributed hydrological model with incompact structure and geomorphological instantaneous unit hydrograph. PhD Thesis, HoHai University, China.

A tracer experiment and numerical simulation of potential nitrogen pollution from agricultural fertilization

LIHUA TANG[1,2], SICONG ZHANG[1,2] & XIUFEN SUN[1]

1 *Department of Hydraulic Engineering, Tsinghua University, Beijing 100084, China*
tanglh@tsinghua.edu.cn

2 *State Key Laboratory of Hydroscience and Engineering, Tsinghua University, Beijing 100084, China*

Abstract Agricultural non-point source pollution is probably the most significant anthropogenic source of nitrogen pollution for water bodies. The excessive fertilizers rushed into surface water bodies by runoff or leached into groundwater by percolation may cause eutrophication or groundwater pollution. It is usually difficult to estimate the non-point source pollution due to its random and distributed nature. In this paper, a winter wheat potted-planting experiment was carried out, and the fertilizer with added ^{15}N as the tracer was added to soil. By examining the nitrogen contents in soil and crop samples during the wheat growth period, the fate of fertilizer nitrogen was analysed, and the potential pollution fertilizer left in soil was estimated. In order to analyse the fertilization and irrigation effects on nitrogen losses, a one-dimensional model, LEACHM (Leaching Estimation And CHemistry Model), was adapted to simulate the nitrogen transport in the agricultural soil profile. The results showed that soil nitrate nitrogen increased significantly after wheat planting, which may result in a high probability of nitrogen losses by runoff or leaching, and therefore excessive fertilization and irrigation may increase nitrogen losses as well as cause a high risk of nitrogen pollution.

Key words tracing experiment; non-point source pollution; agricultural fertilization; LEACHM model

INTRODUCTION

The application of fertilizer has accelerated the increase of crop yields in modern agriculture in recent decades. It was reported that more than 25% of the increase of crop yields is due to the use of fertilizer in the world (Shen, 1992; Wu, 1994). However, the utilization efficiency of fertilizer is rather low. Zeng (1995) pointed out that in China, the utilization efficiency of nitrogen fertilizer was only 30–35%; for phosphorus fertilizer it was about 10–20%. Therefore in order to achieve high crop yields and productivities, large quantities of chemical fertilizers are being put into agricultural fields, and most of them are lost into the environment through runoff, leaching, denitrification, absorption, etc. Leaching of chemicals below the root zone may contaminate groundwater for irrigation or drinking purposes, while excess accumulation of chemical fertilizer in surface water bodies may cause eutrophication problems and even health problems, rendering the water unfit for drinking, recreation, industry, or other common functions. Agricultural non-point source pollution has become the main source of water pollution in many countries (Novotny, 1999; Walton, 2000; Tang, 2003).

Although various studies have proved that chemical fertilizers may cause water pollution, it is difficult to quantify how much pollutant would be discharged into nearby water systems, groundwater or the atmosphere, because of many factors affecting the pollution processes. These factors include the amount and time of fertilizer application, ploughing method, climate condition, topology, etc. (Tsiouris *et al.*, 2002). In general, mathematical simulation is a cheap method to study agricultural pollution. Many numerical models, including one-dimensional (1-D) and 2-D models, have been developed and applied in different areas. Some physically-based models, such as LEACHM, HYDRUS, SWAT, AGNPS, BASINS, can simulate the transport and transformation processes of nutrients and pesticides, considering decomposition, mineralization, crop uptake, etc. Experiments are another means of finding the temporal and spatial rules of fertilizer pollution. Isotope tracers, such as ^{15}N, are good choices for studying the fate of fertilizer in farmland. Many agricultural scientists and hydrologists have used isotopic analysis to understand the fate of agrochemicals; as well as to monitor water quality and hydrochemical behaviour on a catchment scale. Rückauf *et al.* (2004) used the ^{15}N tracer technique to investigate the effect of different soil moisture conditions on nitrate turnover and removal in degraded fen soils. Tetzlaff *et al.* (2007)

collected tracer data, and combined with GIS to understand the catchment's hydrochemical response and hydrological processes. Annable *et al.* (2007) used isotopes, ^{15}N, ^{13}C and ^{37}Cl, to analyse the common agrochemicals for identifying non-point source agricultural contaminants. However, the cost of isotopic experiments on a catchment scale is very expensive, so a pot planting experiment is often used to do some theoretical research.

Nitrogen is a vital nutrient for enhancing crop growth and is also a major pollutant of non-point source pollution (Mehdi & Madramootoo, 1999). Nitrate and ammonia nitrogen are two dominant forms due to their high solubility. The main factors affecting nitrogen transport and transformation in farmland include precipitation (irrigation) intensity, soil type, topography condition, fertilization, plant coverage (Qiao, 2002; Bandaranayake & Arshad, 2006). Among these factors, fertilization and irrigation are what can be effectively controlled by human beings. Therefore agricultural pollution control should be focused on the management of fertilization and irrigation.

In order to quantify the fertilizer pollution in agricultural fields, the aim of the study presented here was to investigate the potential loads of nitrogen after crop planting and fertilization. For this purpose a wheat-potted planting experiment was performed to study the nitrogen transportation and transformation processes during a crop growing period using the ^{15}N tracer technique. Also a 1-D numerical model, LEACHM (Leaching Estimation And CHemistry Model), was used to simulate the transport and losses of nitrogen in the soil profile and the effects of different fertilization and irrigation conditions on nitrogen losses were analysed.

MATERIAL AND METHODS

Experiment description

The experiment of winter wheat-potted planting was carried out from October 2005 to June 2006 at Tsinghua University campus. The experiment pots were made of PVC, about 30 cm in diameter, 40 cm in height, and there was a bottom outlet of 2 cm diameter for each one. The experimental soil, a brown aluvial soil, sandy loam, was obtained from agricultural field in the upper basin of Miyun Reservoir in Beijing. The soil dry density was measured as 1.387 g cm^{-3}. Other chemical contents of initial mixed soil were also measured by the Institute of Soil and Fertilizer in Beijing Academy of Agriculture and Forestry Science, as listed in Table 1.

There were three treatments, ^{15}N fertilizer (the abundance of ^{15}N was 10%), common fertilizer and no fertilizer, and each treatment had three replications. The fertilizer was ammonium sulphate, applied twice during the experiment, once before sowing and then at the wheat jointing stage. The quantities of fertilizer used was 7 g (about 206.2 N kg ha^{-1}), and irrigation amount was about 300 mm. Soil samplings were taken four times during the experiment. The nitrogen contents and ^{15}N abundances of samples were examined and analysed.

Model description

In order to predict the effects of fertilization and irrigation on nitrogen pollution, some scenarios with different fertilization and irrigation patterns were designed, and water and solute movement

Table 1 Initial contents of the experimental soil.

Soil layer (cm)	Total nitrogen (g kg^{-1})	Total phosphorus (g kg^{-1})	NH_4^+-N (mg kg^{-1})	NO_3^--N (mg kg^{-1})	Organic matter (g kg^{-1})
Mixed soil	1.03	1.1	3.31	7.86	18.45
0~5	1.13	1.27	7.30	5.52	20.17
5~10	1.18	1.29	6.83	6.69	21.55
10~20	0.378	0.977	5.89	3.51	8.21
20~30	0.317	0.847	4.86	3.18	8.21

in soil were simulated by adopting a 1-D model, LEACHM (Leaching Estimation And Chemistry Model). This model was developed at Cornell University, USA (Hustson, 2003). The governing equation for water movement in soil is Richards equation, as in equation (1):

$$\frac{\partial\theta}{\partial t}=\frac{\partial}{\partial z}\left[k(\theta)\frac{\partial H}{\partial z}-U(z,t)\right] \quad (1)$$

where θ is volumetric water content (m^3 m^{-3}), H is hydraulic head (mm), K is hydraulic conductivity (mm day^{-1}), t is time (days), z is depth (mm), positive downwards; and U is a sink term, representing water lost per unit time by transpiration (day^{-1}).

Convection–dispersion equations (CDEs) are used in the model, as in equation (2), and the sink term of solute, sorption of solid phase, dispersion of volatile solute in gas phase are also considered.

$$\frac{\partial C_t}{\partial t}=-\frac{\partial J_s}{\partial z}\pm\phi \quad (2)$$

where C_t is the total solute concentration in all phases (liquid, gas, adsorbed) and Φ represents as yet unspecified sources or sinks of solute.

In order to obtain the soil characteristic and model parameters, such as grain-size distribution, initial water content, saturated hydraulic conductivity, breakthrough curve, etc. soil column experiments and some other laboratory tests were performed in combination with the pot planting experiment. Table 2 and Fig. 1 show some comparisons of simulated results and experiment examination, which indicate that the model simulation had a good agreement with the experiment. After model validation, three simulation scenarios were designed as below:

- Different fertilizer amount: by changing the additional fertilizer with 87.6 kg N ha^{-1}, 175.2 kg N ha^{-1} and 262.7 kg N ha^{-1};
- Different nitrogen content in base fertilizer: by changing the nitrogen content with 98 kg N ha^{-1}, 147 kg N ha^{-1} and 196 kg N ha^{-1} in base fertilizer;
- Different irrigation amount: by changing the irrigating water with 300 mm, 330 mm and 360 mm.

RESULTS AND DISCUSSION

Experimental results and analysis

By examining soil and crop samples, the nitrogen contents and distributions in the soil profile during the winter wheat growing period were obtained, and the spatial and temporal distribution of nitrogen in different forms was analysed. From these results the potential pollution loads in soil just before flood season could be estimated.

Table 2 Comparison of NH_4^--N and NO_3^--N between measured and simulated results (unit: mg kg^{-1}).

		Before fertilization		After fertilization	
	Soil layer (cm)	Measured	Simulated	Measured	Simulated
NH_4^--N	0–5	7.3	7.11	40.9	30
	5–10	6.83	7.17	21.7	20.8
	10–20	5.89	5.91	22	20.7
	20–30	4.86	5.39	16.3	22.5
NO_3^--N	0–5	5.52	7.4	101	78.7
	5–10	6.69	5.05	27.6	26.7
	10–20	3.51	3.53	11.9	13.4
	20–30	3.18	3.77	11.8	14.7

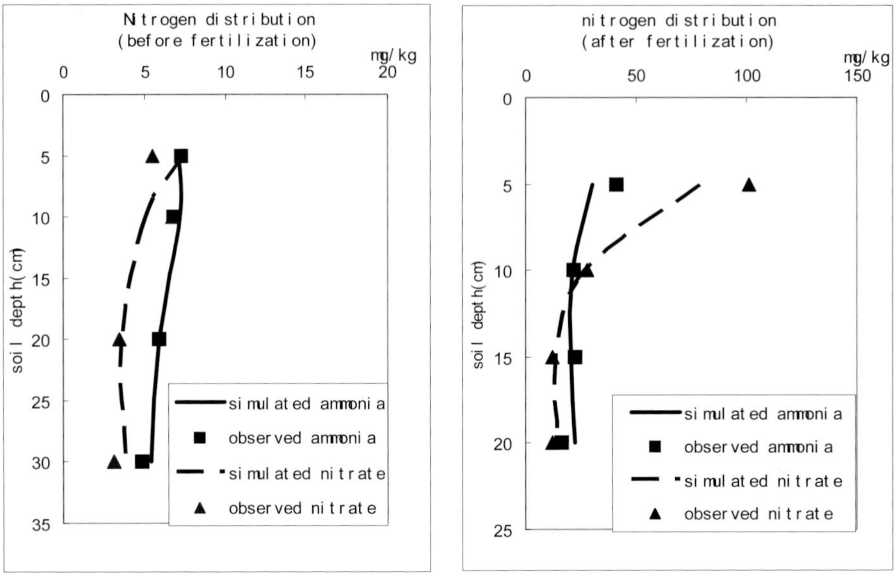

Fig. 1 Nitrogen distribution in soil profile (comparing between simulated and measured values).

Fate of fertilizer nitrogen

Assuming that ^{15}N and ^{14}N uptake by the crop keep a constant proportion, from the measured results of total nitrogen (i.e. TN) content and ^{15}N abundance in soil, the fate of fertilizer nitrogen was analysed. Table 3 shows the fertilizer nitrogen balance in this experiment. The averaged percentages of fertilizer left in soil after the planting period was about 87%; only 8.0–13.6% of fertilizer was taken up by crops. In this experiment there was no leaching because of limited irrigation, therefore the loss of fertilizer nitrogen in this case was by volatilization. Depending on the nitrogen balance analysis, the fertilizer nitrogen loss by volatilization was about 2.45%. For biological nitrogen in different parts of wheat at a mature stage, about 30% of nitrogen in wheat straw came from fertilizer, and in wheat head it was about 37%.

From this experiment it is known that the contribution of fertilizer nitrogen to seasonal crops is very low, and that the large part of fertilizer left in the soil may become potential pollution sources for surface water and groundwater. Because the nitrogen losses in runoff, flow and leaching were not considered in this experiment, the results only present the potential loads for agricultural water pollution.

The distribution of nitrogen in the soil profile

By sampling different layers of soil and examining the nitrogen contents, it is shown in Table 2 and Fig. 1 that both NH_4^--N and NO_3^--N in the soil profile have distributions that decrease with

Table 3 The fate of fertilizer nitrogen.

	Min value		Max value		Average value	
	(g)	(%)	(g)	(%)	(g)	(%)
Left in soil	1.285	86.561	1.292	87.046	1.289	86.803
Plant uptake	0.120	8.063	0.203	13.656	0.161	10.860
Loss	–	–	–	–	0.035*	2.337*
Sum	–	–	–	–	1.485	100

* The fertilizer loss was estimated by mass balance.

depths due to the root system activities and surface fertilization. Also it indicates that after fertilization, nitrogen change is more notable in the top 10 cm soil layer, which is generally the dominant layer for water and soil particles fully mixing under rainfall conditions. If surface runoff is generated, nitrogen in the top layer would be easily flushed into the water system. The model simulated results also had good agreement with measured values, as also shown in Fig. 1.

In general, the fertilizer nitrogen will be re-allocated in different ways after fertilization: some will be taken up by plants, some will be leached into groundwater, some will be washed into runoff, and some will remain in the soil. From the experiment results, it is clear that fertilization makes NO_3^--N increase quickly in the top layer, and then gradually be transported downwards. Due to nitrification and some other chemical reactions, at the end of the planting period the concentration of NO_3^--N in soil obviously increased. At the same time, because of the volatility and instability, the concentration of NH_4^--N and NO_2^--N experienced a small change, and the total nitrogen in the soil stayed at a similiar level to that before planting. The phenomenon implies that the organic nitrogen and hence soil fertility decreased, which may increase nitrogen pollution risk.

The transfer characteristics in different nitrogen forms

In general, the nitrogen in soil can be divided into organic and inorganic. The inorganic nitrogen has three major forms: NH_4^+-N, NO_3^--N and NO_2^--N. Each form can be transferred from one to another under microbal activities. Main chemical processes include decomposition and mineralization (from organic to inorganic), nitrification (NH_4^+-N > NO_2^--N > NO_3^--N) and denitrification (NO_3^--N > N_2O > N_2), and these processes are greatly affected by soil temperature and soil moisture.

In this experiment, some soil samples were taken from each pot and the nitrogen content tested (seen in Table 4). Figure 2 demonstrates the trends of nitrogen contents in soil during the planting period. The total nitrogen (TN) in soil after planting was nearly equal to the initial value; NH_4^+-N and NO_2^--N increased slightly (nitrite nitrogen is omitted in Fig. 2 because of its negligible quantity), while NO_3^--N obviously increased from 7.86 mg kg^{-1} to 66.2 mg kg^{-1} on average. The increase of inorganic nitrogen (which tended to be lost easily) and decreasing organic content may result in low soil productivity and high pollution risk. The possible reason for these changes might be the rising temperature and low soil water content in the main growth period from April to June. Under these conditions mineralization and nitrification actions in soil become stronger, and denitrification action is relatively weak. Therefore, at the end of a wheat planting period, also the beginning of the flood season, inorganic nitrogen concentration in the top soil will increase to a high value, which is easy to be desorbed from soil particles, and then to leach into groundwater by infiltration or to be flushed into surface water bodies by runoff.

Modelling results and analysis

Much previous research indicated that the amounts and patterns of fertilization and irrigation had significant effects on the loss of nitrogen in soil. In this paper the LEACHM model was applied to simulate these effects in an experiment. In such a 1-D vertical simulation, the surface runoff is forbidden, but leaching exists. The losses by leaching, transformation (it presents the nitrification action), and plant uptake are calculated and compared below.

Table 4 Averaged experimental results in different growing period.

Testing items	Total N (mg kg^{-1})	NH_4^--N (mg kg^{-1})	NO_3^--N (mg kg^{-1})	NO_2^--N (mg kg^{-1})
Before planting	1030	3.31	7.86	0.20
Before fertilization	1280	15.97	8.13	0.61
after fertilization(24h later)	1090	22.20	39.50	0.54
After harvesting	1020	12.83	66.20	0.41

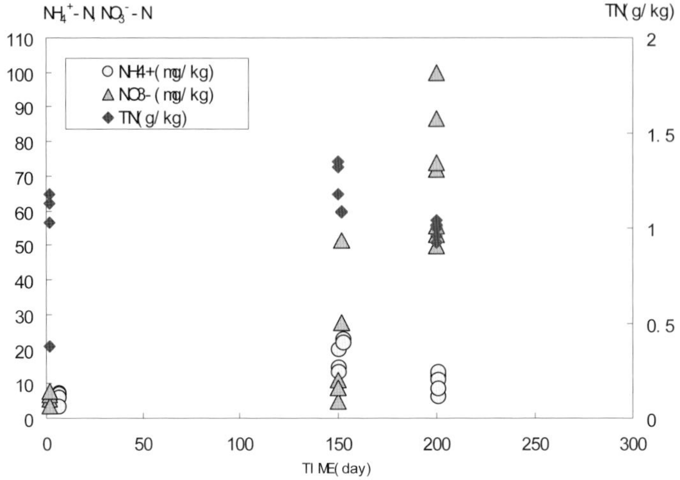

Fig. 2 Trends of nitrogen contents in soil during planting period.

Effects of fertilization level By changing levels of additional fertilizer to 87.6 kg N ha^{-1}, 175.2 kg N ha^{-1} and 262.7 kg N ha^{-1}, respectively, the changes of NH_4^--N and NO_3^--N in soil under different conditions are listed in Table 5. It is shown that both NH_4^--N and NO_3^--N losses by leaching increased by adding additional fertilizer. The amount of NO_3^--N leaching was more than NH_4^--N, which was also proved by other researchers (Qiao, 2004; Zhao, 2006). Losses by transformation of NH_4^--N also increased, while losses by plant uptake were little affected by fertilization when the common fertilization demand was met.

Effect of nitrogen content in base fertilizer The influence of the effect of nitrogen content in base fertilizer is shown in Table 6. Cases 1, 2, 3 represent the nitrogen content with 98 kg N ha^{-1}, 147 kg N ha^{-1} and 196 kg N ha^{-1}, respectively, in base fertilizer. It is indicated that increasing the nitrogen content in base fertilizer had a bigger influence on plant uptake of nitrogen, but fewer losses in drainage and via transformation occurred.

Effect of irrigation amount Irrigation has more notable influences on losses in drainage compared with fertilization. Cases 1, 2 and 3 represent irrigating with 300 mm, 330 mm and 360 mm water, respectively. In Table 7, the results imply that excess irrigation had little influence on plant nitrogen uptake, but caused more losses by drainage and by transformation. Some other studies also approved this conclusion (Tang, 2003). In this case the saving of irrigation water has a significant effect on environmental protection.

Table 5 The losses of NH_4^+-N and NO_3^--N in soil with different fertilization level.

	NH_4^+-N (kg ha^{-1})			NO_3^--N (kg ha^{-1})		
	Case 1	Case 2	Case 3	Case 1	Case 2	Case 3
leaching loss	9.7	11.7	14.2	27.2	28.5	33.6
transformation loss	46.7	53.2	64.8	0	0	0
plant uptake loss	5.4	5.4	5.4	3.5	3.5	3.5

Table 6 The losses of NH_4^+-N and NO_3^--N in soil with different nitrogen content in base fertilizer.

	NH_4^+-N (kg ha^{-1})			NO_3^--N (kg ha^{-1})		
	Case 1	Case 2	Case 3	Case 1	Case 2	Case 3
leaching loss	7.4	7.6	7.9	24.5	25.2	25.8
transformation loss	38.5	39.7	40.6	0	0	0
plant uptake loss	5.5	5.7	5.8	3.6	3.6	3.7

Table 7 the losses of NH_4^+-N and NO_3^--N in soil under different irrigation condition.

	NH_4^+-N (kg ha^{-1})			NO_3^--N (kg ha^{-1})		
	Case 1	Case 2	Case 3	Case 1	Case 2	Case 3
leaching loss	17.6	19.6	22.6	51.0	51.0	51.3
transformation loss	61.5	61.3	61.1	0	0	0
plant uptake loss	5.4	5.4	5.4	3.5	3.5	3.5

SUMMARY AND CONCLUSION

In this study, the tracing experiment shows that the fertilizer utilization ratio was very low, only <15% was taken up by wheat, about 2.35% lost by volatilization, and the majority, >80% of fertilizer applied, remained left in soil after the planting period. Under rainfall or irrigation conditions, the accumulated nitrogen in soil would become a potential source of pollution for groundwater and surface water bodies. Moreover, because chemical fertilizer was surface applied into the top 10 cm of the soil layer, the dominant layer for water and soil particles mixing fully under rainfall conditions, it might add the possibility of surface water pollution. Furthermore, the results indicated that inorganic nitrogen, especially nitrate nitrogen, increased after chemical fertilizer was applied onto farmland. The organic nitrogen content reduced, and resulted in a soil crust. This might also increase the risk of water and nutrient losses. Although the proportions in the different fates of fertilizer nitrogen will vary from the case concerned above, depending on fertilizer chemical content, soil moisture, soil temperature, etc. the risk of surface water and groundwater pollution in an agricultural area is higher than in a non-planting area.

From LEACHM model simulation results, it is shown that both excessive fertilization and irrigation increased nitrogen leaching and caused a greater threat to groundwater quality. Base fertilization had more effects on nitrogen plant uptake than additional fertilization, and consequently improving the fertilizer utilization efficiency and reducing nitrogen loss. Thereby water-saving irrigation and proper fertilization are effective practices of human control on water environmental pollution.

Acknowledgements This work has been supported by a grant from the National Natural Science Foundation of China (no. 50709016), and a grant from the Key Laboratory for River Dynamics and Hydraulic Engineering (no. Sklhse-2006-A-02). The support of samples analysis of the Institute of Atomic Energy Application, Chinese Academy of Agricultural Sciences and Institute of Plan t Nutrients and Resources, Beijing Academy of Agricultural Sciences is gratefully acknowledged.

REFERENCES

Almasri, Mohammad N. & Kaluarachchi, Jagath J. (2004) Assessment and management of long-term nitrate pollution of ground water in agriculture-dominated watersheds. *J. Hydrol.* **295**, 225–245.

Annable, W. K. (2007) ^{37}Cl, ^{15}N, ^{13}C isotopic analysis of common agro-chemicals for identifying non-point source agricultural contaminants *Appl. Geochem.* **22**, 1530–1536.

Bandaranayake, W. M. & Arshad, M. A. (2006) Soil profile distributions of water and solutes following frequent high water inputs. *Soil Till. Res.* **90**, 39–49.

Hustson, J. L. (2003) *LEACHM Computer Program and Documentation.* Cornell University, USA.

Mehdi, B. & Madramootoo, C. A. (1999) Soil nitrate distribution under grain and silage corn using three tillage practices on a loamy sand in southwestern Quebec. *Soil & Tillage Res.* **51**, 81–90.

Novotny, Vladimir (1999), Diffuse pollution from agriculture—a worldwide outlook. *Water Sci. & Technol.* **39**(3), 1–13.

Qiao, Guangjian, Zhang, Junling & Tang, Junzhi (2004) Mechanism of nitrogen pollution in groundwater and measures for its remediation. *Water Resour. Protection* **3**, 9–13.

Rückauf, Ulrike, Augustin, Jürgen, Russow, Rolf & Merbach Wolfgang (2004), Nitrate removal from drained and reflooded fen soils affected by soil N transformation processes and plant uptake. *Soil Biol. & Biochem.* **36**, 77–90.

Shen, Jingwen (1992) study on groundwater pollution by chemical fertilizer, pesticide and waste water irrigation. *Agro-environ. Protection* **11**(3), 137–139.

Tang, Lian, Bai, Dan, Jiang, Renfei & Liu, Zhenzhong (2003) Effect of agriculture non-point source pollution on groundwater pollution and control measures. *Res. Soil & Water Conservation* **10**(4), 212–214.

Tetzlaff, D. Waldron, S., Brewer, M. J. & Soulsby, C. (2007) Assessing nested hydrological and hydrochemical behaviour of a mesoscale catchment using continuous tracer data. *J. Hydrol.* **336**, 430–443.

Tsiouris, S. E., Mamolos, A. P., Kalburtji, K. L. & Alifrangis, D. (2002) The quality of runoff water collected from a wheat field margin in Greece. *Agric. Ecosys. Environ.* **89**, 117–125.

Walton, R. S., Volker, R. E., Bristow, K. L. & Smettem, K. R. J. (2000) Experimental examination of solute transport by surface runoff from low-angle slopes. *J. Hydrol.* **233**, 19–36.

Wu, Zhijie (1994) Chemical fertilizer and biosphere. *Agroenviron. Prot.* **13**(6), 279–282.

Xu, Qian (1996) A review on the status of non-point source pollution of chemical fertilizers and pesticides in China. *Rural Eco-Environment* **12**(2), 39–43.

Zeng, Xiankun (1995) Situation and outlook of fertilizer industry in China. *Acta Pedologica Sinica* **32**(2), 117–125.

Assessing the effects of different cropping patterns on drainage capability of the southern riverine basin

FENGYAN WU[1] & TIESONG HU[2]

1 *Hubei Water Resources Research Institute, Hubei 430070, China*
wufengyan1981@126.com

2 *State Key Laboratory of Water Resources and Hydropower Engineering Science of Wuhan University, Wuhan 430072, China*

Abstract Compared with natural watersheds, only a few studies have been conducted to take the specific characteristics of farmed watersheds into account in hydrological modelling. The purpose of this paper is to develop a conceptual rainfall–runoff model with which both the impact of human activities on flood events and the effects of different cropping patterns on drainage capability of the southern riverine basin could be assessed, especially the existence of drainage practices and ditch networks. Applications of the model are demonstrated for Luoshan drainage basin, which is located in Hubei Province. To analyse the role of cropping patterns, six hypothetical scenarios are discussed. Results show that the drainage system of Luoshan drainage basin can dispose of the excess water from 3-day rainfall totals, with return periods of 6.9 years in five days and with the area of paddy fields being increased, the drainage capability would be strengthened.

Key words cropping patterns; drainage capability; rainfall–runoff model; Luoshan

INTRODUCTION

Most runoff simulation models have been developed at the global-catchment scale or at the local-soil column scale. Few models have been specifically designed for agricultural fields. Compared with natural catchments, farmed catchments, such as irrigation areas and drainage areas, have very different hydrological processes. On farmed catchments, human activities such as agricultural land use, management practices, and crop stages can be significant factors in controlling flood generation. In addition, other factors, including the operating regulations of drainage sluices and pumps and the ditch network, also influence flood routing. These factors must be taken into account in hydrological modelling.

The existing models for simulating hydrological processes in agricultural catchments can be classified into three groups: physically based models (Dunn & Mackay, 1996; Moussa *et al.*, 2002; Jia *et al.*, 2005), conceptual models (Xia & Liu, 1995; Castro *et al.*, 1999), and empirical relations (Chahinian *et al.*, 2005). These models provide a good platform for this study. In this paper, a conceptual hydrological model is designed which takes the effects of crop growth, management practices, and internal drainage systems on hydrological processes into account. It is therefore suitable for evaluating the effects of cropping pattern changes on drainage capability of a large-scale drainage basin.

Model description

A large-scale drainage basin generally consists of a field drainage system and a main drainage system. Internal drainage pumping stations, field ditches and diversion ditches are the part of the field drainage system that first picks up water from the low-lying regions and then removes it to sub-ditches. The main drainage system consists of lakes, main ditches, sub-ditches, exterior drainage pumping stations, and drainage sluices, etc. The function of it is to route water through the basin to the outlet.

Drainage basins can be subdivided into several sub-basins according to the geographical information and the characteristics of the drainage system. The modelling starts with simulating and routing the runoff through the sub-basin, and then simulates the discharge of internal drainage pumping stations according to need. The results are input into the hydrodynamic model, which is adopted to transport flow through the main drainage system as upstream boundary conditions or lateral inflow. A general description of each procedure is given.

Simulation of runoff generation

Since different land uses correspond to different hydrological characteristics, different methods are therefore used. Due to the actual land uses in the study area, four different methods are adopted here.

The water budget on paddy fields is simulated by the monolayer soil-structure Tank model with a surface water layer (Fig. 1). The paddy vertical structure is divided into two layers: the surface water layer and the soil layer. In the surface water layer, the paddy storage depth *HW* in surface water layer rises with rainfall *P* and diminishes with actual evaporation *E*, surface drainage *DW* and infiltration to the soil layer *FW*. *DW* is a function of *HW* and drainage threshold *OH*, which is equal to the highest permissible depth, a function of breeds and stage of growth. *FW* is considered as a constant. A conceptual TANK model is employed to simulate the runoff in the soil layer. The model has two lateral holes and one bottom hole, which represent the outlet of surface runoff, interflow and groundwater runoff, respectively. The storage depth of soil layer *HS* rises with *FW* and diminishes with evapotranspiration *ET*, surface runoff *RS*, interflow *RI*, and groundwater flow *FS*. The maximum tensional storage of soil layer *WM*, controlling coefficient of surface runoff H_1 and controlling coefficient of interflow H_2 are parameters that represent the volume of groundwater, surface runoff and interflow, respectively. The outflow coefficient of surface runoff *KS*, interflow *KI*, and groundwater runoff *KG* are parameters that greatly influence the shape of the stream line.

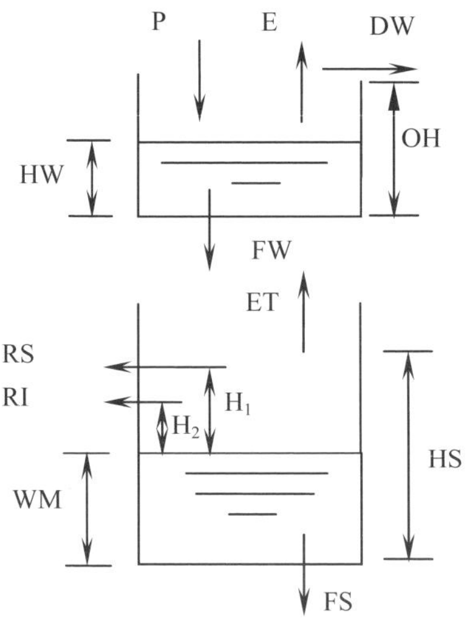

Fig. 1 Vertical water budget of paddy fields.

In this study, QualHYMO modelling is used to simulate the runoff of dry lands and grass lands. QualHYMO (1985) is a continuous hydrological model developed from HYMO by Williams & Hahn (1973), and is especially suitable for rural basins undergoing urbanization or agricultural areas. A detailed discussion of the model can be found in Caterina *et al.* (2001). The model has four calibration parameters, i.e. $S_{\min}$, $S_{\max}$, S_k and API_k. $S_{\min}$ and $S_{\max}$ represent the range of the maximum catchment retention. S_k is a calibration parameter in the API method. API_k is the antecedent precipitation index constant.

Special land uses such as lakes, polders or retention areas are simulated by a single reservoir. Once the water level of the reservoir exceeds the maximum level, all the rainfall is transformed into surface runoff, otherwise no runoff is generated. Evaporation is taken out from the water volume of that reservoir at the beginning. For other impermeable areas such as urban areas, runoff is generated immediately, as long as rainfall rate is greater than evaporation rate. Finally, the total runoff from the sub-basin is the weighted sum, as a function of their respective areas.

The Williams unit hydrograph is used as the transfer function to route surface runoff and interflow to sub-ditches or internal drainage pumping stations in this study. The time to peak t_p and the recession constant x_k are the two parameters to be calibrated. As for the groundwater runoff, a linear reservoir model is adopted to calculate its flow, and the recession coefficient kgg is a calibration parameter.

Discharge simulation of internal drainage pumping stations

The spatial distribution of internal drainage pumping stations is dotted casually, each of them operating independently of the other. For the convenience of the study, based on practice the following assumptions are adopted: (a) all the internal drainage pumping stations in one sub-basin are generalized as one internal drainage pumping station; (b) the discharge of the generalized internal drainage pumping station is the bilinear function of volume of waterlogging. Based on the hypothesis above, at time t, the discharge of the internal drainage pumping station QO_t can be described as:

$$QO_t = QI_t - dV_t/dt = \begin{cases} \sigma V_{\max} & V_t < V_{\max} \\ Q_{\max} & V_t \geq V_{\max} \end{cases} \tag{1}$$

where QI_t represents the inflow output from the runoff generation models. V_t is the volume of waterlogging; σ is the discharge constant, a parameter to be calibrated; $V_{\max}$ is the threshold of the volume of waterlogging, when the actual volume of waterlogging is greater than the threshold, the drainage capability of the internal drainage pumping station reaches its maximum.

Numerical simulation of flow-routing

The flow conditions in a drainage system vary with time, which can be determined by proper hydrodynamic models. In this study, St Venant equations are employed for numerical simulation of one-dimensional (1-D) unsteady open channel flow. The 4-point operators Preissmann implicit finite difference scheme is used here. While St Venant equations are inapplicable to flow conditions since as flow through sluices and pumping stations, special hydraulic equations replaced by the corresponding Taylor's expansions when doing numerical simulation are used; they are called initial boundary conditions. The discharge of a sluice is a function of the number of opened gates, the gate opening, and the upstream and downstream water level. Pumping stations play an important role in reducing the water levels of the main ditch. The discharge of an exterior drainage pumping station is a function of the operating units and lift head. For an intricate network of drainage ditches, there must be many junctions. At each junction, the continuity equation and the dynamic equation should be met.

St Venant equations and junction equations with proper initial and boundary conditions constitute the mathematical model for unsteady flow in the unsteady open-channel network.

APPLICATION

Description of the study area

Luoshan drainage basin is located in Sihu region of Hubei province. The 935-km^2 drainage basin is the lowest region in Jianli County. There are three precipitation stations: Luoshan Station, Chiba Station and Zhuhe Station. The drainage system is formed by man-made sub-ditches, one main ditch named Luoshan Ditch, two exterior drainage pumping stations, two generally closed drainage sluices, and lots of internal drainage pumping stations (Fig. 2). The total design discharges of the two exterior drainage pumping stations are 210 $m^3 s^{-1}$. The total design discharges of internal drainage pumping stations are 126.6 $m^3 s^{-1}$ and the total installed capacities are 13 495 kW, serving a total area of about 378.8 km^2.

Along with the development of the economy and the change of policy, the area of paddy fields increased at the expense of dry lands. The planting area of medium rice also increased at the expense of early and late-season rice. In order to answer how cropping patterns affect the drainage capability of the Luoshan drainage basin, six scenarios were designed, based on data in hand, which can be divided into two sets (Table 1). In Set A, how the drainage capability changes with the change of proportion of the area of paddy fields to the area of dry lands is analysed when all the other factors remain unchanged. In Set B, how the drainage capability changes with the change of the planting area of the medium rice is studied, still supposing all the other factors remain unchanged. At present, the area of paddy fields accounts for 58.7%, and the area of dry lands accounts for 41.3% of the total arable area. The planting area of early-season rice, medium rice and late-season rice is 16.3%, 71.6% and 28.4%, respectively. Therefore, Scenario A2 and B2 correspond to the present.

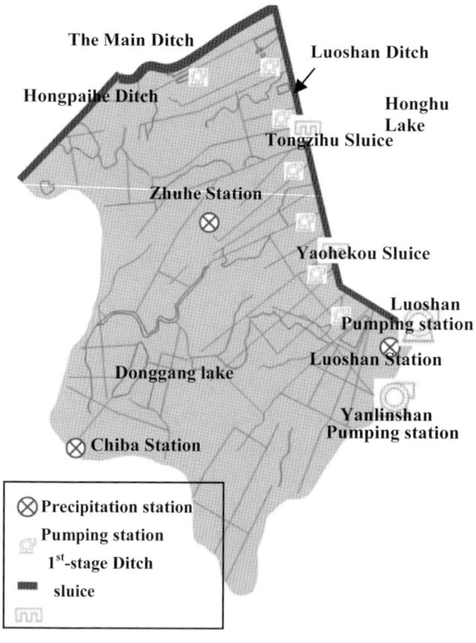

Fig. 2 Diagram of Luoshan drainage basin system.

Table 1 Six Scenarios of cropping patterns (%).

Scenario	Set A paddy fields	dry lands	Scenario	Set B Medium rice	Early-season rice	Late-season rice
A1	20.0	80.0	B1	30.0	59.0	70.0
A2	58.7	41.3	B2	71.6	16.3	28.4
A3	80.0	20.0	B3	100.0	0.0	0.0

Model parameterization and calibration

The model contains many parameters, which reflect geometrical and soil characteristics, land use, etc. Some of them can be obtained by field observations, and the others need to be calibrated. In this study, 14 parameters are calibrated. The adaptive genetic algorithm developed by Zhang (Zhang & Hu, 2002) is used for optimization algorithms. In the algorithms, genetic parameters are regulated adaptively based on population diversity, which is determined according to the similarity of chromosomes. The RMSE between observed and simulated upstream water levels of the Luoshan pumping station and the Yanlinshan pumping station is used as the objective function. The input data is daily hydrological and meteorological data extracted from 1979 to 2004, the water budget model for runoff generation runs with a daily time interval, and the flow routing model runs with a 3-hour time interval.

Due to the absence of detailed measurements of the drainage basin, there is no consistent way to verify the values of the parameters, but the ranges of parameter values can be obtained by analysing the spatial properties and data gathered. The parameter WM represents the maximum tensional storage, and its value is <100 mm and >40 mm. Through the observation in paddy fields, H_1 and H_2 are generally less than 20 mm. The ranges of the parameters S_{min} and S_{max} can be estimated from the SCS curve number. The parameters t_p and x_k define the shape of the Williams unit hydrograph, therefore their ranges can be analysed from typical flood events. Parameter σ represents the relationship between water levels and the discharges of the internal drainage pumping station. By analysing this relationship, the range of its value can be determined. Based on related literature, the other ranges of parameter values are set. Only the value within the range is accepted. The results are listed in Table 2. The Nash-Sutcliffe coefficient used to evaluate model efficiency is generally superior to 0.75, reflecting a relatively good agreement between observed and calculated hydrographs.

Table 2 Calibrated values of parameters.

name	WM (mm)	KS	KI	KG	H_1 (mm)	H_2 (mm)	S_{min} (mm)
value	50	0.32	0.24	0.09	12	8	10
name	S_{max} (mm)	S_k	API_k	t_p	x_k	kgg	σ
value	70	0.02	0.9	28	20	0.96	0.35

RESULTS AND DISCUSSION

Effects of cropping patterns on runoff generation

Criteria for waterlogging control are adopted to reflect the drainage capability of a drainage basin in China. For most Chinese large-scale drainage basins it is a common practice to express the criterion for waterlogging control in draining off all the excess water from 3-day rainfall totals with return periods of certain years in five days for dry lands, or reducing water levels of paddy fields to the depth of submergence tolerance of crops in five days. Therefore, return periods of 3-day rainfall totals were used in this study.

Most of the water in the surface drainage system originates from surface runoff and interflow, therefore the sum of the two kinds of runoff from 3-day rainfall totals with different return periods is analysed in detail. The sum of the two kinds of runoff of 14 3-day rainfall events for the six scenarios are shown in Fig. 3. For the same rainfall event, Scenario A1 produces the highest runoff values and Scenario A3 produces the lowest values. Compared with the present, if the area of paddy fields is decreased by 65.9%, the runoff increases between 2.5% and 26.6% (average 10.5%); if the area of paddy fields is increased by 36.3%, the runoff can be reduced by 5.6% on average. The results indicate that the increase of the area of paddy fields can decrease the runoff. This is due to the flood-retarding affect of the paddy fields. The comparison of the results of Scenario B1 with Scenario B3 in Fig. 3(b) shows that the increase of the planting area of medium rice does not always result in the decrease of the runoff, although for most events this is true. The reason is that the paddy retarding capacity not only relates to the breed, but is also subject to the stages of growth.

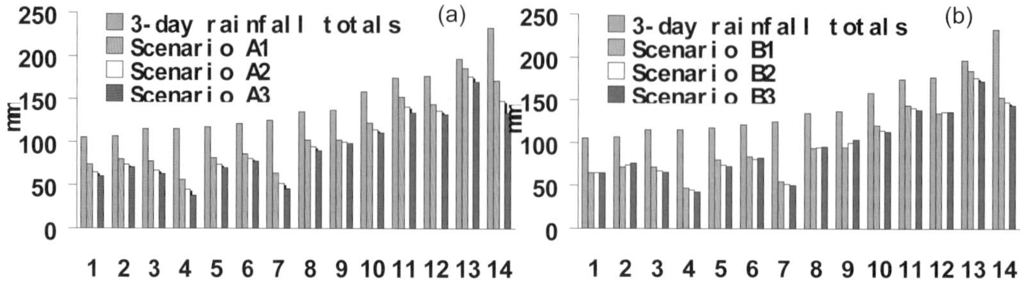

Fig. 3 Comparison of simulated runoffs for the two scenario sets: (a) Scenario Set A and (b) Scenario Set B.

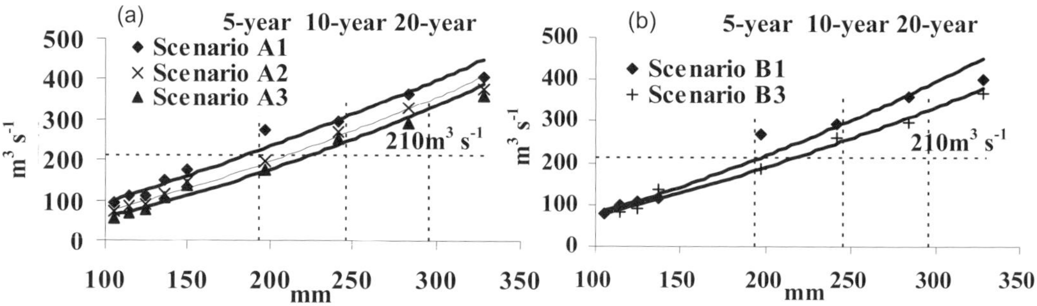

Fig. 4 Relationship between 3-day rainfall totals and required design discharge: (a) Scenario Set A and (b) Scenario Set B.

Effects of cropping patterns on drainage capability

The relationship between 3-day rainfall totals and the required design discharge to control the excess water under the depth of submergence tolerance of crops, or drain off all the excess water in five days for Scenario Set A, is shown in Fig. 4(a). The corresponding results for Scenario Set B are shown in Fig. 4(b).

As shown in Fig. 4(a), in the present conditions, the design discharge of 210 $m^3 s^{-1}$ can only make the volume of waterlogging from 3-day rainfall totals with return periods of 6.9 years under control. Since the result is in accordance with the real situation, it further verifies that the simulation model is effective. For a 10-year return period, the required design discharge should be increased to 259 $m^3 s^{-1}$ and for a 20-year return period, the required design discharge should be 337 $m^3 s^{-1}$. For the same storm event, the required design discharge of Scenario A1 is the highest and the required design discharge of Scenario A3 is the lowest. It indicates that with the increase of the area of paddy fields, the drainage capability of Luoshan drainage basin is enhanced. Increasing the area of paddy fields from 20% to 80%, the drainage capability can be improved from return periods of 4.3 years to return periods of 7.7 years. This is mainly because of the expanding of paddy retarding capacity with the increase of the area of paddy fields.

From Fig. 4(b) it can be seen that for the same storm event, the required design discharge of Scenario B3 is always lower than Scenario B1. The results show that the increase of planting area of medium rice can also improve the drainage capability. The conclusion is inconsistent with the results in Fig. 3(b). This is because the results in Fig. 4(b) are results under the most unfavorable conditions, and the storm event generally occurs in July when the paddy retarding capacity of Scenario B3 is larger than that of the Scenario B1.

CONCLUSIONS

In this paper, a conceptual hydrological model is developed to simulate the hydrological processes of large-scale drainage basins. The model features a simple structure, an explicit concept, and unambiguous parameters, and it can be applied to a wide range of watersheds. Aiming at different land uses, the model adopts different simulating methods, and takes the influence of factors such as the growth, sowing and harvesting on runoff generation into account when modelling paddy fields. As a result, the model is capable of accessing the effects of different cropping patterns on runoff. When conducting the simulation of flow-routing through the network, many human activities, such as the operating regulation of sluices and pumping stations, are taken into account; therefore the effects of cropping pattern changes on drainage capability can be evaluated effectively.

The results demonstrate that the drainage system of Luoshan drainage basin can dispose of the volume of waterlogging from 3-day rainfall totals with return periods of 6.9 years in five days. Increasing the area of paddy fields by 36.6% can improve the drainage capability to return periods of 7.7 years. Increasing the planting area of medium rice can also improve the drainage capability according to the simulation results.

The drainage capability is subject to many factors, but only one factor is analysed in this study, and the other factors remain unchanged. Our future research will extend to cover more factors.

Acknowledgements The authors gratefully acknowledge the helpful comments and suggestions from the anonymous reviewers. This research was partially funded by the Natural Science Foundation of China (Grant no. 50739003).

REFERENCES

Caterina, V., ASCE (2001) Hortonian and variable source area modelling in urbanizing basins. *J. Hydrol. Engng* **8**, 328–335.

Chahinian, N., Moussa, R., Andrieux, P. & Voltz, M. (2005) Comparison of infiltration models to simulate flood events at the field scale. *J. Hydrol.* **306**, 91–214.

Dos Reis Castro, N. M., Auzet, A. V. & Chevallier, P., (1999) Land use change effects on runoff and erosion from plot to catchment scale on the basaltic plateau Southern Brazil. *Hydrol. Processes* **13**, 1621–1628.

Dunn S., M. & Mackay, R. (1996) Modelling the hydrological impacts of open ditch drainage. *J. Hydrol.* **179**, 37–66.

Jia, Y. W., Kinouchi, T. & Yoshitani, J. (2005) Distributed hydrologic mode in a partially urbanized watershed using water and energy transfer process model. *J. Hydrol. Engng* **10**(4), 253–263.

Moussa R., Voltz, M. & Andrieux, P. (2002) Effects of the spatial organization of agricultural management on the hydrological behaviour of a farmed catchment during flood events. *Hydrol. Processes* **16**, 393–412.

QualHYMO (1985) *Ontario Ministry of the Environment (OMOE)*, manual. Environment Canada. Ottawa, Ontario, Canada.

Williams, J. R. & Hahn, R. W. Jr. (1973) HYMO: problem-oriented computer language for hydrologic modelling users manual. *Rep. ARSS-9,* Agricultural Research Service, US Department of Agriculture, Washington, DC, USA.

Xia, Jun & Liu, D. P. (1995) A hydrological system model applied to water transformation and water resources assessment for drainage experiment catchment in plain area in Hubei province. *Shuilixuebao* **11**, 46–55.

Zhang, J. H. & Hu, T. S. (2002) Adaptive genetic algorithm based on population diversity. *J. Computer Engng & Appl.* **38**(9), 49–51.

Possible impact of aerosols on orographic precipitation in eastern China

HANBO YANG[1], DAWEN YANG[1], PINGYUN LI[2] & YAFEI WANG[2]

1 *State Key Laboratory of Hydro-Science and Engineering & Department of Hydraulic Engineering, Tsinghua University, Beijing 100084, China*
yhb99@mails.tsinghua.edu.cn

2 *State Key Laboratory of Severe Weather, Chinese Academy of Meteorological Sciences, Beijing 100081, China*

Abstract To analyse the possible suppression by aerosols of orographic precipitation, three catchments were chosen in eastern China. The suppression was quantified as change in the orographic precipitation enhancement factor (*R*o), defined as the ratio between the hilly and the plains precipitation. The results show a significant downward trend of *R*o in the past 50 years, accompanied by a decrease in visibility (indicating an increase in aerosols concentration). During the period from 1980 to 2002, the three catchments had a decrease in *R*o, with a rate of 22–42% and an increase in the aerosols concentration, with a rate of approx. 0.01–0.02 mg/m^3 (10–30%). It is speculated that the possible cause of *R*o decreasing is the increase in aerosols concentration, and this implies the loss of significant water resources in hilly areas, which may be a reason for aggravation of the water resources shortage, especially in north China and northeast China.

Key words orographic precipitation; enhancement factor; aerosols; visibility

INTRODUCTION

It was recognized 50 years ago, through laboratory experiments in a giant expansion chamber (Gunn & Philips, 1957), that air pollution resulted in more and smaller cloud droplets and suppressed precipitation. Recent studies confirmed that a large quantity of aerosol particles from burning biomass and industrialization had been released into the atmosphere, serving as cloud condensation nuclei (CCN) to initiate cloud formation, and hence participate in the precipitation process. More aerosols result in more, but smaller cloud drops (Ackerman *et al.*, 2000), which suppress the drop-coalescence and warm-rain processes as well as ice precipitation (Rosenfeld, 1999, 2000; Borys *et al.*, 2003; Andreae *et al.*, 2004) and so prolong the time required to convert the small drops into large drops that can precipitate. If the time is longer than the lifetime of cloud drops, less precipitation will occur. Short-lived cloud elements typically form as air rises over topographic barriers, and then evaporate when forced down across the ridge line (Rosenfeld *et al.*, 2007).

To quantify the effect of the orographic component of precipitation, the orographic enhancement factor (*R*o) was introduced, where *R*o is defined as the ratio between the precipitation amounts on the hills and on the upwind plains (Givati & Rosenfeld, 2004). Following this, the effect of aerosols on precipitation was quantified in different countries. Givati & Rosenfeld (2005) analysed the rainfall amounts in northern Israel, and their results show that for the whole period of 1980–2002 *R*o decreased by 15%. Jirak & Cotton (2006) found that along the front range of the Rocky Mountains *R*o had decreased by approx. 30% over the past half-century due to the effect of pollution. In China, Rosenfeld *et al.* (2007) examined *R*o in Mt Hua in the centre of China, and suggested that particulate air pollution caused a decreasing trend of 10–25%.

The historical records of aerosols generally have a very short series, especially in China. Rosenfeld *et al.* (2007) hypothesized that reduced visibility indicated the presence of more aerosols. Yang *et al.* (2007) found that the decrease of visibility in Jinan resulted from the combined influence of PM2.5, PM10 and meteorological conditions. Liang & Xia (2005) proposed that the rapid increase in aerosol loading resulted in visibility deteriorating heavily in eastern China. Analysis of the data of visibility and meteorological variables in Chongqing, China, suggested that the decrease of horizontal visible distance was dominated by the increase of airborne pollution (Peng & Chen, 1992). In general, visibility can be a good indicator of the aerosols concentration in the atmosphere.

The impact of air pollution on orographic precipitation can be quantified with the change ratio of *R*o. Three catchments were chosen for this study, and the objectives were: (1) to analyse the change trends in *R*o and the visibility in the three catchments; (2) to examine the relationship of *R*o with the visibility; and (3) to quantify the impact of aerosols on the orographic precipitation, based on the regression equation between the aerosols concentration and the visibility.

STUDY AREAS, DATA AND METHOD

In the eastern part of China, a rapid growth of industrialization and urbanization, accompanied by a large amount of fossil fuel consumption, induces more aerosol release into the atmosphere. To analyse the possible impact of aerosols on precipitation and water resources, following the method proposed by Givati & Rosenfeld (2004, 2005), three catchments with raingauges in hilly and plain areas were chosen, as well as three stations having observations on visibility and concentration of PM10 (the particle diameter <10 μm) (see the locations in Fig. 1). Three catchments were located from the north to the south in the eastern part of China: Catchment 1 belonging to the Heilongjiang River basin in northeast China, Catchment 2 located in the Hai River basin in north China, and Catchment 3 situated in southeast China. Every catchment has raingauges located in both hilly and plain areas; there are 13 raingauges in total (see Table 1 for more details, data released by the National Meteorological Administration of China). These gauges were sorted into hilly and plain raingauges according to the elevation and topography on the basis of DEM (Digital Elevation Model). The daily precipitation at the 13 gauges was collected from 1961 to 2006.

Fig. 1 Locations of the three catchments chosen in the eastern part of China in this study; the stars represent the adjacent stations with the observation of the concentration of PM10 and visibility.

Table 1 Locations and details of the plains and hilly raingauges in the three catchments.

No.	Hilly stations	Latitude (°)	Longitude (°)	Elevation (m)	Plain stations	Latitude (°)	Longitude (°)	Elevation (m)
1	54287	42.02	128.08	2623	54284	42.10	127.57	774
					54285	42.53	128.25	591
					54276	42.35	126.82	549
2	53588	39.03	113.53	2896	53698	38.03	114.42	81
	53673	38.73	112.72	828	53798	37.07	114.50	77
	53782	37.85	113.55	742				
3	58927	25.10	117.03	342	59126	24.50	117.65	29
					59134	24.48	118.07	139
					59321	23.78	117.50	54

This study also used data collected during 6-hours observation of visibility (in kilometres) at 574 stations from 1980 to 2002 over mainland China, as well as the weather phenomena, and both were released by the National Meteorological Administration of China. If there was fog, precipitation or other special weather phenomena during the observations, the visibility data was excluded from the series. Daily and annual mean of visibility were then both calculated for each station.

For each gauge, annual precipitation was obtained as the sum of daily precipitation. Further, the arithmetic mean of annual precipitation was calculated for the hilly and the plains gauges for each catchment. Then *Ro* was taken as the ratio between the mean precipitation in the hilly areas and that in the plains. The arithmetic mean of visibility was also calculated for each catchment. The relationship between *Ro* and visibility was analysed by linear regression. The relationships of the concentration of PM10 with the visibility in the three stations were also examined by regression analysis on the basis of daily data from 2001 to 2002.

RESULTS AND DISCUSSION

As shown in Fig. 2, all three catchments have significant decrease trends ($p < 0.10$) in annual precipitation in the hilly areas, while Catchments 1 and 2 have slight decreases in annual precipitation and Catchment 3 has a significant increase ($p = 0.01$) in the plains during the period from 1961 to 2006. It shows that annual precipitation in the plains has a downward trend in the northern part, but an upward trend in the southern part; at the same time the precipitation over the hilly areas has a common decrease trend in both northern and southern areas. Additionally, a decrease trend in *Ro* has been detected for the two catchments at a level of $p < 0.01$, except

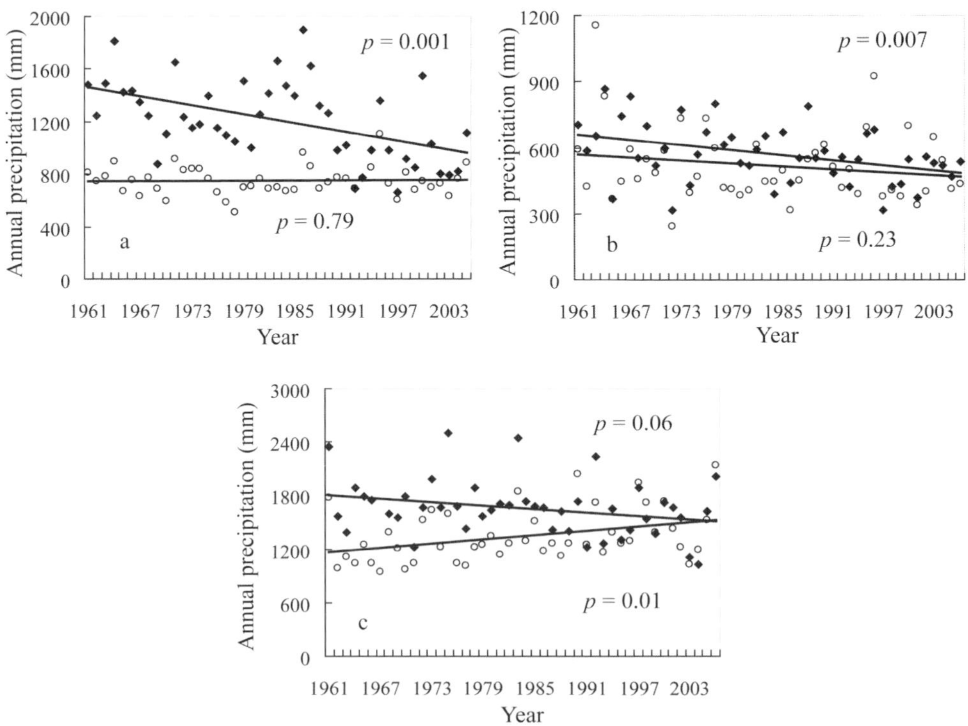

Fig. 2 Trends in annual precipitation of hilly areas and plains in (a) Catchment 1, (b) Catchment 2, and (c) Catchment 3 (○ representing the plains precipitation, and ♦ representing the hills precipitation). The trends are given as two solid lines within each plot. The probability (p) for the lines are given for each plot.

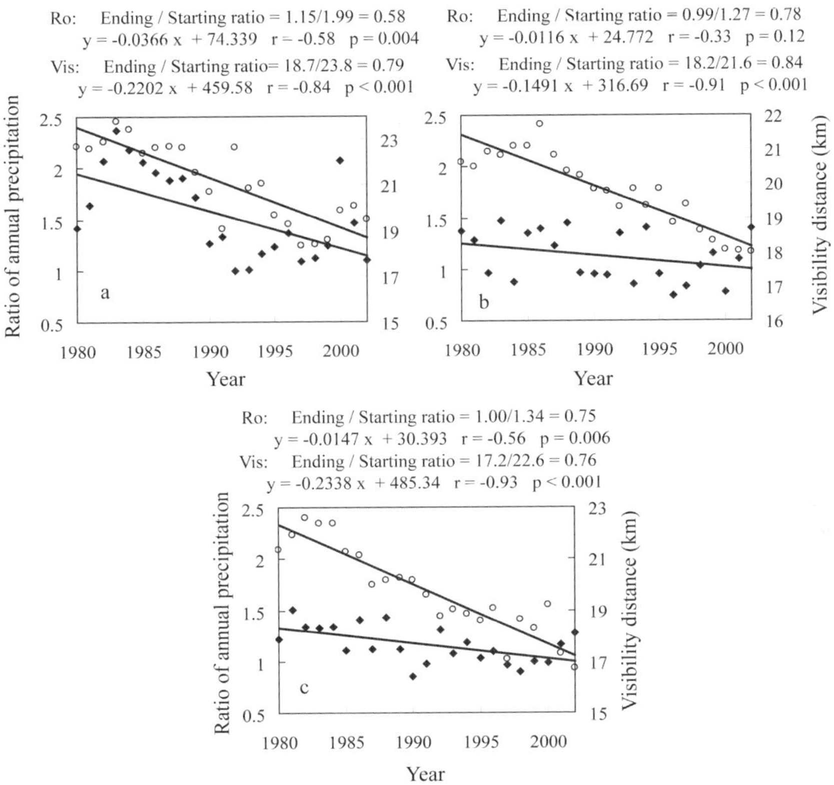

Fig. 3 Trends in *Ro* and visibility in (a) Catchment 1, (b) Catchment 2, and (c) Catchment 3 (○ representing the *Ro*, and ♦ representing the visibility). The trends are given as solid lines within each plot. The regression equations for the lines are given above the top of each plot, as well as the relation coefficient (r) and probability (p).

Catchment 2 at a significant level of $p = 0.12$ (see Fig. 3). As shown in Fig. 3, during the period from 1980 to 2002, *Ro* has an ending/starting ratio of 0.58–0.78; i.e. the hilly areas lost 22–42% precipitation, i.e. 42% in Catchment 1 in northeast China, 22% in Catchment 2 in north China, and 25% in Catchment 3 in southeast China. It implies a loss of water resources, and aggravates the water resources shortage in the arid and semi-arid regions. Similar phenomena were found in the other countries of the world, such as Israel, about 6.5% of the national water consumption was lost due to the suppression of aerosols on orographic precipitation (Givati & Rosenfeld, 2007).

Simultaneously all the catchments have significant decrease trends in visibility at a significance level of $p < 0.001$, with an ending/starting value of 0.76–0.84. The possible cause was speculated to be more aerosols released into the atmosphere due to a rapid growth of industrialization and urbanization in these regions. The relationship between *Ro* and visibility was further examined using regression analysis during the period from 1980 to 2002. The results, as listed in Table 2, show a positive correlation with a correlation coefficient of 0.28–0.66 and a significant level of $p < 0.01$ in two of the catchments, and another with a significant level of $p = 0.20$.

Table 2 Correlation between Ro and the visibility.

Catchment	Correlation coefficient (r)	Significance level (p)
1	0.66	<0.001
2	0.28	0.20
3	0.51	0.01

It can be speculated that the increasing aerosols concentration leads to the decline in visibility, and induces the downward trend in *Ro*. The relationship of aerosols concentration with visibility is shown in Fig. 4. A significant linear correlation ($p < 0.001$) can be found between the concentration of PM10 and the logarithm of the visibility in the three stations. We also assumed that the relationship in the stations can represent the relationship in the adjacent catchment (see Fig. 1). According to the linear regression equation of the concentration of PM10 with the visibility, we can estimate the aerosols concentration in every catchment (see Table 3). The results show that the increase of PM10 concentration is 0.018 mg/m^3 (26.9%) in Catchment 1, 0.016 mg/m^3 (10.1%) in Catchment 2, and 0.011 mg/m^3 (14.1%) in Catchment 3 during the period from 1980 to 2002, corresponding to the decrease rates of *Ro* being 42%, 22% and 25%, respectively. It suggests that *Ro* is sensitive to the aerosols concentration. This was also found by Borys *et al.* (2003), that an addition of 0.001 mg/m^3 of sulphate aerosols to a clean background can decrease the orographic snowfall rate by up to 50% in the Colorado Rocky Mountains.

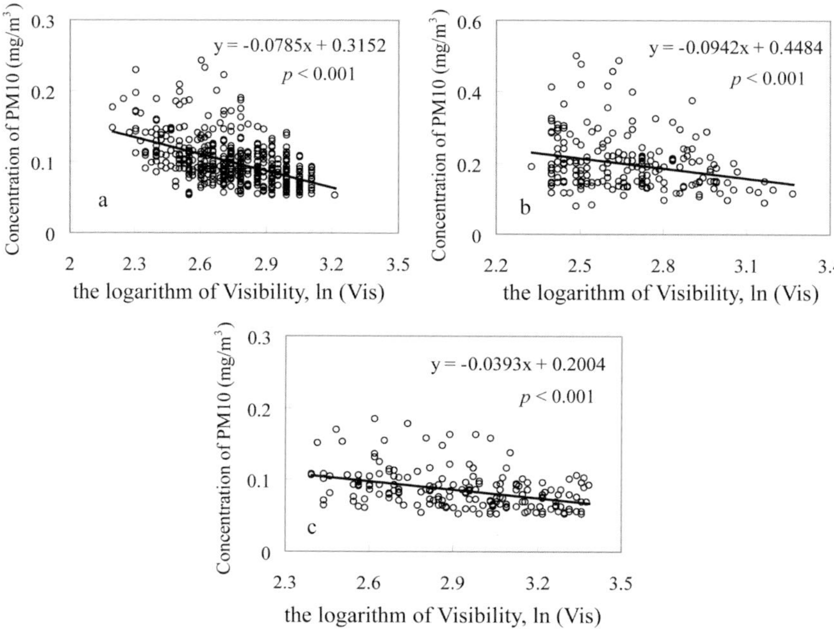

Fig. 4 Relationship between the concentration of PM10 and the logarithm of the visibility. The regression equations for the lines are given, as well as the probability (p).

Table 3 Changes in aerosols (PM10) concentration estimated by regression equation in Fig. 4 for each catchment (the visibility calculated by the regression equation in Fig. 3).

Catchment	PM10 in 1980 (mg/m^3)	PM10 in 2002 (mg/m^3)
1	0.067	0.085
2	0.159	0.175
3	0.078	0.089

Besides the impact on the orographic precipitation, aerosols absorb sunlight and heat air, which lead to higher atmospheric stability (Ramanathan *et al.*, 2005). This results in the depression of the upward motions of the atmosphere, which can reduce cloud formation, and further suppress precipitation and it has been shown in eastern central China (Zhao *et al.*, 2006).

CONCLUSION

In the eastern part of China, the decreasing visibility is possibly as a result of many aerosols released into the atmosphere due to a rapid growth in industrialization and urbanization, accompanied by a large amount of fossil fuel consumption. It was estimated that the increase in aerosols (PM10) concentration was approx. 0.01–0.02 mg/m^3 (10–30%) during the period from 1980 to 2002.

Precipitation in the hilly areas mainly transforms into runoff, supplied to agriculture, cities, and the ecological system in the plains, especially in northeast China and north China. The precipitation enhancement by uplifting on topographic barriers is sensitive to aerosols concentration. This study found that the concentration of PM10 increased by 10–27% in the three catchments, and *R*o decreased by about 22–42% from 1980 to 2002 due to the possible suppression of air pollution on orographic precipitation. It shows a significant loss of water resources in hilly areas, and implies the water resources shortage is aggravated in the plains.

Acknowledgments This research was partially supported by the National 973 Project of China (2006CB403401, 2006CB403405).

REFERENCES

Ackerman, A. S., Toon, O. B., Taylor, J. P., Johnson, D. W., Hobbs, P. V. & Ferek, R. J. (2000) Effects of aerosols on cloud albedo: Evaluation of Twomey's parameterization of cloud susceptibility using measurements of ship tracks. *J. Atmos. Sci.* **57**, 2684–2695.

Andreae, M. O., Rosenfeld, D., Artaxo, P., Costa, A. A., Frank, G. P., Longo, K. M. & Silva-Dias, M. A. F. (2004) Smoking rain clouds over the Amazon. *Science* **303**, 1337–1342.

Borys, R. D., Lowenthal, D. H., Cohn, S. A. & Brown, W. O. (2003) Mountaintop and radar measurements of anthropogenic aerosol effects on snow growth and snowfall rate. *Geophys. Res. Lett.* **30**(10), 1538, doi:10.1029/2002GL016855.

Givati. A. & Rosenfeld, D. (2004) Quantifying precipitation suppression due to air pollution. *J. Appl. Met.* **43**, 1038–1056.

Givati, A. & Rosenfeld, D. (2005) Separation between cloud-seeding and air-pollution effects. *J. Appl. Met.* **44**, 1298–1314.

Givati, A. & Rosenfeld, D. (2007) Possible impacts of anthropogenic aerosols on water resources of the Jordan River and the Sea of Galilee. *Water Resour. Res.* **43**, W10419, doi:10.1029/2006WR005771.

Gunn, R. & Phillips, B. B. (1957) An experimental investigation of the effect of air pollution on the initiation of rain. *J. Met.* **14**, 272–280.

Jirak, I. L. & Cotton, W. R. (2006) Effect of air pollution on precipitation along the front range of the Rocky Mountains. *J. Appl. Met. & Climatol.* **45**(1), 236–245.

Liang, F. & Xia, X. A. (2005) Long-term trends in solar radiation and the associated climatic factors over China for 1961–2000. *Ann. Geophys-Germany* **23**, 2425–2432.

Peng, Z. G. & Chen, J. (1992) Relation of visibility to air pollution. *Chongqing Environ. Sci.* **14**(3), 16–21 (in Chinese).

Ramanathan, V., Chung, C., Kim, D., Bettge, T., Buja, L., Kiehl, T., Washington, W. M., Fu, Q., Sikka, D. R. & Wild, M. (2005) Atmospheric brown clouds: impact on South Asian climate and hydrologic cycle. *Proc. Natl. Acad. Sci.* **102**, 5326–5333.

Rosenfeld, D. (1999) TRMM observed first direct evidence of smoke from forest fires inhibiting rainfall. *Geophys. Res. Lett.* **26**, 3105–3108.

Rosenfeld, D. (2000) Suppression of rain and snow by urban and industrial air pollution. *Science* **287**, 1793–1796.

Rosenfeld, D., Dai, J., Yu, X., Yao, Z. Y., Xu, X. H., Yang, X. & Du, C. L. (2007) Inverse relations between amounts of air pollution and orographic precipitation. *Science* **315**, 1396–1398.

Yang, D. Z., Fang, X. M. & Li, X. S. (1998) Analysis on the variation of sandstorm in northern China. *Q. J. Appl. Met.* **9**(3), 352–358.

Yang, L. X., Wang, D. C., Cheng, S. H., Wang, Z., Zhou, Y. H. & Wang, W. X. (2007) Influence of meteorological conditions and particulate matter on visual range impairment in Jinan, China. *Sci. Total Environ.* **383**, 164–173.

Zhao, C. S., Tie, X. X. & Lin, Y. P. (2006) A possible positive feedback of reduction of precipitation and increase in aerosols over eastern central China. *Geophys. Res. Lett.* **33**, L11814, doi:10.1029/2006GL025959.

Estimating dry matter content for winter wheat using MODIS reflectance data

YONGHONG YI[1], DAWEN YANG[1] & DAOYI CHEN[2]

1 *State Key Laboratory of Hydroscience and Engineering, Department of Hydraulic Engineering, Tsinghua University, Beijing 100084, China*
yiyh05@mails.tsinghua.edu.cn

2 *Department of Engineering, University of Liverpool, Liverpool L69 3BX, UK*

Abstract The accumulation process of dry matter content during the crop growing period is a good indicator of crop condition and yield. This paper investigated the application of remote sensing in estimating the dry matter content. Several vegetation indices (VI) calculated from the MODIS 500 m reflectance data were used for regression analysis. A negative correlation was found between all the VIs and the dry matter content after the VIs reached the peak. This correlation during the later growth period may be more valuable for NPP estimation since the dry matter content was then directly related to the crop yield. However, the daily VI time series were found to be very sensitive to atmospheric conditions and angular effects, which may cover the changes on the VI induced by the crop growth to a certain degree. The Savitzky-Golay filter was employed to smooth the daily NDVI time-series as a test. The smoothed NDVI temporal curves were found closer to the 8-day composite ones. The correlation between the dry matter content and NDVI was also improved after smoothing. Both of these demonstrated that this method could partly remove the noise in the MODIS data, specifically the cloud effect, and disclose the changes in NDVI from the crop growth.

Key words MODIS; dry matter; vegetation index; Savitzky-Golay filter

INTRODUCTION

The accumulation process of the dry mass in the crop vegetative and reproductive stage is closely related to the plant net primary productivity (NPP). Remote sensing can provide effective estimation of dry matter content in a rapid and regional way, which facilitates the prediction of regional crop yield. Vegetation indices (VI) are widely applied in the monitoring of crop growth condition and yield prediction because of the simplicity and computational efficiency (Doraiswamy *et al*., 2004). NDVI, an index combining the absorptive and reflective characteristics of vegetation in the red and near-infrared wavelengths, is sensitive to the vegetation amount (Huete *et al*., 1999). Changes in NDVI time-series indicate changes in vegetation conditions proportional to the absorption of photosynthetically active radiation (Sellers, 1985). NDVI has been widely applied for the estimation of crop yield and end-of-season dry biomass for its availability. However, NDVI is also sensitive to many other factors, e.g. the background reflectance, the changes in the sun and view angles, and the cloud conditions, which could introduce great disturbance in the NDVI time-series. Additionally, NDVI becomes saturated at dense vegetation cover.

As supplements, several indices were proposed to account for these effects. The soil adjusted vegetation index (SAVI, Huete, 1988) and modified soil adjusted vegetation index (MSAVI, Qi *et al*., 1994) were developed to consider the contribution of the soil background through aligning the VI isolines with the greenness isolines (usually expressed in terms of LAI). The Enhanced Vegetation Index (EVI, Huete *et al*., 1999) incorporated both background adjustment and atmospheric resistance concepts for correcting the interactive canopy background and atmospheric influences. Gao (1996) designed a Normalized Difference Water Index (NDWI) using 1240 nm and 860 nm wavelengths, which was proved less perturbed by atmospheric effects than NDVI. The study by Chen *et al*. (2005) proved that NDWI using longer SWIR wavelengths (1640 nm and 2130 nm) showed a lagged saturation in comparison to NDVI for corn and soybeans. In the study of Yi *et al*. (2007), NDWI (1640 nm and 2130 nm) was less sensitive to the atmospheric disturbances and angular effects than NDVI.

Winter wheat is one of the major crops in the northern China plain, so monitoring of its growth and yield has important economic and social impacts. MODIS data shows potential use for retrieval of crop biophysical parameters and for improving accuracy in crop yield assessment, with its merits of moderate spatial resolution (250–500 m) and daily data availability (Doraiswamy *et al.*, 2004). Moreover, MODIS data onboard Terra and Aqua sensors potentially provide twice coverage diurnally, which greatly enhances the ability of monitoring crop condition during the cloudy season.

In this study, four MODIS vegetation indices mentioned above were used to estimate the dry matter content using the linear regression method. The disturbances from the cloud contamination, atmospheric variability and bi-directional effects in VI time-series can greatly affect the monitoring of land cover and terrestrial ecosystems, especially for the daily data (Chen *et al.*, 2004). Therefore, as a test, the Savitzky-Golay filter was employed to reduce these noises in the NDVI time series.

DATA DESCRIPTION

Ground data

Field experiments were conducted around the eddy flux tower (116°3′ 15.3″E, 36°38′ 55.5″N) (see Fig. 1) in the Weishan Irrigation Zone (115°24′–116°30′E, 36°12′–37°00′N). The Weishan Irrigation Zone, with a total area of 36 000 hm^2, is the largest irrigation zone along the downstream of the Yellow River. This area is located in a semi-humid zone with averaged annual precipitation 580 mm. The land cover is mainly constituted of croplands, planted woods, secondary grass and shrubs, and artificial surfaces. The rotation of winter wheat and summer corn is the local primary agriculture practice.

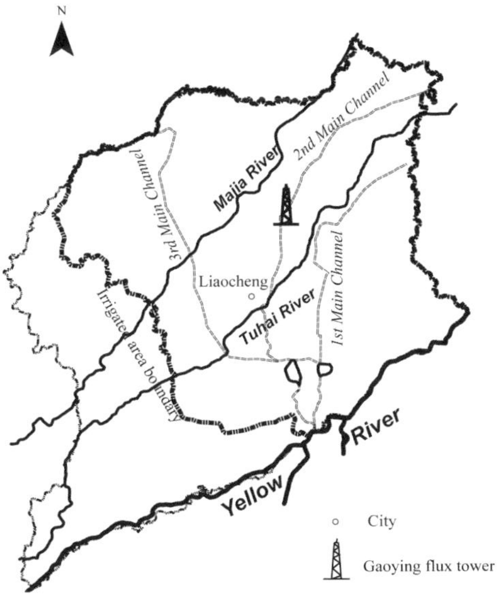

Fig. 1 The Weishan irrigation zone along the downstream of the Yellow River and the experiment site (location of the flux tower).

The ground data were collected in 2006 and 2007. In 2006, the experiment was conducted every other week from 1 April to 3 June. Within a radius of several kilometres around the flux tower, the croplands were almost homogeneously covered by the same wheat crop. Three 1-m^2 blocks were selected randomly around the flux tower, but with nearly uniform row distance. The total number of tillers of wheat in each block was counted, and then the samples taken from the outside of the plot were dried in an oven for 8 h at 70°C to obtain the dry weight. The dry matter of the three samples was averaged and finally the variables in the 1-m^2 area were derived by multiplying the ratio of the tillers in the 1-m^2 area to the sample tillers (in unit of kg m^{-2}). In 2007, the experiment was augmented to six plots located in an area of 1.5 × 1 km^2 around the flux tower. The measurement was conducted weekly from 17 March to 6 June. The same method was taken to measure the dry matter content. The temporal curves of the measured dry matter content in 2006 and 2007 are shown in Fig. 2.

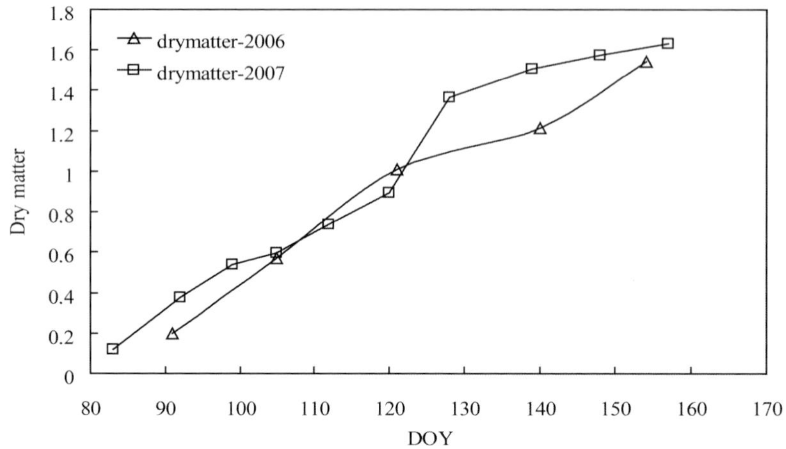

Fig. 2 The temporal curves of dry matter content during wheat growing period in 2006 and 2007.

MODIS data

The MODIS daily and 8-day composite 500 m surface reflectance products were downloaded and used for analysis. The daily products provide an estimation of the surface reflectance for MODIS channels 1 to 7, as it would be measured at the land surface if there were no atmosphere. The 8-day product is a composite over an 8-day period of the daily products. The criteria for qualified pixel screening in the composite are filtering out those pixels with cloudiness, cloud shadows, and low solar zenith angle. When several values pass these quality criteria, the minimum of channel 3 (blue) values are considered, which are mostly contributed by the atmospheric scattering process. The composite process can eliminate the cloud contamination to a certain extent. The data were first re-projected from Sinusoidal Grid (SIN) projection to UTM projection using the MODIS Reprojection Tool (MRT). MODIS Land Data Operational Product Evaluation (LDOPE) software was used to check the cloud status and data quality using the associated quality control and state flags.

METHODOLOGY

Vegetation Index

NDVI has the following form with a normalized transform of the NIR to red reflectance ratio:

$$\text{NDVI}=(\rho_{\text{NIR}}-\rho_{\text{Red}})/(\rho_{\text{NIR}}+\rho_{\text{Red}}) \tag{1}$$

where ρ_{NIR} and ρ_{Red} are reflectances in NIR and red wavelengths, respectively.

SAVI is proposed by Huete (1988) to use vegetation isolines equations to minimize soil background effects. MSAVI is a modified version of SAVI (Qi *et al.*, 1994) and represented by:

$$\text{MSAVI}=\left[2\rho_{\text{NIR}}+1-\sqrt{(2\rho_{\text{NIR}}+1)^2-8(\rho_{\text{NIR}}-\rho_{\text{Red}})}\right]/2 \tag{2}$$

EVI is designed for MODIS, with improved sensitivity into high biomass regions and improved vegetation monitoring through a de-coupling of the canopy background signal and a reduction in atmosphere influence (Huete *et al.*, 1999).

$$EVI = 2.5 \cdot (\rho_{NIR} - \rho_{Red})/(\rho_{NIR} + 6 \cdot \rho_{Red} - 7.5 \cdot \rho_{Blue} + 1) \quad (3)$$

where ρ_{Blue} is blue band reflectance.

NDWI was proposed by Gao (1996) and proved to be less sensitive to atmospheric effects and shows a lag in saturation compared with NDVI. NDWI is calculated as:

$$NDWI = (\rho_{NIR} - \rho_{SWIR})/(\rho_{NIR} + \rho_{SWIR}) \quad (4)$$

where ρ_{SWIR} is reflectance of SWIR wavelengths (e.g. MODIS bands 6, 7).

The Savitzky-Golay filter

The quality of the daily MODIS data was greatly affected by clouds and aerosol contamination. In this study the image was excluded if the cloud and mixed area in the study region exceeded 30%, or the blue band reflectance was above 0.07, to remove undetected cloud effects. However, even after cloud masking, the sub-pixel cloud contamination could not be totally removed, which induced a great variation in the daily reflectance data. This variation may cover the changes in the reflectance due to the crop growth. In this study, the Savitzky-Golay filter was used to reduce the noise caused by cloud contamination and atmospheric variability in the NDVI time-series.

The Savitzky-Golay filter was a simplified least-squares-fit convolution for smoothing and computing derivatives of a set of consecutive values or a spectrum (Savitzky & Golay, 1964). The convolution can be understood as a weighted moving average filter with weighting given as a polynomial of a certain degree. This filter was designed to consecutive data at a fixed and uniform interval. The general equation of this filter can be given as follows:

$$Y_j^* = \frac{\sum_{i=-m}^{i=m} C_i Y_{j+i}}{N} \quad (5)$$

where Y is the original value, Y^* is the resultant value, C_i is the coefficient for the *i*th value of the filter (smoothing window), and N is the number of convoluting integers equal to the smoothing window size (2 m + 1).

This method assumed that the effects of the cloud contamination and atmospheric variability usually depressed the NDVI values and the sudden drops in the NDVI time-series were regarded as noise and were removed. In this study, the cloudy data was first recognized using the above defined cloud mask, and replaced by a linearly interpolated value using the adjacent points. However, if there were over 7 days with cloudy data, the missing data would not be interpolated. Then the Savitzky-Golay filter was used to smooth the NDVI variation and obtain the long-term changing trend. Comparing with the raw data, a new NDVI time-series was generated by replacing the noisy NDVI points with the corresponding points in the smoothed time-series. The new time-series would more closely approach the upper NDVI envelope through an iteration process and best fit the NDVI variations during the full vegetation season. The details of the processing can be found in the study of Chen *et al.* (2004).

RESULTS AND DISCUSSION

Figure 3 presented the daily VI time-series from both Terra and Aqua sensors in 2006 and 2007, respectively, which was only processed using cloud mask. It can be seen that the temporal curves were rather variable, especially for the three indices (NDVI, MSAVI, and EVI). The variations in these indices were found to be closely related to the changes in the blue band reflectance, which was mainly contributed by atmospheric scattering. Especially the NDVI values in both years had

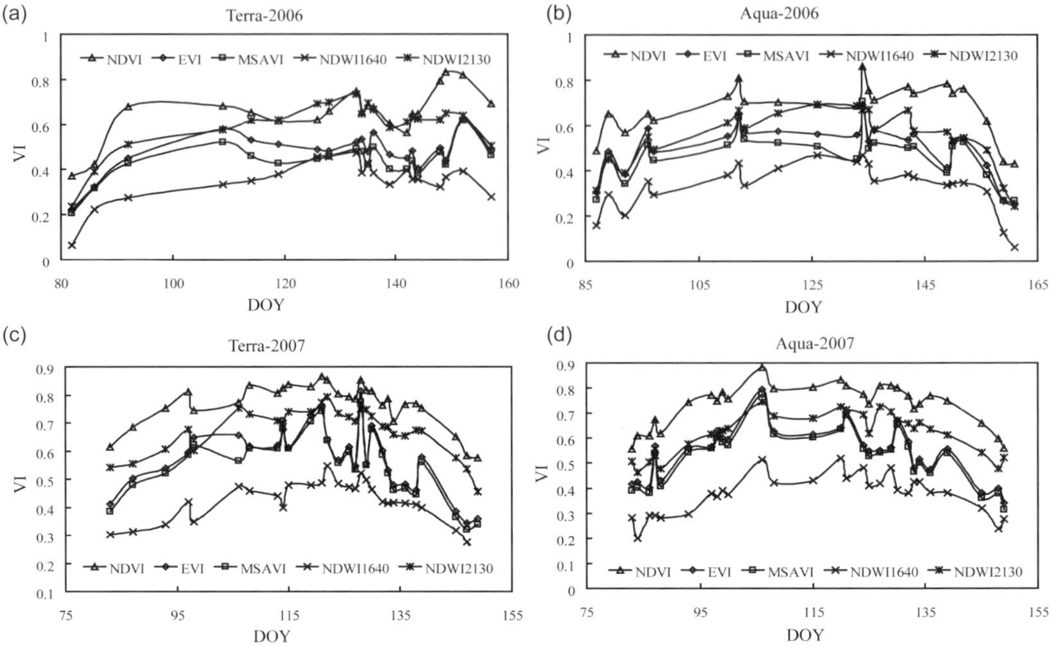

Fig. 3 The temporal curves of daily Terra and Aqua vegetation indices in 2006 and 2007.

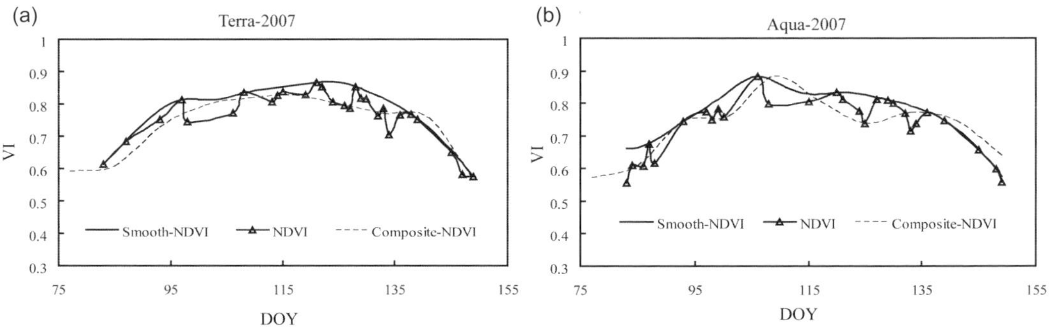

Fig. 4 The comparison of the original, smoothed, and 8-day composite NDVI temporal curves in 2007.

obvious correlations with the blue reflectance. Therefore, it was speculated that these variations were mainly caused by changes in atmospheric conditions. Apart from this effect, the indices were also influenced by the changes in the sun-target-view geometry.

Figure 4 shows the smoothed NDVI time-series in 2007. It was not applied in the 2006 daily NDVI time series because there were too few clear dates from DOY 90 to 110 for both sensors. The filter lost its effect under this condition. The smoothed time series were compared with the 8-day composite ones, in which the influence from the atmospheric variability is partly removed through the composite process. It can be seen from the figure that the smoothed time-series were close to the composite ones, except around DOY 125 for Aqua. However, the valley around this date may be caused by the cloud disturbance and the composite process can not totally remove this effect. Although the smoothed curve was closer to the composite, the smoothed values were generally larger than the composite ones. In the filter method, it was assumed that the cloud always depressed the NDVI values, and the other effects on NDVI were neglected. Therefore, the smoothed NDVI time series were generally matched with the largest values in the time-series.

However, the composite values may choose the values with minimum blue reflectance among the cloud-free dates, not always the date with the largest NDVI value, which could partly remove the angular effects in the reflectance. In consideration of this, the modified Walthall method was applied to the daily NDVI values before filtering to normalize the sun and view angles. The four empirical coefficients in the model were regressed using the modelled reflectance from a 1-D SAIL (Scattering by Arbitrarily Inclined Leaves) model. However, it was found that the BRDF (Bi-directional reflectance) correction did not have much effect on the results. One reason may be that the SAIL model did not provide much reasonable modelling of the canopy reflectance. The other reason may be that the chosen pixels for analysis were not pure pixels, which was limited by the surface heterogeneity in the study area.

Figure 5 shows the scatter plots between several selected indices and dry matter content. For 2006, NDVI and $NDWI_{2130}$ are presented, and for 2007 only NDVI before and after smoothing are shown. In both years, the scatter plots between VI and the dry mass were similar. In the early period, the VI and dry mass showed a positive correlation and after VI arrived at the peak, they showed a negative positive correlation. The different correlation forms may be related to the characteristics of the wheat growth. Before the heading stage (around DOY 120), the growth of the wheat was characterized by the growth in the vegetative components. After the wheat came into the heading stage, the crop grew with notable accumulation of dry mass, but the vegetative parts stopped rapid growth and slowly withered.

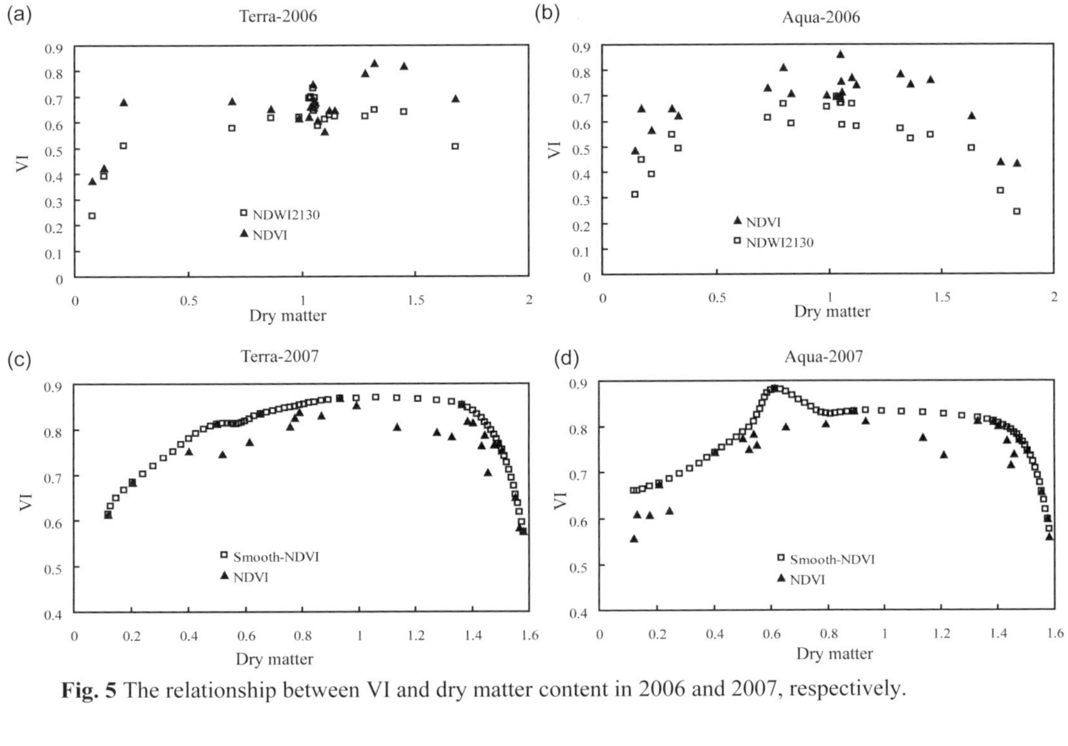

Fig. 5 The relationship between VI and dry matter content in 2006 and 2007, respectively.

Tables 1 and 2 list the determination coefficients (R^2) between all the indices and the dry matter content at the later growth stages in 2006 and 2007, respectively. In 2006, the period was from DOY 119 to 157 for the Terra sensor and from DOY 119 to 161 for Aqua. There were 15 and 14 images during the period for Terra and Aqua, respectively. In 2007, the period was from DOY 115 to 149 for both sensors. There were 19 and 16 images during this period for Terra and Aqua,

Table 1 The summary of R^2 values between dry matter content and daily vegetation indices during the later wheat growth period of 2006.

	NDVI	EVI	MSAVI	$NDWI_{1640}$	$NDWI_{2130}$
Terra	0.29	*0.01*	*0.13*	0.38	**0.41**
Aqua	**0.58**	**0.76**	**0.61**	**0.82**	**0.88**

Bold figures mean that the correlation is significant with $p < 0.01$, and italics mean no obvious correlation with $p > 0.05$. There were 15 samples for Terra from DOY 119 to 157 and 14 samples for Aqua C4 from DOY 119 to 161.

Table 2 The summary of R^2 values between dry matter content and daily vegetation indices during the later wheat growth period of 2007.

	NDVI	NDVI _smooth	EVI	MSAVI	$NDWI_{1640}$	$NDWI_{2130}$
Terra	**0.48**	**0.51**	**0.40**	**0.44**	**0.43**	**0.48**
Aqua	**0.42**	**0.51**	**0.51**	**0.52**	**0.46**	**0.42**

Bold figures mean that the correlation is significant with $p < 0.01$. There were 19 samples for Terra and 16 samples for Aqua both from DOY 115 to 149 in 2007.

respectively. The difference in the chosen dates between the two years was due to the difference in the crop phenology, which can be observed from the dry matter accumulation curve in Fig. 2.

From Table 1, it can be seen that Terra and Aqua VIs had very different performance in the correlation with the dry matter content in 2006. For Terra, only $NDWI_{2130}$ showed significant negative correlation with the dry matter content. EVI and MSAVI showed no correlation with the dry matter content at all. However, for Aqua, significant correlation was found between all the indices and the dry matter content. The two NDWI indices performed best in both sensors. This may be because NDWI, employing longer wavelengths, was less sensitive to the disturbance from the cloud and atmospheric conditions (see Fig. 3).

In 2007, the indices performed similarly for both sensors. A significant correlation was found between all the indices and the dry matter content. The correlation between the smoothed NDVI and the dry matter content did not improve much compared with the original NDVI and the other indices. From Fig. 5, it can be found that the observed dry matter content did not increase much, whilst the VI values kept decreasing after DOY 139. The change in the dry matter content at the last period in 2007 was much different from that in 2006. In 2006, the dry matter content had a sharp increase after DOY 140. If only considering the dates before DOY 139, the correlation between the smoothed NDVI and dry matter content was greatly improved. The R^2 values were increased from 0.51 to 0.85 for Terra and from 0.51 to 0.93 for Aqua. As we knew, the wheat did not experience serious water stress in 2007, and the difference in the dry matter accumulation at the later stages should be investigated. One of the possible error sources may be from the measurements. In 2007, the measurement was averaged from six plots and made once a week, and the error from the multiple measurements was only defined arbitrarily by the growth process of each sample.

CONCLUSIONS

The MODIS derived vegetation indices were used to estimate the dry matter content during the winter wheat growing period in 2006 and 2007.

The daily VI time-series, especially the three indices (NDVI, EVI and MSAVI) employing shorter wavelengths, showed great deviation, which may be caused by the cloud and atmospheric variability and the changes in the sun-target-view geometry. Comparatively, NDWI was more resistant to these disturbances and may potentially be more reliable for monitoring crop growth conditions, especially when the uncertainty in the data set was large, e.g. the Collection 4 Terra data in 2006.

After smoothing, the daily NDVI time-series was closer to the 8-day composite ones, which showed the success of the algorithm in removing the cloud contamination and other disturbances. However, the smoothed NDVI values were typically larger than the composite values. This was because the filter method ignored the angular effects and only considered the depressing effect of the cloud on NDVI values. The sun and view angles in the NDVI time-series should be normalized before the filtering, especially for sensors with a large Field of View (FOV) like MODIS.

The VIs showed a positive correlation with the dry matter content before the VIs arrived at the peak, and after that, a negative correlation was found between them. Except for the Terra indices in 2006, all the indices showed significant correlation with dry matter content during the later growth period, which was characterized by the notable accumulation of the dry mass in the crop. The correlation at the later growth stages may be much more valuable for NPP estimation since the dry matter content is then directly related to the crop yield.

Acknowledgements This research was supported by the National Natural Science Foundation of China (50679029).

REFERENCES

Chen, D., Huang, J. & Jackson, T. J. (2005) Vegetation water content estimation for corn and soybeans using spectral indices derived from MODIS near- and short-wave infrared bands. *Remote Sens. Environ.* **98**(2–3), 225–236.

Chen, J., Jonsson, P., Tamura, M., Gu, Z., Matsushita, B. & Eklundn, L. (2004) A simple method for reconstructing a high-quality NDVI time-series data set based on the Savitzky-Golay filter. *Remote Sens. Environ.* **91**, 332–344.

Doraiswamy, P., Hatfield, J., Jackson, T., Akhmedov, B., Prueger, J. & Stern, A. (2004) Crop condition and yield simulation using Landsat and MODIS. *Remote Sens. Environ.* **92**(4), 548–559.

Gao, B. (1996) NDWI—a normalized difference water index for remote sensing of vegetation liquid water from space. *Remote Sens. Environ.* **58**(3), 257–266.

Huete, A. (1988) A soil adjusted vegetation index (SAVI). *Remote Sens. Environ.* **25**, 295–309.

Huete, A., Justice, C. & van Leeuwen, W. (1999) MODIS vegetation index (MOD13). Algorithm Theoretical Basis Document (ATBD). http://modis.gsfc.nasa.gov/data/atbd/atbd_mod13.pdf.

Qi, J., Chehbouni, A., Huete, A. R., Kerr, Y. H. & Sorooshian, S. (1994) A modified soil adjusted vegetation index. *Remote Sens. Environ.* **48**(2), 119–126.

Savitzky, A. & Golay, M. J. E. (1964) Smoothing and differentiation of data by simplified least squares procedures. *Analytical Chem.* **36**, 1627–1639.

Sellers, P. J. (1985) Canopy reflectance, photosynthesis and transpiration. *Int. J. Remote Sens.* **6**, 1335–1372.

Yi, Y., Yang, D., Chen, D. & Huang, J. (2007) Retrieving crop physiological parameters and assessing water deficiency using MODIS data during winter wheat growing period. *Canadian J. Remote Sens.* **33**(3), 189–202.

2 Modelling approaches

A large-scale routing scheme for stream simulation and its application to river basins in China

FENG CHEN[1,2] & ZHENGHUI XIE[1]

1 *Institute of Atmospheric and Physics, Chinese Academy of Sciences, Beijing 100029, China*
zxie@lasg.iap.ac.cn

2 *Graduate School of Chinese Academy of Sciences, Beijing, China*

Abstract This paper presents a routing scheme for large-scale river basin stream simulation, with a grid flow direction scheme and the Muskingum-Cunge method considering lateral flow into the river. The grid flow direction scheme is to identify the grid flow directions by incorporating the GTOPO30 Digital Elevation Model (DEM) data, the multi-branch tree method, and the "stream burn" method. The routing scheme was applied to route runoff at river basins derived from the land surface model VIC-3L with the regionalized parameters at the resolution 0.5°×0.5° in continental China into streamflow at hydrological stations. The results show that the created grid flow directions are reasonable, and the continental simulation results agree well with the observations on a large river basin scale in China.

Key words DEM; flow direction; large-scale routing scheme; Muskingum-Cunge method; Pfafstetter basin numbering system

INTRODUCTION

The land component of climate models represents many processes that control exchanges of energy, momentum and water between soil, vegetation and the atmosphere, and has long been recognized as important for hydrological forecasting such as streamflow prediction and climate change studies (Dickinson *et al.*, 1993; Bonan *et al.*, 1996; Nijssen *et al.*, 1997; Zeng *et al.*, 2002; Dai *et al.*, 2003). Streamflow is derived from runoff with a routing scheme, which can be used in evaluating model efficiency by meteorologists and predicting water resources and disasters by hydrologists (Saunders, 1999). Therefore, a suitable large-scale routing scheme for streamflow simulation by a land surface model is necessary for streamflow simulation.

A routing scheme consists of grid flow directions identification, overland flow concentration and river routing. To identify the flow directions, there has been a strong research effort toward resolving a satisfactory method for determining flow directions in large-scale models (O'Donnel *et al.*, 1999; Fekete *et al.*, 2001; Turcotte *et al.*, 2001; Hao *et al.*, 2002; Olivera *et al.*, 2002; Shaw *et al.*, 2003, 2005; Ye *et al.*, 2005). There have been two main methodologies used in up-scaling hydrological routing information (Shaw *et al.*, 2005). The first method attempts to consider all of the topographic information available for each sub-grid from a more detailed DEM in up-scaling routing patterns. Although this method can produce very good results with flow directions reflecting the actual drainage pattern of the basin, it is very computationally expensive and can require costly software (Shaw *et al.*, 2003). The second approach uses average elevations, determined from a more detailed DEM, for each grid cell to route flow between neighbouring grid cells. This approach is computationally efficient, but does not produce adequate representation of the channel network of the basin (Arora & Boer, 1999; Shaw *et al.*, 2003).

This paper presents a new routing scheme for large-scale river basin stream simulation, with a grid flow direction scheme and the Muskingum-Cunge method considering lateral flow into the river. The flow direction scheme is to identify the flow directions based on the principle of multi-branch tree generation (Weiss, 1997), which is widely used in computer sciences, and the stream burn method (Hutchinson, 1989; Hao, 2002) in which the actual river information is combined with the DEM data in order to reduce the bias between the channel structure of a basin derived from DEM and the actual drainage patterns. Streamflow simulation for river basins in continental China by the routing scheme and the land surface model VIC (Liang *et al.*, 1994, 1996; Liang & Xie, 2001) with the regional parameters at the resolution 0.5° × 0.5° (Xie *et al.*, 2007) is given.

METHODOLOGY

As mentioned above, a grid flow direction identification method and a routing algorithm are the two key components of a routing scheme. We identify the flow directions and extract the drainage network for a region by the multi-branch tree method and the stream burn method. Based on the extracted flow directions and drainage network, the large-scale basin is sub-divided into small sub-basins and numbered following the Pfafstetter basin numbering system (Yang *et al.*, 2000). The daily runoff at a grid produced by a land surface model is convolved with a triangular unit hydrograph as the overland flow concentration, and the river routing is calculated by the extracted channel network and Muskingum-Cunge method considering lateral flow into the river. The schematic diagram of the routing scheme is shown in Fig. 1.

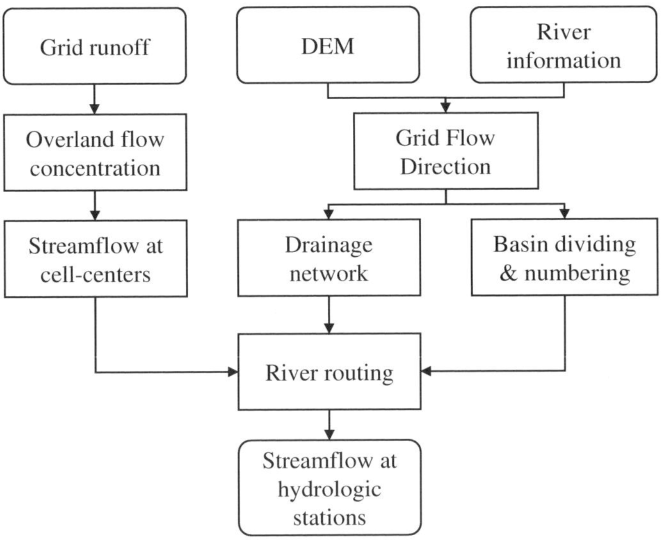

Fig. 1 The schematic diagram of the routing scheme.

Grid flow direction identification

By using the DEM data, many important hydrological characteristics can be gained, such as slope rate, slope direction, drainage network, basin boundary (Saunders, 1999). Since DEMs only describe the cell centre's elevation or the cell average elevation, the channel structure of a basin derived from a DEM may not adequately reflect actual drainage patterns on a large-scale, which is from the discontinuous river net, the sinks in the DEM, and the parallel river net in smooth regions. For each cell in a basin, flow out direction at a cell is assigned to the lowest neighbour cell by examining the elevation of itself and the eight surrounding cells according to the D8 algorithm (O'Callaghan, 1984). It performs well at a sufficiently fine resolution; however, the drainage network extracted by the D8 algorithm is less reliable when the resolution of the DEM becomes coarser. We use the multi-branch tree method and stream burn ideology (Hutchinson, 1989; Hao, 2002) to extract a continuous, no-sink and unparallel drainage network.

The streamflow at the outlet of a basin comes from runoff at grid cells for the basin. Each grid cell in the basin can be considered as a node in the multi-branch tree; the multi-branch tree contains the entire cells in the basin, and the outlet of the basin as the root node. Every father node contains a subset of the cells whose flow is assigned to the father node cell. Once the flow out direction at a cell is identified, the node associated with the cell is added into the multi-branch tree at its father node. The grid flow direction identification for a basin is reduced to construct the multi-branch tree. For the outlet of a basin i.e. the root node of the multi-branch tree for the basin, is called the first layer node. The flow out directions for neighbours of the first layer node are identified by D8 algorithm, and those neighbour nodes whose flow out directions are assigned to the first layer node are added into the tree as the son nodes of the root node. For constructing the (m + 1)th layer nodes of the tree, the flow out directions for the nodes, which are not included in the m layers but at least one of their neighbours are identified by D8 algorithm, and those nodes whose flow out directions are assigned to the mth layer nodes are added into the tree as the son nodes of the mth layer nodes. There may still be cells whose flow directions are not identified after searching all the nodes in the tree, which means that the DEM has sinks. In order to solve this problem, we find the lowest cell among the cells whose neighbour has been in the tree, and assign the flow to its lowest neighbour that has been in the tree. When all the sinks are well treated, the grid flow directions can be identified. For all the nodes in the multi-branch tree connected with the root tree, the drainage network is continuous. Figure 2 shows the drainage network (the resolution of the DEM is 0.1°×0.1°). It is seen from Fig. 2(b) that the river is discontinuous in the circles by comparing the actual drainage network. Figure 2(c) shows the continuous drainage network extracted by using the multi-branch tree method.

However, the extracted drainage network is incorrect in the circle in Fig. 2(c). The main reason for this problem is that there are little differences between the cell elevations. In order to reduce the bias between the drainage network derived from DEM and the actual one, we use the stream burn method (Hutchinson, 1989; Hao, 2002) to include the natural river information into the DEM data. We can get the river DEM based on GIS, and the new DEM by the original DEM minus the river DEM. The drainage network, including the natural river information, is better than the original one (Fig. 2(d)).

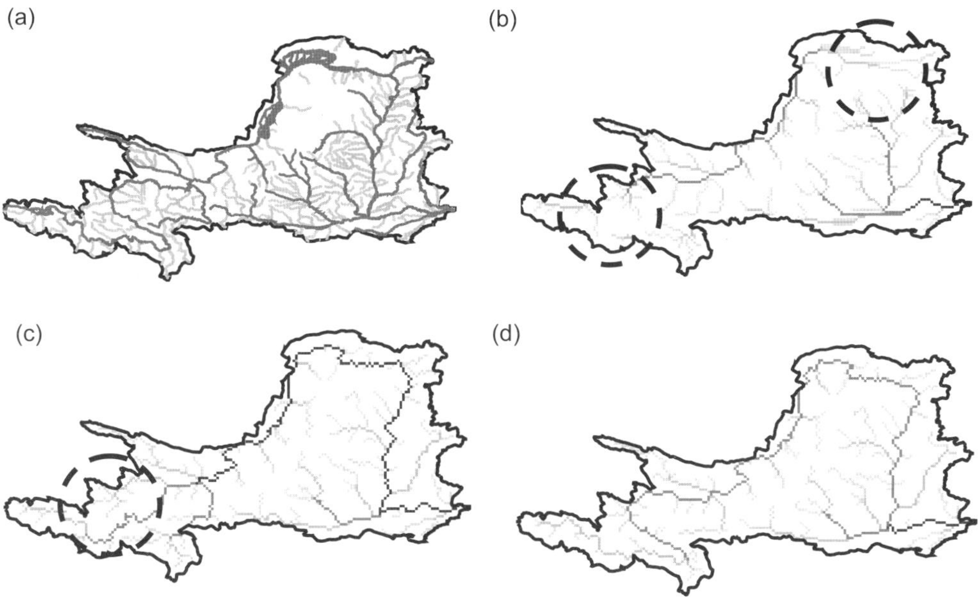

Fig. 2 The drainage network of the Yellow basin (above the Huayuankou station): (a) the actual drainage network; (b) the drainage network extracted from the original DEM by using Arc/Info analytical software; (c) the drainage network extracted from the original DEM by using the multi-branch tree method; (d) the drainage network extracted from the new stream-burned DEM by using the multi-branch tree method.

Based on the multi-branch tree method and the stream burn method, the grid flow directions for all grid cells in the region are identified, the extracted drainage network is continuous, and every cell is connected with the outlet. Otherwise, the water in sinks always choses the lower way to flow out.

River and basin creating

The river and basin created is based on the flow directions and the accumulated flow matrix, which is calculated by the number of cells flowing into the cell (Ye *et al.*, 2005). We suppose that the river channel is created in the cells whose accumulated flow values are bigger than other cells nearby. The whole river (or basin) can be identified by searching the flow direction matrix and the accumulated flow matrix when the outlet cell of a river (or basin) is found.

Basin partition

To represent the hydrological process well, the large basin is sub-divided into several sub-basins by the method which always divides the basin into nine sub-basins, and numbering them with the Pfafstetter basin numbering system (Yang *et al.*, 2000). According to the Pfafstetter basin numbering system, the whole basin can be divided into several sub-basins, and the relationship between any two sub-basins can be easily found from the numbering system. For example, the sub-basin labelled XYZ, where XY is the upper basin-level's serial number, and Z is the serial number of the current basin-level. Figure 3 shows the relationship between the Yellow River basin (above the Huayuankou station) and its sub-basins.

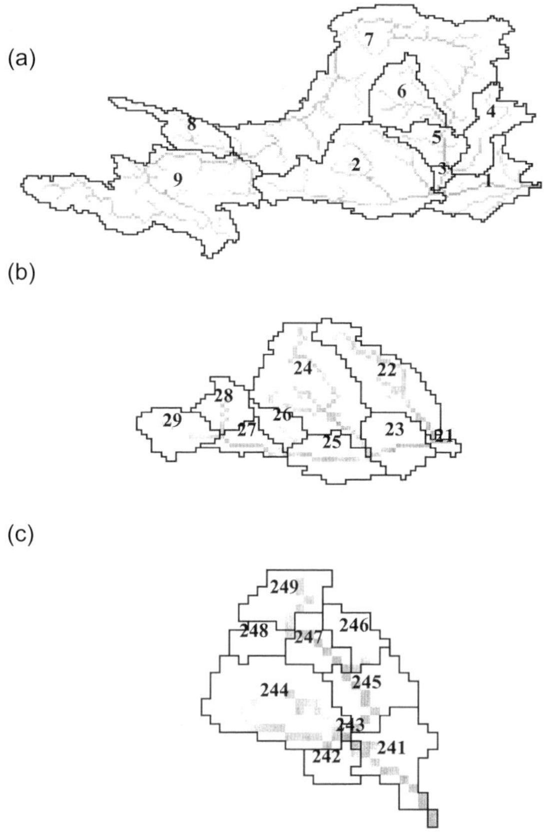

Fig. 3 The relationship between Yellow river basin (above the Huayuankou station) and its sub-basins: (a) the first level coding; (b) the second level coding of the sub-basin 2 in the first level; (c) the third level coding of the sub-basin 24 in the second level.

Overland flow concentration

Although the grid-based land surface model produces only one time series of runoff values for each grid cell, this runoff should not be considered as being produced at a single point, but as being distributed non-uniformly over the area of that grid cell (Nijssen *et al.*, 1997). To account for differences in travel time of runoff produced in different parts of the grid cell, the daily runoff produced by a land surface model is convolved with a triangular unit hydrograph, which simulates routing within the grid cell.

River routing

We suppose that the hydrographs produced for each grid cell in a basin are then summed together and distributed uniformly along the trunk stream and routed to the basin outlet. The Muskingum-Cunge method considering lateral flow into the river (Cunge, 1969; Price *et al.*, 1975; Rui *et al.*, 1995; Rui *et al.*, 2000; Yuan *et al.*, 2005) is applied to flood routing to obtain the outlet's discharge.

APPLICATION

Based on the runoff produced by Xie *et al.* (2007), which presented a methodology for regional parameter estimation of the VIC-3L land surface model and simulated the streamflow in continental China, we chose eight hydraulic stations located in five large basins for verification.

The detailed information for the stations and the verification results can be seen in Table 1. Compared with the actual control area, the relative errors of the simulated control area of the eight basins are –3.1%, 3.9%, 0.9%, 6.5%, 2.5%, 7.5%, 11.0% and 1.5%, respectively. The main reason for the bias is that the resolution is too coarse in our study, and the boundary of basins could not be so regular. Generally, large basins will have a small relative error while the small basins will have a large one.

Table 1 The information of eights stations and the verification results.

River basin	Station	Actual control area (10^4 m^2)	Simulated control area (10^4 m^2)	Area bias[a] (%)	Bias[b] (%)	RRMSE[c] (%)	Ce[d]	Simulation period
Pearl River	Wuzhou	32.7	31.7	–3.1	21.0	2.7	0.82	1/1/1980–31/12/1990
Yangtze River	Yichang	100.6	104.5	3.9	0.8	2.1	0.89	1/1/1980–31/12/1990
	Datong	170.5	169.0	–0.9	6.2	1.4	0.88	1/1/1980–31/12/1990
Huaihe River	Bengbu	12.1	12.9	6.6	41.4	5.9	0.66	1/1/1980–31/12/1990
Yellow River	Tangnaihai	12.2	11.9	2.5	–14.6	3.5	0.81	1/1/1980–31/12/1989
	Guide	13.4	14.4	7.5	0.8	4.3	0.67	1/1/1980–31/12/1989
Songhuajiang River	Jiangqiao	16.3	18.1	11.0	–17.8	5.4	0.74	1/1/1980–31/12/1990
	Haerbin	39.0	39.6	1.5	–8.2	5.1	0.54	1/1/1980–31/12/1990

[a] Area bias – the bias of the control area between simulated and actual: $bias = (A_s - A_o)/A_o \times 100\%$

[b] bias - defined as: $bias = (\overline{Q}_s - \overline{Q}_o)/\overline{Q}_o \times 100\%$

[c] RRMSE – the relative root mean squared error, defined as: $RRMSE = \frac{1}{n}\sqrt{\sum_{i=1}^{n}(Q_{s,i} - Q_{o,i})^2} / \overline{Q}_o \times 100\%$

[d] CE - the Nash-Sutcliffe model efficiency coefficient, defined as:

$CE = (\sum_{i=1}^{n}(Q_{o,i} - \overline{Q}_o)^2 - \sum_{i=1}^{n}(Q_{s,i} - Q_{o,i})^2)/\sum_{i=1}^{n}(Q_{o,i} - \overline{Q}_o)^2 \times 100\%$, with $Q_{s,i}$ and $Q_{o,i}$ the simulated and observed flows in month *i*.

Figures 4 and 5 show the observed and simulated monthly streamflow time series and the mean monthly hydrographs for the eight large basins, respectively. The routing scheme performed well based on the runoff produced by the VIC-3L model with the transferred parameters, and the simulated streamflows matched the observed values well. Although the relative errors of some stations are not very good, four in eight basins' relative errors are less than 10%, and the smallest is 0.8%. All the relative root mean squared errors are <6%, which shows small amplitude of the difference between the simulated and observed streamflows. Five in eight basins' Nash-Sutcliffe coefficients are larger than 0.7, and four basins' coefficients are larger than 0.8.

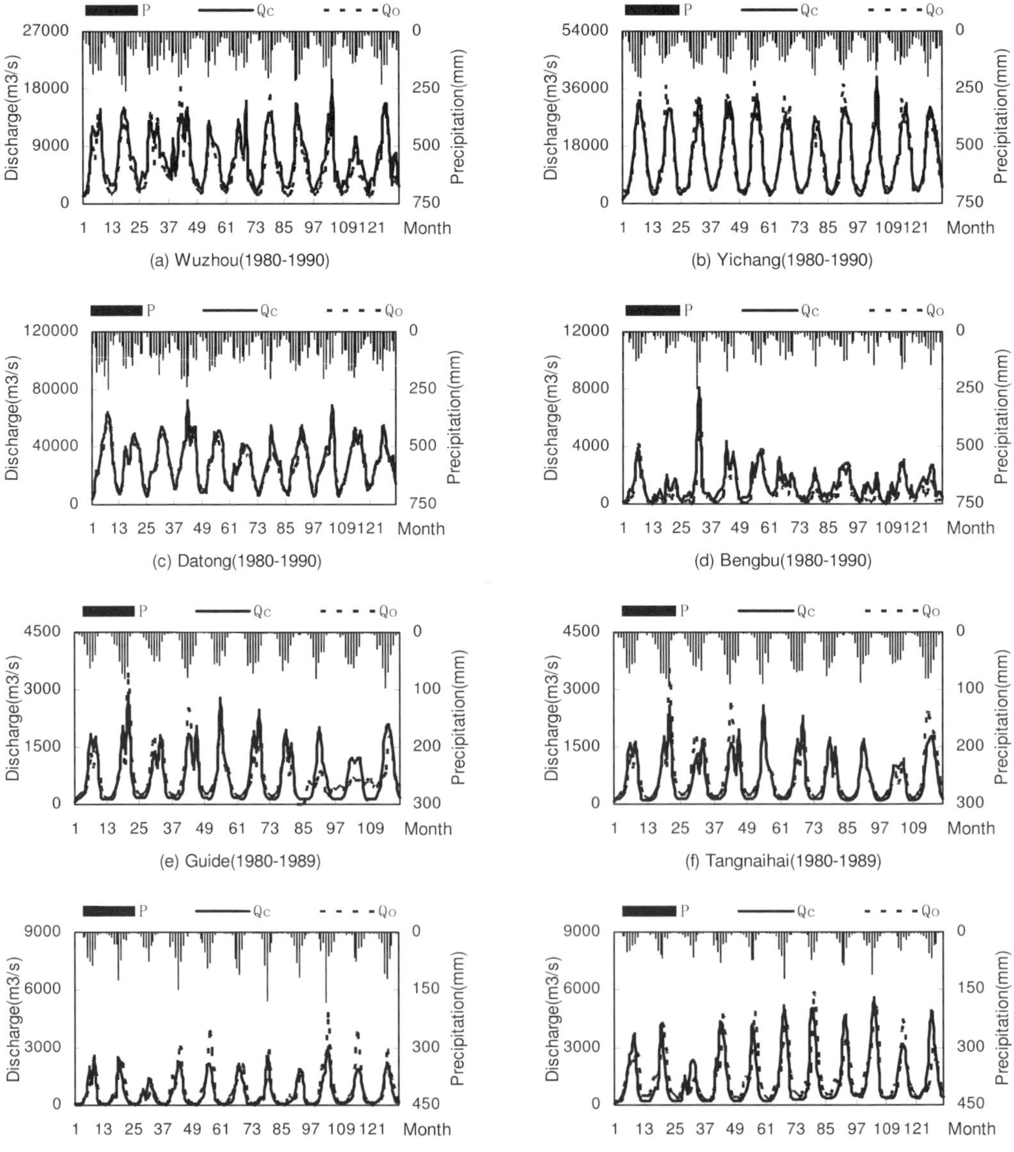

Fig. 4 The observed and the simulated monthly streamflow time series for the eight verification basins.

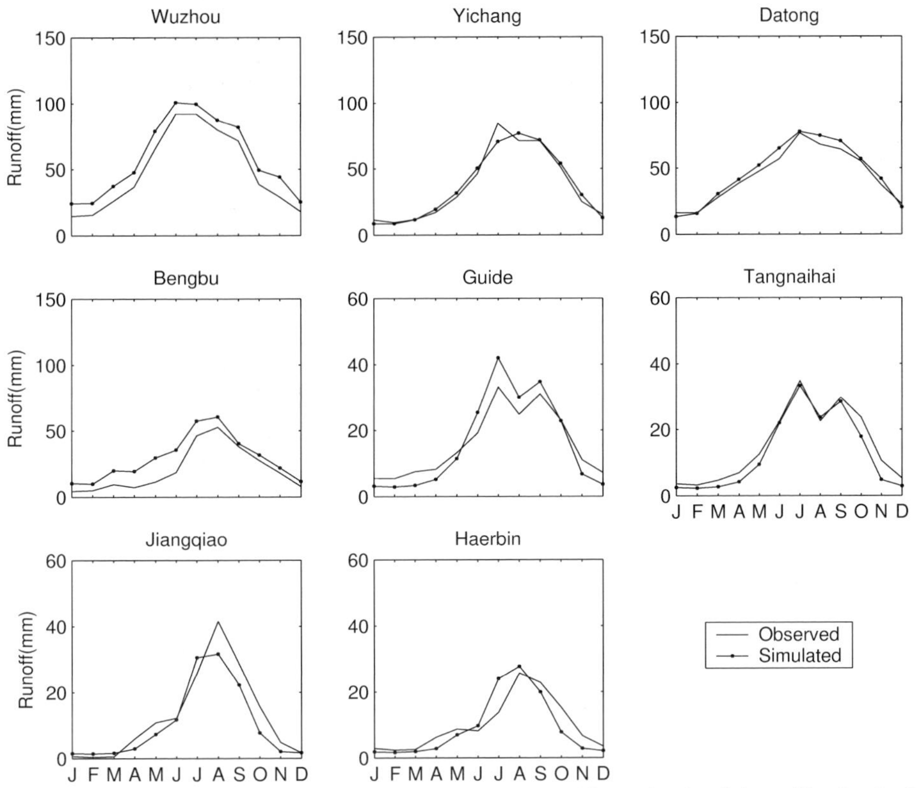

Fig. 5 Mean monthly hydrographs of observed and simulated flows for the eight verification basins.

CONCLUSIONS

In this paper, a routing scheme for large-scale river basin streamflow simulation was presented. Based on the D8 algorithm, the multi-branch tree method and the stream burn method were combined to identify the grid flow directions. The grid cell daily runoffs produced by the VIC-3L model with transferred parameters, were routed to the hydrological stations by this routing scheme, and then compared to the monthly-observed streamflows at the corresponding basins.

The results show that the drainage network extracted from DEM matches up well to the actual one, and the simulated streamflows matched the observed values well. Although the relative errors of some stations are not very good, the bias may not only be introduced by the routing scheme, but also by the model uncertainty and the parameter uncertainty. In general, the routing scheme performs well in a large-scale stream simulation and can be applied to the large-scale routing.

Acknowledgements This work was supported by the National Natural Science Foundation of China under Grant no. 90411007, the Knowledge Innovation Project of Chinese Academy of Sciences under Grant nos. KZCX2-YW-217 and KZCX2-YW-126-2, the National Basic Research Program under the Grant 2005CB321704, and CAS International Partnership Creative Group "The Climate System Model Development and Application Studies".

REFERENCES

Arora, V. K. & Boer, G. J. (1999) A variable velocity flow routing algorithm for GCMs. *J. Geophys. Res.* **104**(D24), 30965–30979.

Bonan, G. B. (1996) A Land Surface Model (LSM version 1.0) for ecological, hydrological, and atmospheric studies: technical description and user's guide. NCAR Technical Note NCAR/TN-417+STR. National Center for Atmospheric Research. Boulder, Colorado, USA.

Cunge, K. A. (1969) On the subject of a flood propagation method (Muskingum Method). *J. Hydraul. Res.* **7**(2), 205–230.

Dai, Y. J., Zeng, X. B., Dickinson, R. E., Baker, I., Bonan, G. B., Bosilovich, M. G., Denning, A. S., Dirmeyer, P. A., Houser, P. R., Niu, G. Y., Oleson, K. W., Schlosser, C. A. & Yang Z. L. (2003) The common land model. *Bull. Am. Met. Soc.* **84**(8), 1013–1023.

Dickinson, R. E., Henderson S. A. & Kennedy P. J. (1993) Biosphere–Atmosphere Transfer Scheme (BATS) version 1e as coupled to the NCAR Community Climate Model. NCAR technical note NCAR/TN-387+STR. National Center for Atmospheric Research, Boulder, Colorado, USA.

Fekete, B. M., Vörösmarty, C. J. & Lammers, R. B. (2001) Scaling gridded river networks for macroscale hydrology: development, analysis, and control of error. *Water Resour. Res.* **37**(7), 1955–1967.

Hao, Z. C. & Li, L. (2002) The method of creating of the net of river based on DEM. *Hydrology* **22**(4), 8–10 (in Chinese).

Hutchinson, M. F. (1989) A new procedure for gridding elevation and stream line data with automatic removal of spurious pits. *J. Hydrol.* **106**, 211–232.

Liang, X. (1994) A two-layer Variable Infiltration Capacity land surface representation for General Circulation Models. *Water Resour. Series* TR140, 208.

Liang, X., Lettenmaier, D. P. & Wood, E. F. (1996) One-dimensional statistical dynamic representation of subgrid spatial variability of precipitation in the two-layer Variable Infiltration Capacity model. *J. Geophys. Res.* **101**(D16), 403–422.

Liang, X. & Xie, Z. H. (2001) A new surface runoff parameterization with subgrid-scale soil heterogeneity for land surface models. *Adv. Water Resour.* **24**(9–10), 1173–1193.

Nijssen, B., Lettenmaier, D. P., Liang X., Wetzel, S. W. & Wood, E. F. (1997) Stream simulation for continental-scale river basins. *Water Resour. Res.* **33**(4), 711–724.

O'Callaghan J. F. & Mark D. M. (1984) The extraction of drainage networks from digital elevation data. *Computer Vision, Graphics, and Image Processing* **28**, 323–344.

O'Donnel, G., Nijssen, B. & Lettenmaier, D. P. (1999) A simple algorithm for generating streamflow networks for grid-based, macroscale hydrological models. *Hydrol. Processes* **13**(8), 1269–1275.

Olivera, F., Lear, M. S., Famiglietti, J. S. & Asante, K. (2002) Extracting low-resolution river networks from high-resolution digital elevation models. *Water Resour. Res.* **38**(11), 1231.

Price, R. K. & Harrison, A. J. M. (1975) *Flood Routing Studies*, 33–34. Natural Environment Research Council, London, UK.

Rui, X. F. (1995) *The Theory for Runoff Producing and Concentration*, 85–87. Water Resources and Electric Power Press Beijing, China (in Chinese).

Rui, X. F. & Wang, L. L. (2000) A research on flood routing method with prediction. *Adv. Water Sci.* **11**(3), 291–295 (in Chinese).

Saunders, W. (1999) Preparation of DEMs for use in environmental modeling analysis. *ESRI User Conference*, 24–30 July.

Shaw, D., Martz, L. W. & Pietroniro, A. (2003) A methodology for preserving channel flow networks and connectivity patterns in large-scale distributed hydrological models. *Hydrol. Processes* **19**(1), 149–168.

Shaw, D., Martza, L. W. & Pietroniro, A. (2005) Flow routing in large-scale models using vector addition. *J. Hydrol.* **307**, 38–47.

Turcotte, R., Fortin, J. P., Rousseau, A. N., Massicotte, S. & Villeneuve, J. P. (2001) Determination of the drainage structure of a watershed using a digital elevation model and a digital river lake network. *J. Hydrol.* **240**(3–4), 225–242.

Weiss, M. A. (1997) *Data Structures and Algorithm Analysis in C.* Addison-Wesley, Menlo Park, California, USA.

Xie, Z. H., Yuan, F., Duan, Q. Y., Liang, M. L., Zheng, J. & Chen, F. (2007) Regional parameter estimation of the VIC Land Surface Model: methodology and application to river basins in China. *J. Hydromet.* **8**(3), 447–468.

Yang, D. W., Musiake, K., Kanae, S. & Oki, T. (2000) Use of the Pfafstetter basin numbering system in hydrological modeling. In: *Proc. 2000 Annual Conference, Japan Society of Hydrology and Water Resources*, 200–2001.

Ye, A. Z., Xia, J., Wang, G. S. & Wang, X. N. (2005) Drainage network extraction and subcatchment delineation based on digital elevation model. *J. Hydrol. Engng* **36**(5), 531–537 (in Chinese)

Yuan, F., Ren, L. L., Yu, Z. B. & Xu, J. (2005) A river flow routing model based on digital drainage network. *J. Hydrodynamics* 17(4), 483–488.

Zeng, X. B., Muhammad, S., Dai, Y. J., Dickinson, R. E. & Myneni, R. (2002) Coupling of the common land model to the NCAR community climate model. *J. Climate* **15**(14), 1832–1854.

Investigation of organized hydrological heterogeneity from a spatial analysis perspective

HONGYI LI[1] & MURUGESU SIVAPALAN[1,2]

1 *Department of Civil and Environmental Engineering, University of Illinois at Urbana-Champaign, Illinois, USA*
hli23@uiuc.edu

2 *Department of Geography, University of Illinois at Urbana-Champaign, Illinois, USA*

Abstract Spatial analysis as a framework for exploring spatial patterns is well known in social-geography, econometrics and other fields, but has seldom been applied in catchment hydrology. This paper presents a new method for the investigation of the spatial patterns of hydrological variables from a *spatial analysis* perspective. Using the topographic index (TI) as an example, the preliminary yet promising results obtained from this analysis suggest that the Global Moran's I index constructed from an asymmetric inverse flow distance weight matrix is an effective bulk signature of hydrological spatial structure, and the Local Indicator of Spatial Association (LISA) captures well the transition from hydrological heterogeneity to hydrological homogeneity with increasing spatial scale. These results suggest that complementary to existing statistical methods, spatial analysis could be a promising framework for quantifying and modelling spatial patterns exhibited by hydrological variables.

Key words spatial analysis; TI; Global Moran's I; asymmetric inverse flow distance weight matrix; LISA

INTRODUCTION

The significance of spatial structure within hydrological variables has been extensively appreciated and studied (Western *et al.*, 2001; Milzow *et al.*, 2004; Marani *et al.*, 2006; Valdes *et al.*, 2006), and it has also been realized that standard statistical methods and even geostatistical methods are not sufficient to detect and quantify such spatial structures (Western *et al.*, 2001; Cai & Wang, 2006). The existence of spatial structure violates the assumption of standard statistical hypothesis tests (Cai & Wang, 2006), which assume that spatial variables are independent, thus automatically neglecting the close correlation that exists in space. Standard geostatistics is able to represent spatial continuity, but not spatial structure/organization. Western *et al.* (2001) convincingly demonstrated that the hydraulic connectivity among spatial hydrological variables, e.g. soil moisture, could not be captured and quantified by standard geostatistical methods.

Spatial analysis, however, is an alternative and effective framework for studying spatial patterns within hydrological variables, besides standard statistics and geostatistics. Cai & Wang (2006) used Moran's I to quantify the global autocorrelation of the Topographic Index (TI), and examined the impacts of spatial scale and resolution on the Moran's I of TI. However, the Moran's I in their work is constructed from an Inverse-Euclidean-Distance weight matrix without considering the hydrological connection among spatial locations, and therefore can actually only capture the spatial *continuity* of TI. We extend their work by constructing Moran's I from an asymmetric-Inverse-Flow-Distance weight matrix, which has more significant hydrological meaning than the Inverse-Euclidean-Distance weight matrix, and enables us to investigate spatial *connectivity*. Moran's I as a bulk indicator integrates the whole spatial structure into one single value. This is convenient for comparison among various basins (Cai & Wang, 2006) after standardization, but is not suitable for exploring the detailed spatial structures which might shed light on the underlying physical mechanisms. We therefore also apply another indicator, the Local Indicators of Spatial Association (LISA), to the TI of the same watershed. Interestingly, we find that LISA is able to effectively capture the apparent transition from hydrological heterogeneity to homogeneity that occurs with increasing spatial scale.

In this paper we first introduce Global Moran's I and LISA. We then apply these two indicators on TI from a natural watershed, and present the results along with some discussion of the results. Finally we draw a few preliminary conclusions.

METHODOLOGY

Spatial autocorrelation refers to the correlation of the values of one single variable through a spatial region, i.e. the value of a variable at one location is related to the values of the same variable at other neighbouring locations. Global Moran's I and LISA are the most widely used spatial statistical tools for quantifying spatial autocorrelation (Anselin, 1995, 2004). The mathematical description of Global Moran's I is as follows:

$$I = \frac{n\sum_{i=1}^{n}\sum_{j=1}^{n} W_{ij}(X_i - \overline{X})(X_j - \overline{X})}{S_0 \sum_{i=1}^{n}(X_i - \overline{X})^2} \tag{1}$$

where X is the variable of interest, i and j indicate different locations, $\overline{X}$ is the average of X through the whole study area, W_{ij} is the so-called spatial weight between X_i and X_j, which quantifies the level of relation between these two values. The larger the value of W_{ij}, the more closely correlated X_i and X_j are. $S_0 = \sum_{i=1}^{n}\sum_{j=1}^{n} W_{ij}$ is used here as a normalization value such that the range of I is from –1 to 1. Positive values indicate spatial similarity, i.e. clustering of similar values, and negative values indicate spatial dissimilarity, i.e. clustering of dissimilar values. Zero value indicates no spatial autocorrelation, i.e. spatial randomness. Considering "similarity" might have a different meaning in hydrology. We hereafter refer to spatial similarity as spatial homogeneity, and spatial dissimilarity as spatial heterogeneity.

W_{ij} could be flexibly defined according to the types of spatial relation, for example, W_{ij} can be defined as inverse Euclidean distance D_{ij}, i.e. $W_{ij} = 1/D_{ij}$. This definition is consistent with the First Law of Geography, which states that "every thing is related to something else, but nearer things are more related than distant things" (Tobler, 1970). However, the global Moran's I based on this definition of spatial weight matrix is just a measure of spatial continuity. In this paper, W_{ij} is defined as inverse flow distance, and flow distance is the distance between any two locations along a natural flow path. The difference between Euclidean distance and flow distance is illustrated in Fig. 1. Note that $[W_{ij}]$ is a spatial weight matrix, and hence $[W_{ij}]$ constructed from inverse asymmetric flow distance is an asymmetric matrix, which is the one utilized in this work.

It is often assumed in spatial analysis that the values at two locations will be independent if the distance between them exceeds a certain value, i.e. threshold distance (hereby noted as $D_{threshold}$). That is, if the two locations are far enough, they are not related, thus the corresponding weight is automatically zero. Therefore, for each value of threshold distance $D_{threshold}$ there will be a corresponding weight matrix, and thus a corresponding value of Global Moran's I.

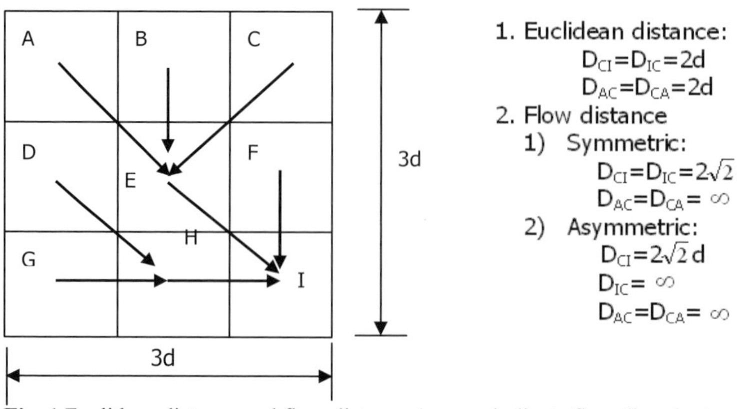

Fig. 1 Euclidean distance and flow distance (arrows indicate flow direction).

Notice that although Global Moran's I is a dimensionless number, there is no theoretical foundation to directly compare the values of Global Moran's I (for different variables within the same region or the same variable within different regions). A more rigorous way is to use standardized Global Moran's I (Luc Anselin, personal communication), calculated as follows:

$$z = E(I)/SE(I) \tag{2}$$

where $E(I)$ and $SE(I)$ are the average and standard deviation of Global Moran's I, respectively. A positive standardized Global Moran's I means spatial similarity, whereas a negative value means spatial dissimilarity.

According to its definition, Global Moran's I is a bulk indicator of spatial correlation integrated over the whole region, and cannot help detect detailed information on spatial processes within a region. Local Indicators of Spatial Association (LISA) are able to represent local spatial autocorrelation for each location, and thus are a powerful tool to explore spatial processes within the region of interest (Anselin, 1995). Notice that LISA is a generic name of a series of local spatial statistics. Local Moran's I, as a special case of LISA, can be defined as:

$$I_i = (X_i / m_2)\sum_{j=1}^{n} W_{ij} X_j \tag{3}$$

where $m_2 = \sum_i X_i^2$ does not vary with location *i*. Global Moran's I is the spatial mean of Local Moran's I.

RESULTS AND DISCUSSION

Global Moran's I of TI

A small watershed with a drainage area of about 3 km^2 in Brown County, Indiana, USA, was selected for this illustrative study. Topographic Index (TI) (Beven & Kirby, 1979) has significant hydrological meaning and exhibits spatial structure, and is therefore selected as the variable under study. But there is no doubt that these spatial analysis tools can be applied to any other hydrological variable that exhibits spatial structure. TI of this watershed is calculated based on a 30 × 30 DEM, as shown in Fig. 2.

Table 1 shows Global Moran's I calculated from different distance types and different threshold distances. Generally speaking, the values of standardized Global Moran's I from inverse

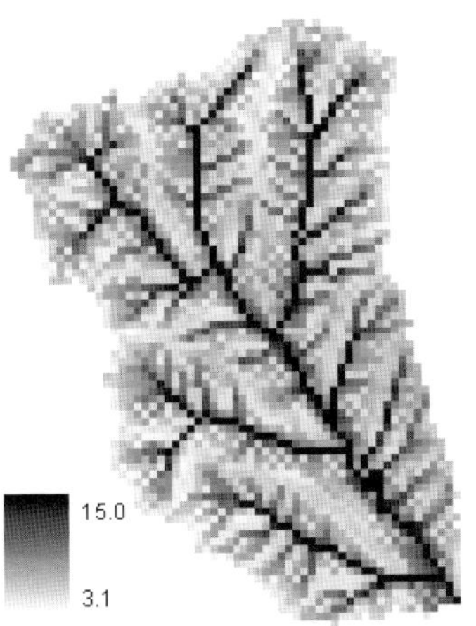

Fig. 2 Distribution of TI in the watershed under study.

Table 1 Moran's I and Z values under different weight matrices.

$D_{threshold}$	Distance type	Global Moran's I	Standardized Global Moran's I
30 m	Euclidean distance	0.4406	32.34
30 m	Flow distance	0.7284	44.06
150 m	Euclidean distance	0.1380	41.94
150 m	Flow distance	0.6045	80.37

flow distance weight matrices are usually greater than those from inverse Euclidean distance weight matrices. Although not fully presented here, the results under other threshold values also confirm this conclusion. The reason for this phenomenon is that the probability of two locations on the same flow path having similar values (inverse flow distance weight) is larger than that of two locations close to each other, but not on the same flow path (inverse Euclidean distance weight), while the probability of two locations on the same flow path having dissimilar values is smaller than that of two locations close to each other while not on the same flow path. This would be more easily understood by considering two locations, one in the river, and another on a hillslope close to the previous one, but with the two locations not being on the same flow path. The values of TI will be significantly different between them. The weight from inverse Euclidean distance is a fairly large value due to short Euclidean distance, which leads to smaller Moran's I due to the dissimilarity between the values of TI. The weight from inverse flow distance is zero since they are not hydraulically connected, and thus will not affect the value of Moran's I.

Impact of threshold distance

Threshold distance is another important factor affecting Moran's I. There are, in fact, two methods to define neighbouring area of a location with the help of threshold distance (Luc Anselin, personal communication). One is called the cumulative neighbourhood method, which defines all locations having an effective distance smaller than the given threshold distance as neighbours. Such a neighbourhood is like a dish. The previous results are based on this method. The other one is called the band neighbourhood method, which defines all locations having an effective distance larger than a minimum distance $D_{threshold,\min}$ while smaller than a maximum distance $D_{threshold,\max}$ as neighbours. Such a neighbourhood is like a circular band, or a donut. The difference between $D_{threshold,\max}$ and $D_{threshold,\min}$ is called the width of the band, and is usually a constant. The sensitivity of standardized Global Moran's I on threshold flow distance is examined in this work with the cumulative neighbourhood method and the band neighbourhood method, respectively.

Figure 3(a) shows that, under the cumulative neighbourhood method, standardized Global Moran's I of TI first increases quickly, reaches the maximum at a threshold flow distance of about 500 m, and then decreases relatively slowly with increasing threshold flow distance. This is consistent with Fig. 3(b), where the band width in Fig. 3(b) is 30 m in this work, and the threshold distance is estimated as the mean of $D_{threshold,\max}$ and $D_{threshold,\min}$. In Fig. 3(b), under the band neighbourhood method, the standardized Moran's I of TI starts as a positive value (similarity among nearby neighbours), decreases to zero at a threshold distance of 500 m, and then flips to negative. The abrupt increase of the standardized Moran's I of TI from about 2600 m is probably caused by error since the longest flow distance in the study area is about 2600 m, and can thus be neglected. A possible hypothesis which could be inferred from Fig. 3 is that the values of TI at two locations having a certain distance (here about 500 m) might not be related, i.e. they are totally independent.

Local Moran's I

As we have discussed previously, the values of local Moran's I are the local components of Global Moran's I, and a map of local Moran's I is a detailed description of the spatial structure of the hydrological variable of interest.

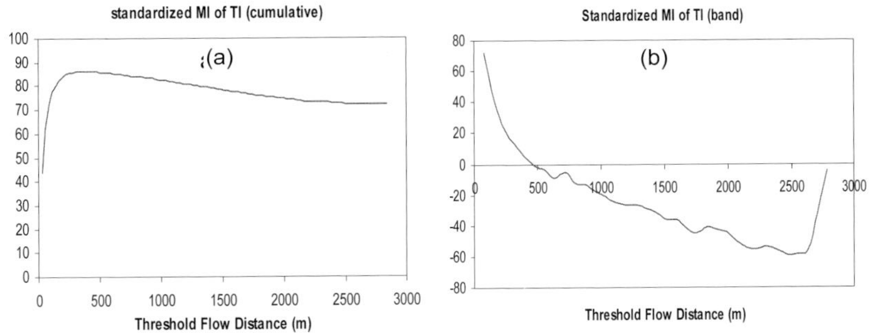

Fig. 3 Standardized Global Moran's I (under asymmetric inverse flow distance weight matrix) changing with threshold distance.

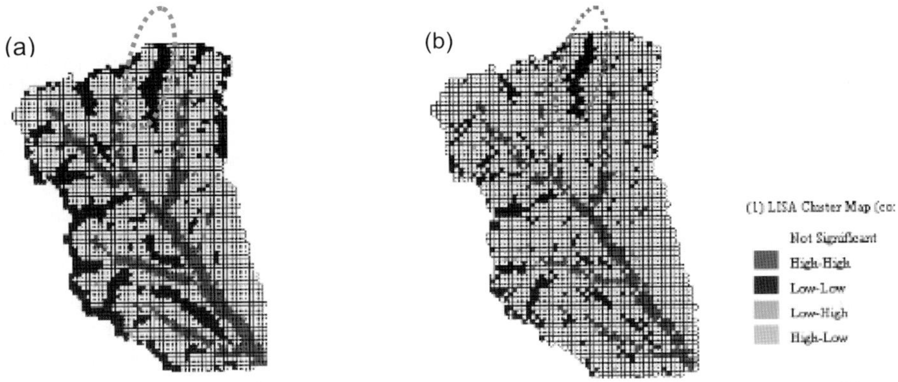

Fig. 4 LISA cluster maps under traditional weight matrices: (a) queen contiguity weight, (b) inverse Euclidean distance weight.

Figure 4 shows the maps of local Moran's I calculated from traditional weight matrices: queen contiguity weight matrix and inverse Euclidean distance weight matrix. Queen contiguity is a type of contiguity (the other two are rook contiguity and bishop contiguity). If two spatial units share a common boundary or a common vertex, then we define their relationship as queen contiguity. Notice that our work is based on 30 × 30 DEM, i.e. grid areas. These maps are calculated with software *GeoDa* (Anselin, 2004a) and *R* (Anselin, 2004b). The High-High clusters are the locations with high TI values surrounded by high TI values (similarity), Low-Low clusters are the locations with low TI values surrounded by low TI values (similarity), Low-High clusters are the locations with low TI values surrounded by high TI values (dissimilarity), and High-Low clusters are the locations with high TI values surrounded by low TI values (dissimilarity). The common feature of Fig. 4(a) and (b) is that the grids near ridges are defined as Low-Low clusters as highlighted by dashed circles, i.e. they are defined as similar neighbours.

However, some grids near ridges are not real neighbours from a hydrological perspective. Although some grids at the ridges are adjacent, they may not be hydrologically connected, since water will disperse in different directions. The probability of them being on a common flow path is considerably low. So according to the definition of inverse flow distance weight, the weight between such two locations at the ridges is most probably zero. This fact is well reflected by the maps of standardized local Moran's I shown in Fig. 5. It can thus be concluded that LISA from inverse flow distance weight matrix is more reasonable than the LISA from other weight matrices, in the sense that it effectively removes the hydrologic-pseudo-neighbours.

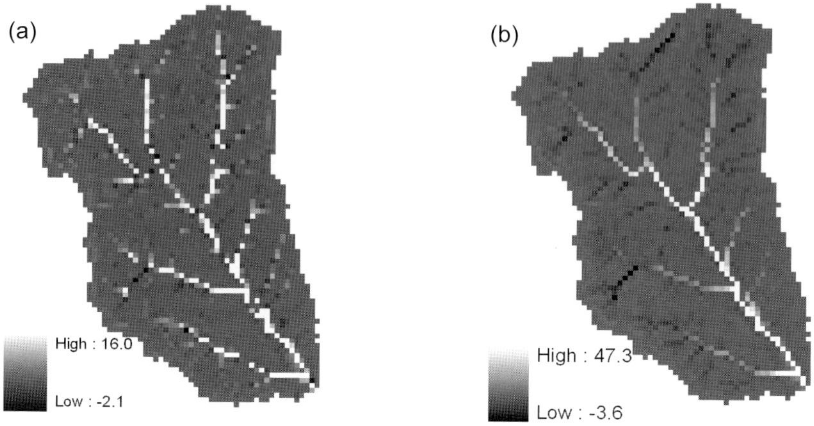

Fig. 5 Standardized Local Moran's I under inverse flow distance weight matrix: (a) threshold flow distance = 30 m, (b) threshold flow distance = 3000 m.

Another interesting pattern from the inverse flow distance weight matrix method (Fig. 5), which is not imposed by the results of the other two methods (Fig. 4), is that the LISA values increase along the streams in the downstream direction, i.e. from negative (significant heterogeneity) at the headwater cells to positive (significant homogeneity) at the downstream cells. Note that the grid cells within the river reaches are **river cells**. According to the definition of inverse flow distance weight matrix, and thus the corresponding definition of hydrological neighbours, for those river cells at the headwaters, their neighbours are mainly the cells on the hillslope (hereafter noted as **hillslope cells**). Noticing that the river cells have high TI values while the hillslope cells have low TI values, it is not difficult to expect dissimilarity among the most upstream river cells and their neighbours. However, for any cell on the main stream, river cells consist of a large part of its neighbours, and because the river neighbour cells have apparently larger weights (due to shorter flow distances) than hillslope neighbour cells, i.e. the river neighbour cells have more impacts than the hillslope cells, so it is reasonable to expect similarity among the cells on the main stream. The increasing of LISA values along the channel in the downstream direction is especially clearly shown in Fig. 5(b). Together with the increasing of LISA values, the drainage area, i.e. spatial size, is also increasing in the downstream direction. That is, a river cell with a larger drainage area is also having a larger LISA value, which implies more spatial homogeneity.

CONCLUSIONS

From the above discussions, a few preliminary conclusions can be drawn:

- Spatial analysis is a promising framework to detect and quantify spatial pattern/structure besides standard statistics and geostatistics, as illustrated here with standardized Global Moran's I and Local Moran's I.
- From a hydrological perspective, inverse flow distance weight matrix has more hydrological meaning than inverse Euclidean distance proposed by Cai & Wang (2006), and thus has more potential to be applied for hydrological modelling on a catchment scale.
- Comparing with Global Moran's I, LISA, here referring to local Moran's I based on inverse flow distance weight matrix, is more able to detect local spatial pattern and help us to detect spatial trends in a region.
- LISA of a hydrological variable nicely represents the transition from spatial heterogeneity (dissimilarity) to homogeneity (similarity) from the headwater to the watershed outlet, with

increasing drainage area. It is the authors' feeling that, although further theoretical clarification is necessary, this result could possibly be used as one of the ways to estimate the reasonable size of the Representative Elementary Watershed (REW) (Reggiani *et al.*, 1998) for distributed hydrological modelling.

Acknowledgements The first author wants to thank Luc Anselin, Dingbao Wang and Ximing Cai for the valuable discussions and suggestions.

REFERENCES

Anselin, L. (1995) Local indicators of spatial association–LISA. *Geograph. Anal.* **27**(2), 93–115.

Anselin, L. (2004a) Exploring spatial data with GeoDa. *A Workbook, Spatial Analysis Laboratory*, 129–147. Spatial Analysis Laboratory, Dept. of Agricultural and Consumer Economic and CSISS, University of Illinois at Urbana-Champaign, Illinois, USA.

Anselin, L. (2004b) Spatial regression analysis in R: *A Workbook, Spatial Analysis Laboratory*, 13–34. Spatial Analysis Laboratory, Dept. of Agricultural and Consumer Economics and CSISS, University of Illinois at Urbana-Champaign, Illinois, USA.

Cai X. & Wang, D. (2006) Spatial autocorrelation of topographic index in catchments. *J. Hydrol.* **328**, 581–591.

Marani, M., Silvestri, S., Belluco, E., Ursino, N., Comerlati, A., Tosatto, O. & Putti, M. (2006) Spatial organization and ecohydrological interactions in oxygen-limited vegetation ecosystems. *Water Resour. Res.* **42**, W06D06, doi:10.1029/-2005WR004582.

Milzow, C., Molnar, P., McArdell, B. W. & Burlando, P. (2006) Spatial organization in the step-pool structure of a steep mountain stream (Vogelbach, Switzerland). *Water Resour. Res.* **42**, W04418, doi:10.1029/2004WR003870.

Reggiani, P., Sivapalan, M. & Hassanizadeh, S. M. (1998) A unifying framework for watershed thermodynamics: Balance equations for mass, momentum, energy, entropy and the 2nd law of thermodynamics. *Adv. Water Resour.* **22**(4), 367–398.

Tobler, W. R. (1970) A computer movie simulating urban growth in the Detroit region. *Econom. Geography Suppl.* **46**, 234–240.

Valdes, D., Dupont, J.-P., Laignel, B., Ogier, S., Leboulanger, T. & Mahler, B. J. (2007) A spatial analysis of structural controls on Karst groundwater geochemistry at a regional scale. *J. Hydrol.* **340**, 244–255.

Western, A. W., Blöschl, G. & Grayson, R. B. (2001) Towards capturing hydrologically significant connectivity in spatial patterns. *Water Resour. Res.* **37**(1), 83–97.

Appropriate contributing area threshold of a digital river network extracted from DEM for hydrological simulation

LI LI[1,2], JIAHU WANG[1,2] & ZHENCHUN HAO[1]

1 *State Key Laboratory of Hydrology-Water Resources and Hydraulic Engineering, Hohai University, No. 1 Xikang Road, Nanjing, China*
llwjh@163.com

2 *National Weather Center, University of Oklahoma Norman, Oklahoma 73019, USA,*

Abstract Contributing area threshold is an important parameter of extracting digital river networks (DRNs) from digital elevation models (DEMs). Impact of its change on runoff concentration, methods of determining it, and its spatial distribution are studied in this article. Conclusions showed that with the increase of contributing area threshold, basin mean concentration time of flood became longer and flood peak decreased, while the time of peak was almost unchanged. It meant that the runoff concentration of the hydrological model was sensitive to the contributing area threshold. A new approach was developed, in which the river system in digital maps was considered as the correct river network (CRN) while that extracted from DEM was called the digital river network (DRN). Drainage density and fractal dimension of DRNs derived from different thresholds were compared with that of CRN to choose the best one. Contributing area thresholds of 60 catchments were estimated using this method. It was found that the spatial distribution of them followed some rules, e.g. it changed approximately with power function of upslope gradient in areas with the same vegetation. It would be of benefit to take into account its spatial variation during extracting catchment properties and simulating hydrological processes.

Key words digital river network; DEM; contributing area threshold; catchment properties

INTRODUCTION

Distributed hydrological models have been widely used in hydrological simulating and studies of impact of climate change on water resources. Extracting catchment properties from DEM is usually the beginning of developing a distributed hydrological model based on DEM. A variety of algorithms (Band, 1986; Tribe, 1992; Garbrecht & Martz, 1997) have been reported in the literature and the most pertinent problems have been highlighted. These are related to the treatment of flat areas and to the location of the channel heads. The former has been researched extensively and some methods have been developed (Hutchinson, 1989; Martz & Garbrecht, 1998; Turcotte & Fortin, 2001). However, studies about the latter are far fewer than the former. In the following channel heads are discussed, including sensitivity of flow concentration to them and the methods of determining it.

In geomorphology, the head location of the channel is a concept related to river erosion, so it is described as the result of sediment transport and erosion processes. The contributing area of all grids can be obtained after pit filling and flow direction calculation during the period of extracting the river network from DEM. Grids with a larger contributing area more possibly belong to channels. Based on this concept, a common approach of extracting drainage networks from a DEM is to consider a grid cell as being part of a channel if its contributing area is larger than a defined contributing area threshold (O'Callaghan *et al.*, 1984). The threshold is also called a critical contributing area. It is considered as a parameter of digital river network extraction and distinguishes stream processes from flow concentration of hillslopes.

The structure of stream networks, such as river density, classification of streams and partition of sub-basins, may be different with different threshold (Baxter *et al.*, 2004). Impacts of its change on landscape properties (length of streams, main channels and drainage paths) and scale properties (Horton law and Strahler law) can also not be ignored (Roger *et al.*, 1996). Some studies showed that its change would influence width function (Yang *et al.*, 2000) and basin geomorphologic instantaneous unit hydrograph (Liu *et al.*, 2003). So the change of contributing area threshold would affect the proportion of overland and channel processes, and as a consequence, would affect the basin flow concentration in hydrological modelling.

The contributing area threshold varies with local climate, soil, geology, vegetation, and landscape (Martz *et al.*, 1992). A lot of literature (Giorgio *et al.*, 1996; Karea *et al.*, 1996) has declared that there was a relationship of power function between contributing area (A) and local slope (S). The threshold value can be evaluated from the plot of slope *vs* area. This approach is based on the premise that there will be a single inflection point in the longitudinal profile near the head of a channel. Moreover, it is expected that the slope will reach its maximum value at this inflection point. So a single value is usually applied due to the lack of more detailed information, ignoring its spatial distribution with change of hydrology, geology, relief and climate (Carpenter *et al.*, 2001). The underlying hypothesis is applied to find the location of channel heads, that channel heads are located in zones where fluvial transport becomes dominant over diffusive transport, corresponding to the spatial transition from convex to concave slope profiles (Jürgen *et al.*, 2003). Some authors get it from a log–log plot of local slope against contributing area (Tarboton *et al.*, 1992) or relationship between drainage density and corresponding assumed area threshold (Kong, 2003). Others get it by photointerpretation techniques (Dennis & Arthur, 1997; Lopez & Camarasa, 1999).

Unfortunately, there is a considerable scatter in these slope/area plots and furthermore the network identified in this way shows a constant drainage density. Forced by the need to obtain a non-uniform drainage density in the different parts of the catchment, as expected in natural basins, some authors assumed contributing area as a surrogate variable for discharge (Rodriguez-Iturbe *et al.*, 1992) and used a threshold of AS^k to identify channel network (Giannoni *et al.*, 2000).

In order to study impact of contributing area threshold on basin runoff concentration, two small catchments (named as Sta1 and Sta22, respectively; their details are listed in Table 1) lying in the middle reach of the Yellow River Basin were selected in this study. Some approaches in common use to evaluate contributing area threshold are discussed, taking the two catchments as examples. A new method was proposed to estimate the threshold according to drainage density of the correct river network. Finally, the contributing area thresholds of 60 catchments were evaluated and analysed to obtain their distribution rules. Both the GTOPO30 USGS DEM and digitized river network from 1:250 000 topographic maps of China were used during the study.

Table 1 Information of the two catchments.

ID	Information of outlets			Information of catchments	
	Long. (East)	Lat. (North)	Altitude(m)	Number of grids	Area of catchment (km^2)
Sta1	112.363	36.413	1006	285	197
Sta22	110.329	34.171	1013	937	664

SENSITIVITY OF BASIN FLOW CONCENTRATION TO CONTRIBUTING AREA THRESHOLD

The digital river system, flow path network and their slope describe the properties of basin runoff concentration if ignoring the non-uniformity of the underlying surface. Variation of them implies the changed response of the basin to flow concentration process in distributed hydrological simulation. The reservoir action of the watershed can be presented as the translation and flattening-out of the flood hydrograph. Consequently, different contributing area thresholds may result in change of concentration time and flood hydrograph.

Change of basin mean concentration time

The concentration time of each water drop in one watershed is different from others because of their varied flow velocity and path. Similarly, the concentration time of flow in each grid is different to others in distributed hydrological simulations because grids are considered as minimum units. In the study, concentration time of each grid from its centre to the outlet was evaluated by equation (1):

$$T_i = \sum_{j=1}^{M_i-1} t_j \,;\, t_j = \frac{l_j}{v_j} = \frac{l_j}{K_V S_j^{\frac{1}{2}}} \tag{1}$$

where, T_i is the concentration time of flow from centre of the ith grid to the outlet, M_i is the number of grids it passes through, t_j is the time it needs from the jth grid to the next downstream grid, l_j is the surface distance, v_j is average flow velocity, S_j is gradient, and K_V is defined as the velocity constant, implying the effect of hydraulic factor, such as roughness and hydraulic radius, on flow velocity.

Basin mean concentration time is defined as the average concentration time of all water drops. In distributed hydrological modelling, it is defined as that of all grids. In order to study its universal change, define:

$$Kv_{channel} = \beta Kv_{slope};\ \tau = \frac{1}{Kv_{slope}} \tag{2}$$

where, $Kv_{channel}$ and Kv_{slope} are the velocity constant of channel and overland runoff, respectively, and β is a constant. They change with different stream and underlying surface. In this study, they are assumed to be invariable for the same basin. So t_j for channel grids and slopes may be expressed as:

$$t_j = \frac{l_j}{Kv_{channel} S_j^{\frac{1}{2}}} = \frac{l_j}{\beta Kv_{slope} S_j^{\frac{1}{2}}} \text{ and } t_j = \frac{l_j}{Kv_{slope} S_j^{\frac{1}{2}}} \tag{3}$$

So basin mean concentration time can be calculated with the following equations:

$$T_i = \tau \left(\sum_{j=1}^{m_i} \frac{l_j}{S_j^{\frac{1}{2}}} + \sum_{j=1}^{n_i-1} \frac{l_j}{\beta S_j^{\frac{1}{2}}} \right) = \tau \alpha_i \,;\ T = \frac{1}{N} \sum_{i=1}^{N} T_i = \frac{1}{N} \sum_{i=1}^{N} \alpha_i \tau = \left(\frac{1}{N} \sum_{i=1}^{N} \alpha_i \right) \tau \tag{4}$$

where, N is the number of grids in the basin, m_i is the number of slope-grids passed through by runoff from the ith grid to the outlet-grid, n_i is that of channel-grids. The concentration time of each grid to the outlet is the function of τ. It is convenient to calculate mean basin concentration time and compare it with that of different conditions.

Impact of surface roughness on overland runoff is larger than that of channel runoff because flow in the channel is deeper. So the velocity constant of channel, $Kv_{channel}$, is larger than that of overland Kv_{slope}. They would vary with the change of conditions. For example, the value of Kv_{slope} is about 0.21 for forested surface and 1.5 for farmland (Li *et al.*, 2003). While the value of $Kv_{channel}$ may reach 89.9 in the Xiaoli River, which is a branch of the Yellow River, according to its flow velocity of 6.67 m/s and mean gradient of 5.5‰. Upon that, the value of constant β was assumed as 10, 50 and 100, respectively, in the study. The change of river basin mean concentration time with critical contributing area in the two studied catchments is shown in Fig. 1, in which the y-coordinate means the relative increment of basin mean concentration time *vs* that when critical contributing area was one grid. It shows that basin mean concentration time becomes longer with the increasing contributing area threshold. With the increase of β, basin mean concentration time is also longer. Furthermore, the sensitivity is related to basin area. Its change is more apparent when basin area is smaller because of the larger percentage of slope grids.

Change of flow hydrograph

The reservoir action of the basin would also change with the variation of contributing area threshold. Assume that net precipitation (R) occurs on a basin uniformly in one time-step Δt and the runoff concentration of each grid is independent of others. Linear reservoirs are applied to simulate runoff concentration from each grid to the next downstream grid. These reservoirs for channel or overland runoff both assume no change along the flow path. The difference of runoff

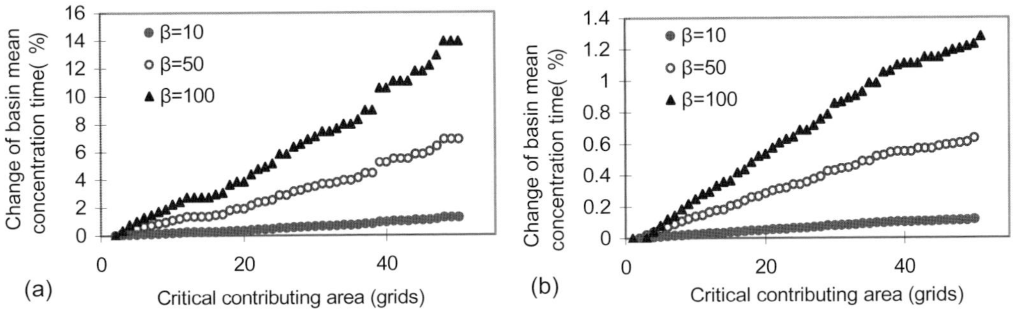

Fig. 1 Relative change of basin mean concentration time with critical contributing area: (a) Sta1 basin area, 197 km^2, (b) Sta22 basin area, 664 km^2.

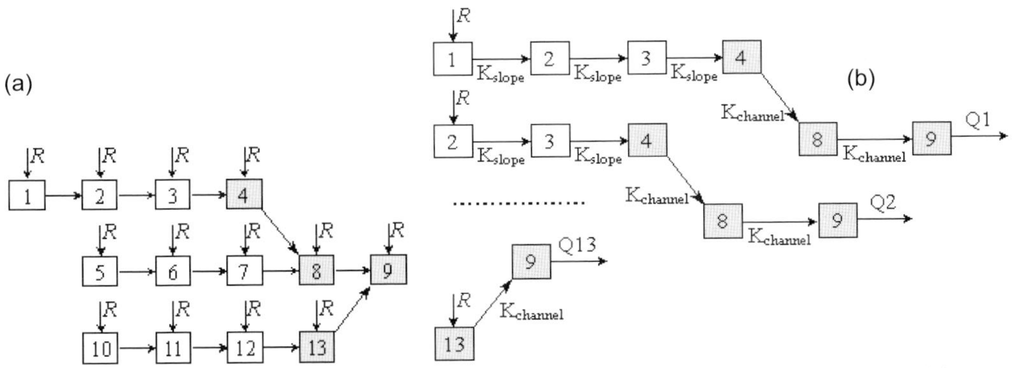

Fig. 2 Sketch map of basin flow routing; the number in grids is the ID of each grid, white grids mean slope grids and grey grids mean channel grids: (a) flow path network and net precipitation, (b) flow routing simulation of each grid.

concentration process between channels and hillslopes lies on the value of outflow coefficient K. Figure 2 illustrates this, in which, $K_{channel}$ is the outflow coefficient of channel and K_{slope} is that of hillslopes. So the flood hydrograph at the outlet can be obtained by adding up that of each grid:

$$Q_{out} = Q_1 + Q_2 + \cdots\cdots + Q_N \tag{5}$$

where, Q_{out} is the total runoff at outlet, Q_i is the outlet runoff contributed from the ith grid.

The percentage of overland and channel runoff is different, with the change of critical contributing area, which results in the change of flow hydrograph and flood peak. Flow hydrographs of Sta1 and Sta22 with different outflow coefficients were calculated to investigate its sensitivity to contributing area threshold. Figures 3 and 4 show when the value of K_{slope} and $K_{channel}$ were 0.2 and 0.6, respectively. The threshold is 10, 15, 20, 25, 30, 35 or 45 grids. With increasing contributing area threshold, the outflow of basins became slower, discharge of flood peak was decreasing and the recession period became longer, while the time of flood peak changed slightly.

NEW APPROACH TO THRESHOLD EVALUATION

The impact of the digital river network on basin runoff concentration is due to its characteristics, such as contributing area, channel length and slope, drainage density and morphology. Change of river network caused by different contributing area thresholds embodied in its statistical properties, such as density, and morphology properties. In this study, drainage density was applied to evaluate threshold and river system in digital maps was considered as the correct river network

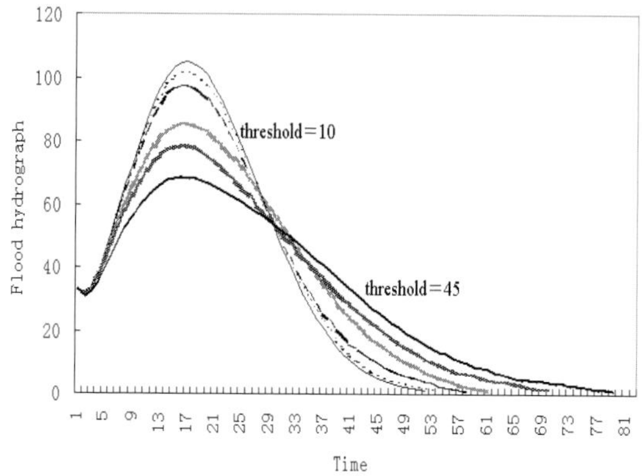

Fig. 3 Change of flood hydrograph in Sta1.

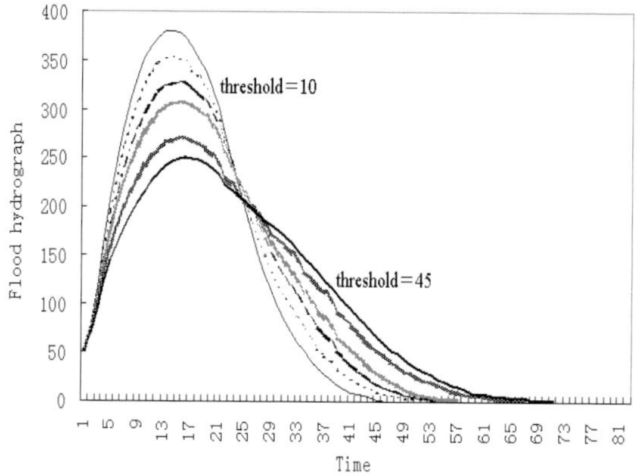

Fig. 4 Change of flood hydrograph in Sta22.

(CRN). Drainage density of correct and digital river networks with different contributing area thresholds were calculated and compared, selecting the one closest to that of the correct river system as the right one. The relevant contributing area threshold is also considered as the reasonable value.

The river system in digital maps was preprocessed first as shown in Fig. 5, in order to calculate its characteristics automatically. In the figure, "gridding" means partitioning the channel of the river system into grids according to cells of DEM, which can be accomplished by some GIS software. "Vectoring" means determining flow direction of grids with channel segments based on their elevation (Wang *et al.*, 2005). "Single-gridding" means making channels be single-grid width, which makes characteristics of the correct river system comparable with that of the digital river.

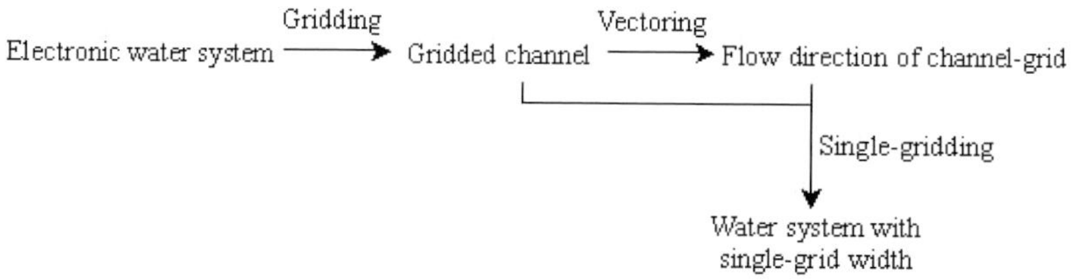

Fig. 5 Preprocessing of correct river system.

Table 2 Characteristics comparison of correct and digital river system.

Characteristics	Sta1			Sta22		
	Correct	Digital	Error	Correct	Digital	Error
Drainage density (km/km^2)	0.222	0.226	1.80	0.387	0.397	2.58
Fractal dimension	1.089	1.137	4.41	1.460	1.478	1.23

Drainage density of CRNs in Sta1 and Sta22 were 0.222 km/km^2 and 0.387 km/km^2, respectively, the evaluated threshold with this method being 21 and 8 grids. Drainage density and fractal dimension were applied to judge the similar degree between correct and digital river network for the former showed the density of channel, and the latter described its morphology and growth (Jin *et al.*, 2000). Some of their characteristics are listed in Table 2. Table 2 shows that fractal dimension of digital and correct river network were close when making thresholds 21 and 8 in Sta1 and Sta22, respectively. It implies that they were similar in morphology characteristics and development, as well as in density of channels.

SIMPLE ANALYSIS ABOUT SPATIAL DISTRIBUTION OF THRESHOLD

Contributing area threshold locates the point where overland runoff becomes channel runoff. As mentioned before, it is different not only with landscape and relief, but also with geology, hydrology and climate. Spatial distribution of contributing area threshold exhibits that of river characteristics. It would benefit to improve the precision and rationality of digital river networks to take into account its distribution. The spatial distribution of threshold was discussed after evaluating the basin mean contributing area threshold. Thresholds of 60 small catchments located in the Sanmenxia-Huayuankou region of the Yellow River were evaluated with the above method based on the assumption of contributing area threshold being equal in the same catchment. It is the reason for choosing small catchments in this study. Density of channel head and fractal dimension of digital river system in each catchment were also calculated at the same time. Compare them with that of the correct river network and similarly estimate thresholds corresponding to channel head density and fractal dimension. Basins in which the three thresholds being similar were sought out for later study while they were different, sometimes due to errors coming from preprocessing of correct river network, digital channel extraction and resolution of DEM. Twenty five catchments were sought out and their thresholds were analysed to find the spatial distribution rules. Conclusion of a simple analysis showed that thresholds of these catchments were related to vegetation and mean gradient of hillslopes. The main vegetation of the study area is woods and farmland. The relationships between contributing area threshold and mean gradient of hillslopes with different vegetation are shown in Fig. 6. They are both in power law. But there were three

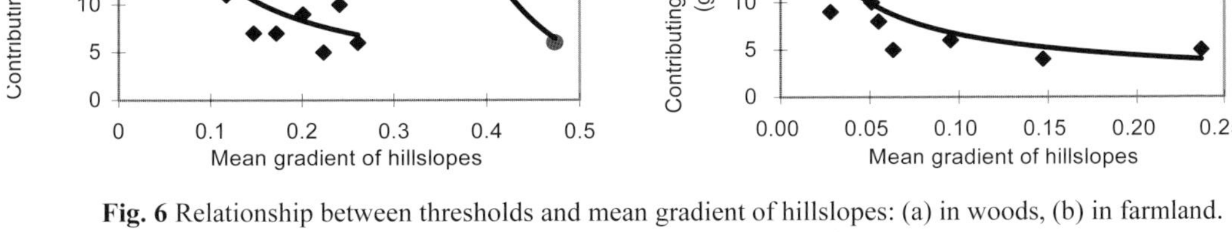

Fig. 6 Relationship between thresholds and mean gradient of hillslopes: (a) in woods, (b) in farmland.

catchments deviated from the main relationship in Fig. 6(a), which may be related to other factors, such as geology and soil.

CONCLUSION

Contributing area threshold is an important parameter during the extraction of the digital river network, which determines the basin runoff concentration of distributed hydrological models by partitioning the proportion of overland and channel runoff processes. Studies of the sensitivity of watershed runoff concentration on contributing area threshold showed that flood peak and hydrograph changed, apparently with different threshold. Appropriate thresholds will describe the obtained digital channel and flow path characteristics of basin runoff concentration better and help to decrease the uncertainty of hydrological simulation. A new method of estimating thresholds based on correct river networks was suggested after testing some methods in common use. The result obtained using the new method was similar, to the correct river network in morphology characteristic and development. Simple analysis of its spatial distribution showed that a power law existed between contributing area threshold and mean gradient of hillslopes. Furthermore, the relationship was related to vegetation and other factors.

REFERENCES

Band, L. E. (1986) Topographic partition of watersheds with digital elevation models. *Water Resour. Res.* **22**(1), 15–24.

Baxter, E. V., Zhengtao C. & Anubhav, G. (2004) Evaluation of a physics-based distributed hydrologic model for flood forecasting. *J. Hydrol.* **298**, 155–177.

Carpenter, T. M., Georgakakos, K. P. & Sperfslage, J. A. (2001) On the parametric and nexrad-radar sensitivities of a distributed hydrologic model suitable for operational use. *J. Hydrol.* **253**, 169–193.

Dennis, L. J. & Arthur, C. M. (1997) A spatially distributed hydrologic model utilizing raster data structures. *Computers & Geosci.* **23**(3), 267–272.

Garbrecht, J. & Martz, L. W. (1997) The assignment of drainage direction over flat surfaces in raster digital elevation models. *J. Hydrol.* **193**, 204–213.

Giannoni, F., Roth, G. & Rudari, R. (2000) A semi-distributed rainfall–runoff model based on a geomorphologic approach. *Phys. Chem. Earth (B)* **25**(7–8), 665–671.

Hutchinson, M. F. (1989) A new procedure for gridding elevation and stream line data with automatic removal of spurious pits. *J. Hydrol.* **106**, 211–232.

Jin, D., Chen, H. & Guo, Q. (2000) An experimental study on influence of materiel component to non-linear relation between sediment yield and drainage network development. *Acta Geographica Sinica* **55**(4), 439–448.

Jürgen, V. V., Roberto C. & Francesca B. (2003) Deriving drainage networks and catchment boundaries: a new methodology combining digital elevation data and environmental characteristics. *Geomorphology* **53**, 281–298.

Kong, F. (2003) Distributed watershed hydrological model and flow concentration based on digital technique. PhD Thesis, Hohai University, Nanjing, China.

Li, L. (2003) Extraction of catchment properties from DEM and its using in hydrologic simulation. MSc Thesis, Hohai University, Nanjing, China.

Liu, Y. B., Gebremeskel, S. & De Smedt, F. (2003) A diffusive transport approach for flow routing in GIS-based flood modelling. *J. Hydrol.* **283**, 91–106.

Lopez, G. M. J. & Camarasa A. M. (1999) Use of geomorphological units to improve drainage network extraction from a DEM: Comparison between automated extraction and photointerpretation methods in the Carraixet catchment (Valencia, Spain). *Int. J. Applied Earth Observation & Geoinformation* **1**(3/4), 187–195.

Martz, L. W. & Garbrecht, J. (1992) Numerical definition of drainage network and subcatchment areas from digital elevation models. *Comput. Geosci.* **18**(6), 747–761.

Martz, L. W. & Garbrecht, J. (1998) The treatment of flat areas and closed depressions in automated drainage analysis of raster digital elevation models. *Hydrol. Processes* **12**, 843–855.

O'Callaghan, J. F. & Mark, D. M. (1984) The extraction of drainage networks from digital elevation data. *Computer Vision, Graphics & Image Processing* **28**, 323–344.

Rinaldo, A., Rigon, R., Vogel, G. K. & Rodriguez-Iturbe, I. (1995) Can one gauge the shape of a basin? *Water Resour. Res.* **31**(4), 1119–1127.

Rodriguez-Iturbe, I., Rinaldo, A., Rigon, R., Bras, R. L., Marani, A. & Ijjasz-Vasquez, E. (1992) Energy dissipation, runoff production, and the three-dimensional structure of river basins. *Water Resour. Res.* **28**(4), 1095–1103.

Roger, M. & Claude, B. (1996) Fractal analyses of tree-like channel networks from digital elevation model data. *J. Hydrol.* **187**, 157–172.

Roth, G., La Barbera, P. & Greco, M. (1996) On the description of the basin effective drainage structure. *J. Hydrol.* **187**, 119–135.

Tarboton, D. G., Bras, R. L. & Rodriguez-Iturbe, I. (1992) A physical basis for drainage density. *Geomorphology* **5**, 59–76.

Tribe, A. (1992) Automated recognition of valley lines and drainage networks from grid digital elevation models – a review and a new method. *J. Hydrol.* **139**, 263–293.

Turcotte, R., Fortin, J.-P., Rousseau, A. N., Massicotte, S. & Villeneuve, J.-P. (2001) Determination of the drainage structure of a watershed using a digital elevation model and a digital river and lake network. *J. Hydrol.* **240**, 225–242.

Vandaele, Karel, Poesen, Jean, Govers, Gerard & van Wesemael, Bas (1996) Geomorphic threshold conditions for ephemeral gully incision. *Geomorphology* **16**, 161–173.

Wang, J., Hao, Z. & Li, L. (2005) Method for automatic determining the flow direction of raster river network. *J. Hydraul. Engng* **36**(8), 972–977.

Yang, D., Herath, S. & Musiake K. (2000) Spatial resolution sensitivity of catchment geomorphologic properties and the effect on hydrological simulation. *Hydrol. Processes* **15**, 2085–2099.

A statistically-based runoff-yield model

ZHONGMIN LIANG, JUN WANG, YANG LEI & YE SHI

State Key Laboratory of Hydrology-Water Resources and Hydraulic Engineering, Hohai University, Nanjing 210098, China

zmliang@hhu.edu.cn

Abstract This paper describes a statistical-based rainfall–runoff model considering spatial variation of rainfall and other factors that control the runoff-yield process. Both infiltration excess and saturation excess runoff-yield mechanisms are coupled in this model. The spatial variation of a rainfall event is described by introducing a probability density function (pdf) constructed by combining a simple linear function with a negative exponential function. Meanwhile, the soil infiltration capacity over a basin is simulated by a parabolic type pdf. According to the joint probability distribution of rainfall and soil infiltration capacity, the probability distribution of the surface runoff can be derived, a quasi-analytical solution for surface runoff is obtained, and furthermore the surface runoff and soil infiltration are computed by adopting the infiltration excess mechanism. Part of the infiltration supplements the soil moisture, and the rest recharges to subsurface and groundwater flow through using a water storage capacity curve. The total runoff is composed of infiltration excess and saturation excess runoff. As an example, the model was applied to Lushi basin, a sub-basin at the middle reach of the Yellow River. Results were compared with those obtained by TOPMODEL; it is shown that the model presented in this study is applicable to such a basin with semi-humid or semi-arid hydrological characteristics.

Key words statistically-based runoff-yield model; spatial variation of rainfall; infiltration excess mechanism; saturation excess mechanism; TOPMODEL

INTRODUCTION

Hydrological models are useful tools to study the mechanism of rainfall–runoff processes. Based on runoff-yield mechanisms, hydrological models are categorized into infiltration excess and saturation excess runoff-yield mechanisms. The former is typically applicable in arid or semi-arid areas, while the latter is applicable in humid or semi-humid areas. The conceptual model integrating with the two mechanisms was developed in the 1990s. Sivapalan *et al.* (1987) provided a theoretical framework considering both infiltration excess and saturation excess runoff-yield mechanisms. These integrated models are expected to be used in semi-humid and semi-arid areas. Based on whether they consider runoff-yield spatial variation or not, hydrological models are classified into distributed semi-distributed and lumped models. With the development of RS and GIS techniques, distributed models with physical explanation were studied and quickly developed. In the process of runoff yield, these models take the inhomogeneous distribution of runoff-yield factors into account, and model parameters with theoretical basis can be explicitly and physically explained. Between the lumped and distributed models, is the semi-distributed model, among which TOPMODEL (Beven, 1979) is the most representative. However, the determination of parameters in distributed or semi-distributed models is rather complicated and the requirement for data is much stricter, thus limiting the applications of these models.

In order to simplify the model structure and keep the accuracy, a statistical approach is sometimes applied to describe the spatial variation of runoff-yield. Two major points should be considered for the runoff-yield influencing factors (Xu *et al.*, 1996). The first type of factor is the spatial variability of the underlying surface, including geology, topography and soil structure. Another factor is certain hydrological and meteorological conditions, such as rainfall and evaporation. Among these factors, the spatial variation of rainfall is important in influencing the soil moisture and runoff-yield. Warrilow *et al.* (1986) used a negative exponential-type function to express the distribution of the rainfall over a basin. Sivapalan & Wood (1986) studied the effect of spatial heterogeneity of soil and rainfall; they derived quasi-analytical solutions for the ponding time and infiltration rate. These studies demonstrated that using a probability density function (pdf) could offer a valuable alternative for characterizing the spatial heterogeneity of a rainfall event.

In this paper, a new rainfall pdf constructed by combining a simple linear function with a negative exponential function is put forward to reflect the rainfall spatial distribution. According to the joint spatial probability distribution of rainfall and soil infiltration, the empirical Horton model (Horton, 1933) was adopted to separate surface runoff and infiltration. Part of the infiltration water supplements the soil water, and the rest is taken as the subsurface and groundwater flow, which is computed by a water storage capacity curve (Zhao *et al.*, 1980) based on saturation excess mechanism. Therefore, the total runoff is composed of infiltration excess and saturation excess runoff. It also means that this model can be applied to areas where runoff-yield is controlled by both infiltration and saturation excess runoff-yield mechanisms.

As an example, the statistically-based model was applied to simulate floods for Lushi basin. In addition, results were compared with those of TOPMODEL. It indicated that the simulation accuracy of the model is acceptable, which may make the model applicable to such semi-arid and semi-humid areas.

MODEL DESCRIPTIONS

Model structure

The runoff of this statistical-based model is divided into two parts: the surface runoff and the subsurface and groundwater runoff. The former is calculated by the infiltration excess runoff-yield mechanism through using the joint distribution of rainfall and infiltration capacity. The latter is calculated by the saturation excess runoff-yield mechanism through using a water storage capacity curve.

Surface runoff

The spatial variation of rainfall in any time step of a rainfall event can be described by probability distribution functions or probability density functions (pdf). In this study, the net rainfall at any point within a basin is supposed to comply with:

$$f(P_i) = \frac{1}{m^2} P_i \exp(-\frac{P_i}{m}) \tag{1}$$

where, P_i is the net rainfall at point i in the basin; m is a parameter which can be determined by the rainfall data from all the gauges in the basin by using goodness-of-fit. Figure 1 illustrates an example of the pdf from equation (1).

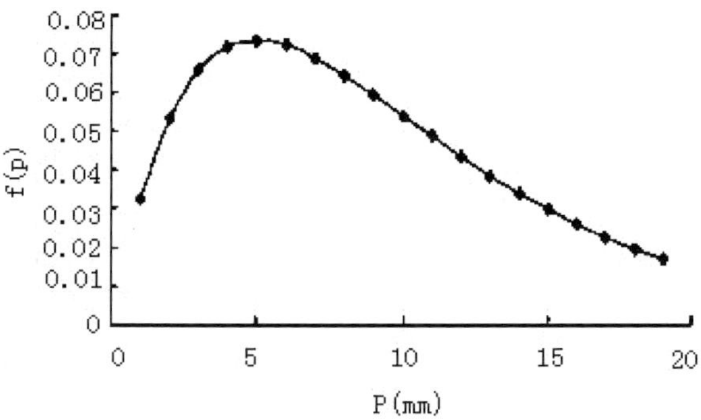

Fig. 1 Illustration of a probability density function of rainfall.

Due to the dissimilarity of soil type and soil structure of the basin, the water retention capacity and initial water storage are heterogeneous, so that the infiltration capability of each point is not the same at each time interval. In order to reflect the spatial variation of soil infiltration capacity, a cumulative distribution function (cdf) is introduced as:

$$F(F_i)=1-(1-\frac{F_i}{F_{mm}})^n \tag{2}$$

where F_i is the soil infiltration capacity at point i. $F_{mm}=(n+1)F_m$ is the maximum soil infiltration capacity of the basin, while F_m is the mean of F_i. n is an index which reflects the unevenness of infiltration distribution. Based on equation (2), the pdf of soil infiltration capacity can be described as:

$$f(F_i)=\frac{dF}{dt}=\frac{n}{F_{mm}}(1-\frac{F_i}{F_{mm}})^{n-1} \tag{3}$$

If $P_i > F_i$, the surface runoff generated at point i is $P_i - F_i$, and the rest infiltrates to the soil to supplement soil moisture or join in the generation of subsurface and groundwater flow. If $P_i \le F_{i,}$ all rainfall infiltrates to the soil moisture. As both the rainfall and infiltration capacity are treated as random, the surface runoff is also a random variable, therefore the cdf of surface runoff $R_s = P - F$ can be expressed as:

$$F(R_{si})=\iint_{P_i-F_i<R_{si}} f(F_i,P_i)dF_i dP_i \tag{4}$$

where $f(F_i,P_i)$ is the joint distribution density function of dualistic variables (F_i,P_i). R_{si} is the surface runoff at point i. Soil infiltration capacity F_i and rainfall P_i are considered as two independent stochastic variables, thus equation (4) can be converted to:

$$F(R_{si})=\int_0^{F_{mm}}\left[f(F_i)\int_{P_{\min}}^{R_{si}+F_i} f(P_i)dP\right]_i dF_i \tag{5}$$

where P_{min} is the minimum precipitation. Substituting equations (1) and (2) into equation (5):

$$F(R_{si})=\int_0^{F_{mm}}\left[\frac{n}{F_{mm}}(1-\frac{F_i}{F_{mm}})^{n-1}\int_{P_{\min}}^{R_{si}+F_i}\frac{1}{m^2}P_i\exp(-\frac{P_i}{m})dP_i\right]dF_i \tag{6}$$

Solving the above integral, the distribution function of R_{si} can be deduced as:

$$F(R_{si})=\frac{n}{F_{mm}}\left[k-\frac{k_{r1}}{m}\cdot R_{si}\cdot\exp(-\frac{P_{\max}}{m})-(k_{r1}+\frac{1}{m}k_{r2})\exp(-\frac{R_{si}}{m})\right] \tag{7}$$

$$k=\frac{F_{mm}}{n}\cdot(\frac{1}{m}P_{\min}+1)\cdot\exp(-\frac{P_{\min}}{m}) \tag{8}$$

$$k=\frac{F_{mm}}{n}\cdot(\frac{1}{m}P_{\min}+1)\cdot\exp(-\frac{P_{\min}}{m}) \tag{9}$$

$$k_{r1}=\int_0^{F_{mm}}\exp(-\frac{F_i}{m})\cdot(1-\frac{F_i}{F_{mm}})^{n-1}dF_i \tag{10}$$

where k_{r1} and k_{r2} are temporary variables having no relation to R_{si}, and can be solved by numerical integration. So that the computation of surface runoff is essentially a quasi-analytical solution, the average or mean surface runoff can be calculated by:

$$R_s=E(R_{si})=\int_0^{+\infty}R_{si}dF(R_{si}) \tag{11}$$

Generally, the upper limit of the integral is the maximum precipitation $P_{\max}$ of the rainfall event at the computation time step, then R_s can be computed as:

$$E(R_s)=\frac{n}{F_{mm}m}(k_{r2}+2k_{r1}m)\left[m-(P_{\max}+m)\exp(-\frac{P_{\max}}{m})\right]-\frac{nk_{r1}}{F_{mm}m}\cdot P_{\max}{}^2\cdot\exp(-\frac{P_{\max}}{m}) \tag{12}$$

After the computation of surface runoff, the soil infiltration F can be computed by deducing surface runoff from rainfall P.

Subsurface and groundwater flow

When precipitation reaches the ground, some of it converts to surface runoff R_s, which can be calculated by equation (12), while the rest of it infiltrates to the soil. Part of the infiltration is consumed as soil evaporation, and part of it supplements soil moisture. Here we assume, at certain points within the basin, that subsurface and groundwater flow will occur when the soil water storage achieves field water capacity.

In order to reflect the unevenness of the infiltration capacity, a water storage capacity curve was brought in (shown as Fig. 2). From the infiltration and the curve, the subsurface and groundwater flow can be calculated. An empirical parabola function was adopted to compute runoff-yield area:

$$\alpha = 1-(1-\frac{W}{W_{mm}})^b \tag{13}$$

where, α is the ratio of runoff-yield area to the total basin area; W represents the water storage of a point within the basin; $W_{mm} = (1+b)W_m$ is the maximum water capacity of all points, while W_m is the mean areal water capacity; b is an index which reflects the uneven distribution of water storage capacity over the basin.

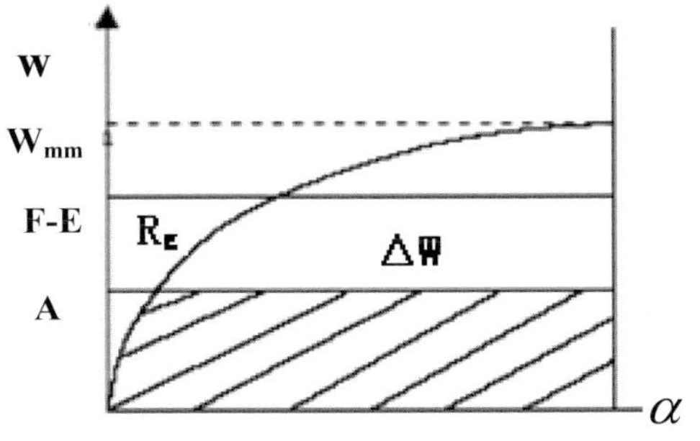

Fig. 2 Water storage capacity curve.

According to the initial mean areal water storage W_0, the corresponding point water storage capacity A (as shown in Fig. 2) can be computed as:

$$A = W_{mm}\left[1-(1-\frac{W_0}{W_m})^{\frac{1}{b+1}}\right] \tag{14}$$

The initial mean areal water storage W_0 at each time step is valued between 0 and W_m, which can be computed based on the water balance law.

$$W_0^t = W_0^{t-1} + F^{t-1} - E^{t-1} - R^{t-1} \tag{15}$$

where, W_0^t, W_0^{t-1} are denoted as mean areal water storage at time t and $t-1$ respectively; F^{t-1} is the soil water infiltration, while E^{t-1} is soil evaporation and R^{t-1} just represents the saturated runoff generated at time $t-1$. In addition, a coefficient k is introduced to compute actual evaporation E by $E = kE_p$, in which, E_p represents the soil evaporation potential.

If $F - E < 0$, the infiltration is used up for evaporation and no runoff, while $F - E > 0$, runoff is generated in following way:

when,

$$F - E + A < W_{mm} \qquad R_g = F - E - W_m + W_0 + W_m[1 - (F - E + A)/W_{mm}]^{1+b} \tag{16}$$

$$F - E + A \geq W_{mm} \qquad R_g = F - E - (W_m - W_0) \tag{17}$$

R_g is denoted as subsurface and groundwater runoff. When the water storage of all points in the basin reaches their field water capacities, the total infiltrated water will convert to subsurface and groundwater runoff after the deduction of evaporation.

Parameter estimation

The model parameters can be classified into three groups: parameters of rainfall pdf, parameters of infiltration capacity pdf, and those of the water storage capacity curve. Parameter m can be determined by the goodness-of-fit approach. The mean areal maximum infiltration capacity F_m in the infiltration capacity distribution function can be computed by an empirical Horton model. Finally, index n, mean areal maximum water storage W_m, and the parabola index of water storage capacity curve b are normally determined through calibration.

APPLICATION

Study area

The Lushi basin, a sub-basin at the middle reach of Yellow River, is studied in this paper. Figure 3 is a draft map showing the location of Lushi basin and rainfall gauges within it. The control area of it is about 4716 km^2 and the mainstream length is about 180 km.

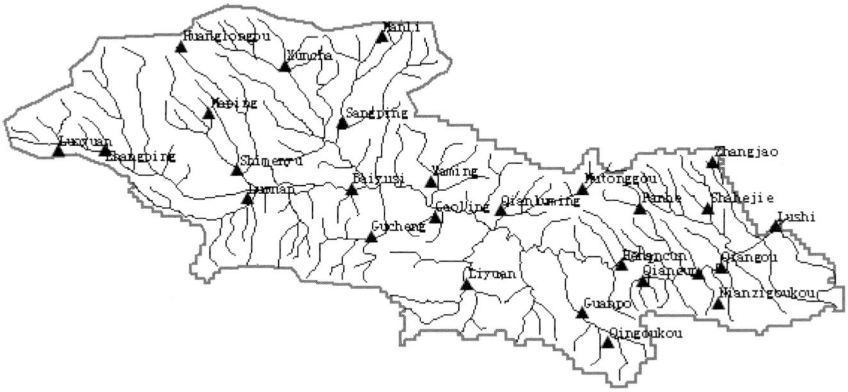

Fig. 3 Location of the rainfall gauges within Lushi basin.

The topographical elevation is higher in the west and lower in the east. The mean annual precipitation is around 600–800 mm. Rainstorms usually occur in July or August, and rainfall amount in these two months takes up 38% of the total yearly amount. Floods are mainly formed by storms, with characteristics of high peak flow and short duration.

Data

DEM data In this study, DEM data with resolution of 30″ × 30″ provided by USGS.

Rainfall gauges 26 rainfall gauges in the basin were used, which are shown in Fig. 3.

Evaporation Daily evaporation data at Lushi gauge were used to represent the aerial evaporation of the study basin.

Storm data Six storm events were selected with rainfall data at the 26 rainfall gauges and corresponding discharge data at Lushi gauge. The time step of each of the rainfall–runoff events is 1 hour.

Model application

The model runs for each of the time steps (1 hour) of a rainfall event, thus obtaining the simulated runoff process. The parameter m in the pdf of rainfall is estimated by rainfall data through using the goodness-of-fit method, and varies for each calculation time interval. n is an index to reflect the uneven feature of soil infiltration capacity over the basin and is fixed to 0.3 in this study. Generally, in the mountainous area, b is about 0.1 in small basins, 0.2–0.3 in medium scale basins and 0.3–0.4 in large basins. Here, b is given by 0.25. W_m is empirically defined to 130 mm. The model was used to simulate six flood events and results were compared with that of TOPMODEL.

RESULT ANALYSIS

Relative errors of the total runoff (in mm) for the six simulated flood events are listed in Table 1 for both the statistical-based runoff-yield model and TOPMODEL. For the statistical-based

Table 1 Comparison of observed and computed runoff depths.

				Statistical-based model		TOPMODEL	
Flood number	Start time	Mean areal precipitation (mm)	Observed runoff depth (mm)	Simulated runoff depth (mm)	Relative error (%)	Simulated runoff depth (mm)	Relative error (%)
1	1980070110	57.7	20.2	18.5	9	22.9	–13
2	1983072810	63.3	26.5	22.0	13	23.5	11
3	1984070410	72.1	23.2	23.7	–2	23.5	–1
4	1985091410	74.0	34.1	36.2	–6	29.8	12
5	1989070910	71.8	28.1	22.7	19	34.1	–21
6	1996080210	14.0	4.8	4.6	6	4.4	9

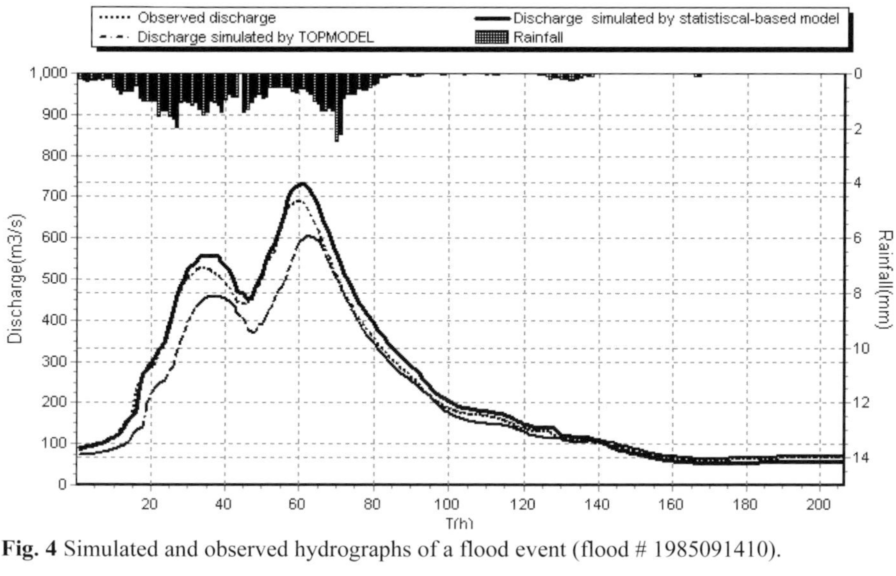

Fig. 4 Simulated and observed hydrographs of a flood event (flood # 1985091410).

model, the absolute value of the relative error for each flood event is less than 20%, with a minimum value of 2% and maximum value of 19%. For TOPMODEL, the absolute value of the relative errors for the six flood events are between 1 and 21%. As a whole it is indicated that the model is comparable to TOPMODEL, and could provide better simulations for rainfall–runoff events for the studied cases. As an example, the simulated and observed hydrographs of the flood event with start time 1985091410 is shown in Fig. 4.

SUMMARY AND CONCLUSION

A statistical-based runoff-yield model coupling the saturation and infiltration excess runoff-yield mechanisms is studied. A new pdf is introduced to describe the spatial distribution of rainfall, a quasi-analytical solution for runoff-yield calculation is obtained. The model was applied to simulate six flood events in Lushi basin, and results were compared with those obtained from TOPMODEL. It indicates that the model could achieve acceptable accuracy and is comparable to TOPMODEL for the studied flood events.

This statistically-based model considers the spatial variation of rainfall, soil infiltration capacity and water storage capacity that are the main factors controlling the runoff-yield process. Comparing the traditional approaches for calculation of areal rainfall such as arithmetic mean and Thiessen polygon, the application of a pdf possesses the advantage to describe the spatial heterogeneity of a rainfall event.

Acknowledgements This study was supported by the National Natural Science Foundation of China (grant no. 50779013), Specialized Research Fund for the Doctoral Program of Higher Education of China (grant no. 20070294018) and The National Basic Research Program of China ("973" Program, grant no. 2007CB7141973). We acknowledge two anonymous reviewers, whose suggestions have improved the paper.

REFERENCES

Beven K. J. & Kirkby, M. J. (1979) A physically based variable contributing area model of basin hydrology. *Hydrol. Sci. Bull.* **27**(1), 43–69.

Horton, R. E. (1933) The role of infiltration in the hydrologic cycle. *Eos (AGU)* **14**, 446–460.

Liang, X., Lettenmaier, D. P. & Wood, E. F. (1996) One-dimensional statistical dynamic representation of subgrid spatial variability of precipitation in the two-layer variable infiltration capacity model. *J. Geophys. Res.* **101**(16), 403–422.

Sivapalan, M. & Wood, E. F. (1986) Spatial heterogeneity and scale in the infiltration response of catchments. In: *Scale Problems in Hydrology* (ed. by V. K. Gupta, I. Rodriguez-Iturbe & E. F. Wood), 81–106. Reidel, Dordrecht, The Netherlands.

Sivapalan, M., Beven, K. & Wood, E. F. (1987) On hydrologic similarity 2. A scaled model of storm runoff production. *Water Resour. Res.* **23**(12), 2266–2278.

Warrilow, D. A., Sangster, A. B. & Slingo, A. (1986) Modeling of land-surface processes and their influence on European Climate. *Dyn. Clim. Tech.* Note **38**, 92 pp., Meteorological Office, Bracknell, Berkshire, UK.

Zhao, R. J., Zhuang, Y., Fang, L. R., Lin, X. R. & Zhang, Q. S. (1980) The Xinanjiang model. In: *Hydrological Forecasting to the Operation of Water Resource Systems* (Proc. Symposium, Oxford, April 1980), 351–356. IAHS Publ. 129, IAHS Press, Wallingford, UK.

Uncertainty estimation for the Xin'anjiang model parameters

FANGGUI LIU[1,2], XIAOFANG RUI[1,2], HEHAI XIE[1,2], JIE FENG[1,2] & JIYU YU[1,3]

1 *State Key Laboratory of Hydrology-Water Resources and Hydraulic Engineering, Nanjing 210098, China*
qufuliufanggui@163.com or liufanggui@hhu.edu.cn

2 *College of Hydrology and Water Resources, Hohai University, Nanjing 210098, China*

3 *Shandong Water Conservancy Vacational College, Rizhao, 276826, China*

Abstract Due to the uncertainties in hydrological forecasting mainly derived from hydroclimatic input data, hydrological model structure and model parameters, the investigations on the uncertainty of model parameters are crucial to improve the precision of flood forecasting. The Xin'anjiang (XAJ) model developed by Renjun Zhao is considered an effective conceptual watershed hydrological model, and it has been extensively employed for hydrological modelling problems. In this study, the generalised likelihood uncertainty estimation (GLUE) methodology for XAJ model identification allowing for equifinality was proposed to identify the uncertainty of model parameter sets. Additionally, the GLUE analysis was utilized to determine the Yanduhe catchment, one tributary of the Yangtze River watershed, and the uncertainty of XAJ model parameters. Based on the view of two storm events, we observed *SM* in the XAJ model is very sensitive. For example, its likelihood values displayed the peak value area and its little change will have a large influence on the simulated results; while *K*, *B*, *EX* are not sensitive. Further, we found the observed discharge hydrograph can not be located wholly within the upper and the lower limits of the simulated discharge hydrograph, and some discharge values may fall outside of the 90% uncertainty bounds. This indicates that the XAJ model can not be used to simulate the discharge hydrograph well because of the uncertainty of the model.

Key words uncertainty estimation; GLUE methodology; XAJ model; Yanduhe catchment

INTRODUCTION

Predictions in ungauged basins (PUB) is one of the important and difficult problems in hydrology arousing general concern at home and abroad (Rui *et al.*, 2007). As such, the IAHS decade of PUB (2003–2012) (Sivapalan *et al.*, 2003), focusing on the effective methodology of hydrological simulation to decrease the uncertainty of hydrological prediction and increase its forecasting precision, has been organized, including the establishment of the China PUB Working Groups in 2004 (Yang *et al.*, 2004). According to investigations by Sivapalan *et al.* (2003), Yang *et al.* (2004), etc. the uncertainty of hydrological prediction is mainly caused by the uncertainties of the input data, model structure and model parameters. Therefore, it is definitely necessary to investigate the uncertainty of parameters in hydrological models.

It may be endemic to mechanistic modelling of watershed hydrological systems that there are many different parameter sets within a chosen model structure that may be behavioural or acceptable in reproducing the observed behaviour of that system. This has been called the equifinality concept (Beven & Freer, 2001). The points that produce equifinality include (Rui *et al.*, 2007): (1) The objective function is multivalued; (2) There are numerous interactions among model parameters; (3) Model parameters are stochastic. As such, equifinality makes it uncertain to ultimately find one "optimal" parameter set, so a novel modality should be analysed to fully evaluate such uncertainty. In recent years, although some advances in the uncertainty research of the watershed hydrological model structure and parameter sets have been achieved (Guo *et al.*, 1995; Beven & Freer, 2001), the GLUE methodology of Beven & Binley (1992) is by far one of the most effective uncertainty analysis methods in such a research field (Beven & Binley, 1992; Beven & Freer, 2001).

In this study, the uncertainty of parameters sets in XAJ model was analysed using the GLUE method to investigate the equifinality phenomenon so that the XAJ model can be better applied in hydrological modelling.

XAJ MODEL DESCRIPTION

The XAJ model, first built in 1973, is a conceptual watershed model. Its basic feature is the concept of runoff formation on repletion of storage, which means that runoff is not produced until the soil moisture content of the aeration zone reaches field capacity, and thereafter runoff equals the rainfall excess without further loss. The XAJ model is mainly used for hydrological forecasting in agricultural, forested and pastural land in humid or semi-humid regions (Zhao, 1992).

The schematic diagram of XAJ model is shown in Fig. 1. It demonstrates that the outflow has four major components including the flow routing, the runoff production, the evapotranspiration and the runoff separation.

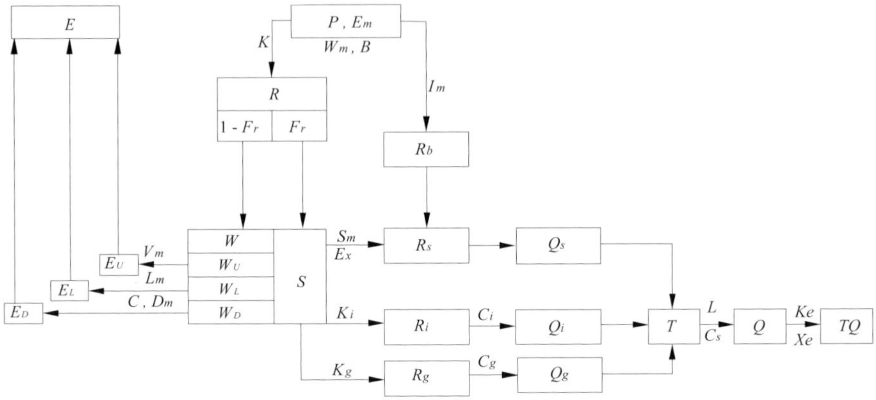

Fig. 1 Flowchart of XAJM (Zhao, 1992).

According to the model structure, the runoff was divided into three components, namely surface runoff, interflow and groundwater flow. The definition of the parameters (Zhao Renjun, 1992) is given in Table 1. Values of a large number of XAJ model parameters cannot be obtained from field measurements and need to be determined through a model calibration procedure. The trial-and-error procedure is used for the optimization of model parameters.

Table 1 Parameters of XAJ model.

notation	definition	notation	definition
K	coefficient for potential evapotranspiration	E_x	exponent of the free water capacity distribution curve
W_U	upper zone tension water capacity	S	mean areal free water storage capacity of the surface soil layer
W_L	lower zone tension water capacity	K_i	daily interflow coefficient
W_D	deep zone tension water capacity	K_g	daily groundwater coefficient
W	average areal tension water capacity	C_i	daily interflow recession coefficient
C	coefficient of deep evapotranspiration	C_g	daily groundwater recession coefficient
I_m	ratio of impervious area to the total area	C_s	recession constant in the "lag and route " method for routing
B	exponent of the tension water curve	L	corresponding "lag"

THE GLUE METHODOLOGY

Rationale of GLUE

The background of the GLUE methodology has been an attempt to recognize more explicitly the fundamental limitations of hydrological models as simulators of catchment rainfall–runoff processes (Beven, 1989, 1993; Grayson, 1992). One implication of such a recognition is that it should not be assumed that there is one "optimal" model parameter set which can be found to represent a catchment (whether a lumped or distributed representation). For the GLUE methodology, parameter set combinations within the given parameter ranges are calculated using Monte-Carlo simulation. Then the likelihood function is used, and a function value is computed by utilizing both the simulated results and the observed values. As such, once the likelihood weight is computed, the likelihood values of different parameter set combinations are captured. Further, a marginal value is chosen for all of the likelihood values, which of course is a subjective choice to a certain extent. If the likelihood values are smaller than the marginal value, they are specified as zero, since these parameter sets can not reasonably describe the function characteristics of the model. Conversely, if the likelihood values are larger than the marginal value, they are rescaled and sized. So the uncertainty bounds in model prediction below a certain confidence limit are calculated (Beven & Binley, 1992; Beven & Freer, 2001; Xiong & Guo, 2004).

Analysis steps of GLUE

Based on the rationale of the GLUE methodology mentioned above, its estimation steps are described as follows:

(a) The formal definition of the likelihood measure. The presented results employ a coefficient of determination as the basic likelihood measure and basic likelihood measurement is written in the form:

$$L(\theta_i \mid Y) = 1 - \sum_{j=1}^{n}(Q_{ij} - Q_{Oj})^2 \Big/ \sum_{j=1}^{n}(Q_{ij} - \overline{Q_O})^2 \qquad (1)$$

in which $L(\theta_i \mid Y)$ is the likelihood measure for the ith parameter set, Q_{ij} the simulated value for jth time step, Q_{Oj} the observed value for jth time step, $\overline{Q}_O$ the average value of the observations and n the sequence number.

(b) Determination of the parameter ranges and an *a priori* distribution. Initially the ranges are assumed as wide as possible based on the rational physical properties. For most of the applications of GLUE *a priori* distribution is replaced by a uniform distribution and uniform random sampling across the specified parameter range is adopted.

(c) Uncertainty estimation of model parameters. Scatter plot of likelihood values for selected model parameters is plotted and uncertainty of model parameters is computed and analysed.

(d) Calculation of uncertainty bounds. In this study, 5% and 95% accumulative likelihood distributions are used as uncertainty limits in the predictions.

CASE STUDY

Description of the Yanduhe catchment

The Yanduhe catchment, located upstream of the Yangtze River watershed, is a first-order tributary of the Yangtze River. It occupies an area of 601 km^2, with the Yanduhe stream gauging station at its outlet. Its exact location is at latitude 31°12'N and longitude 110°18'E. For the catchment, the mean annual rainfall is 1222.2 mm, so it is a typical humid region and has rich precipitation. The soil in the catchment has a strong infiltration capacity and excess rainfall is difficult to generate. Further, as the excess-infiltration runoff is the main mode of runoff generation, the XAJ model can be used to simulate both its runoff-generating process and runoff concentration process. Five rainfall gauges named as Banqiao, Xiagu, Duizi, Songziyuan and Yanduhe, and one evaporation station are located in the catchment.

On the basis of 60 × 60 m DEM data of 1:50 000, the Yanduhe catchment is divided into 540 000 grid cells with 600 rows and 900 columns (see Fig. 2). In this investigation, the distribution curve of the topographic index $\ln(\alpha/\tan\beta)$ is calculated using the method developed by Xie & Huang (2006).

Fig. 2 3-D map of the Yanduhe catchment extracted from the DEM.

THE RESULTS ANALYSIS

The XAJ model is selected as the rainfall–runoff model of the Yanduhe catchment. In this study, only the runoff generation parameters in the XAJ model are considered. The concentration parameters in the XAJ model will be further investigated in the future. Previous studies indicated that *K*, *B*, *SM* and *EX* are very sensitive, and they have large impacts on the simulations. These parameters are therefore used to investigate the uncertainty of the simulated results.

Employing 19810625 and 19860909 storm events of the Yanduhe catchment and the deterministic factor as the likelihood measure, 2 × 5000 parameter sets were chosen to estimate the uncertainty based on uniform random sampling within the specified parameter ranges (see Table 2). The scatter plot of likelihood values for four selected XAJ model parameters of the Yanduhe catchment are displayed in Figs 3 and 4.

Table 2 *a priori* distribution of parameters.

Parameters	Minimum value	Maximum value	Mean value
K	0.4	1.0	0.7
B	0.1	0.4	0.25
SM	20 (mm)	110 (mm)	65 (mm)
EX	1.0	1.7	1.35

According to Figs 3 and 4, it is found that the value ranges of *SM* reduce greatly after its rating through the discharge hydrograph at the catchment outlet. It is also noted that *SM* shows peak value ranges, while *K*, *B*, *EX* are not sensitive at all. We also observed that high likelihood values exist in the whole parameter distribution range. In addition, many parameter value sets are displayed in high likelihood value ranges, namely equifinality, indicating that there are many equivalent parameter sets.

In this study, for parameter sets' deterministic factor is greater than zero, their likelihood values are rescaled and sized. In the following, the 90% uncertainty bounds for simulations of XAJ model, including the observations and the upper bound and the lower bound of uncertainty, are calculated and shown in Figs 5 and 6. The uncertainty bounds change with the discharge process, due to its larger values in the higher discharge zone and fewer values in the lower discharge zone. The simulated discharge limits can not totally encompass the observed discharge hydrograph,

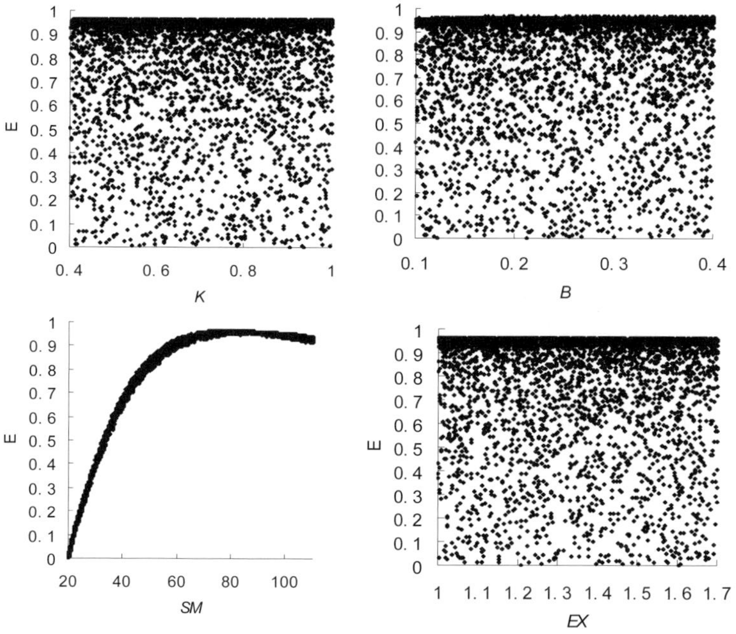

Fig. 3 Scatter plot of likelihood values for four selected parameters of the Yanduhe catchment for the 25 June 1981 storm event..

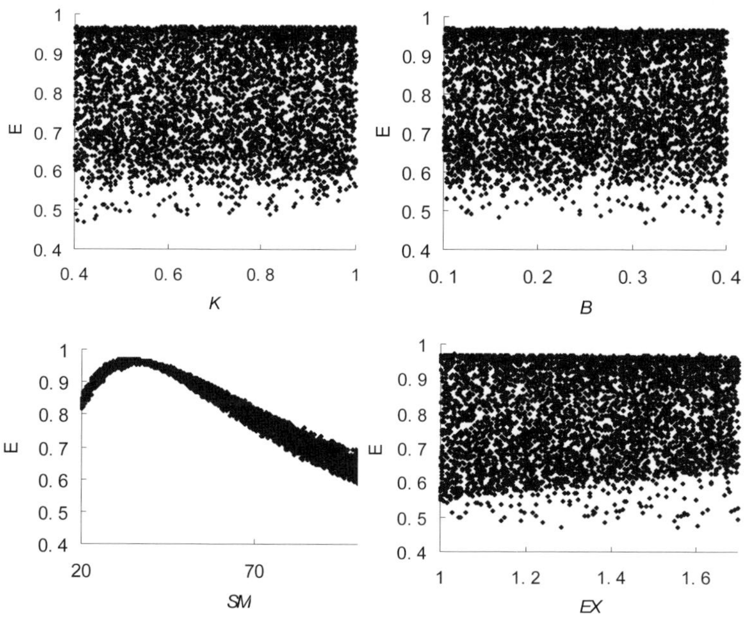

Fig. 4 Scatter plot of likelihood values for four selected parameters of the Yanduhe catchment for the 9 September 1986 storm event.

because some discharge values always fall outside of the 90% confidence interval. This shows that XAJ model can not simulate the discharge hydrograph well due to the uncertainty of the model.

The 90% uncertainty bounds are not wide enough to encompass all of the observed discharges during the calibration period and, the possibilities are listed as follows:

(a) The influence of *a priori* parameter distribution. The uniform distribution is simply used to generate model parameters. However, *a priori* distribution of model parameters under study is not well determined.

(b) The influence of the sampling method for Monte Carlo simulation. Although Monte Carlo simulation may overcome some disadvantages of automatic scan, random scan and trial-and-error scan in higher dimensional parameter space of hydrological models, many combinations of parameters for the model structure with multi parameters require several ten thousand or tens of thousands, even up to more than one million parameter samplings. The computational efficiency is very low. For simulation of the flood process with a long duration, it is time consuming with large samplings so it is better if the samplings are not too large.

(c) The influence of hydrological data. The rainfall interpolation time interval is specified as 1 h for this catchment, while the time interval for the initial step is longer. If the rainfall is divided into the same interpolation time interval, the simulated discharge hydrograph does not agree well with the observed discharge hydrograph.

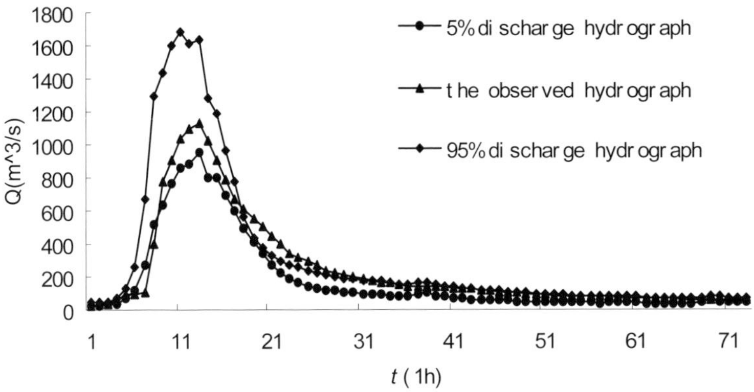

Fig. 5 Ninety percent uncertainty bounds for simulations for the 25 June 1981 storm event in the Yanduhe catchment

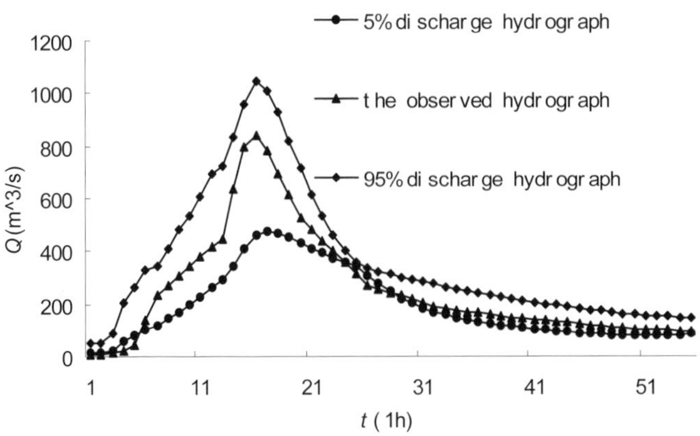

Fig. 6 Ninety percent uncertainty bounds simulations for the 9 September 1986 storm event in the Yanduhe catchment.

DISCUSSION AND SUMMARY

When uncertainty and equifinality are concisely defined, the rationale and the analysis steps of the GLUE methodology are proposed and the methodology is applied to the Yanduhe catchment, located in the Yangtze River watershed. Additionally, in our computation *a priori* distribution of XAJ model parameters is specified via estimation, and the code for XAJ model algorithm was run with 2 × 5000 randomly generated parameter sets. It is noted that the equifinality is easy to get by. *SM* in XAJ model is very sensitive, it shows peak value ranges, while *K*, *B*, *EX* are not sensitive at all. We also observed that high likelihood values exist in the whole parameter distribution range. According to assumed rules, the upper and the lower limits of 90% uncertainty bounds are calculated. Given the parameter distribution and the parameter space, the observed discharge hydrograph is not able to totally fall inside the upper and the lower limits, which also demonstrates that the parameter ranges can not encompass all the values. As a watershed hydrological model is a complex system, the influencing factors causing the uncertainty should be investigated in future. Additionally, using the GLUE methodology is easy to find the equifinality phenomenon, but how to figure out the equifinality problem by using it should be further studied in future.

Acknowledgements This project was supported by the National Natural Science Foundation of China (50609005), the Fok Ying-Tong Education Foundation for Young Teachers in the Higher Education Institutions of China (101075) and the Specialized Research Fund for the Doctoral Program of Higher Education of China (20050294002).

REFERENCES

Beven, K. J. (1989) Changing ideas in hydrology: The case of physically based models. *J. Hydrol.* **105**, 157–172.

Beven, K. J. (1993) Prophecy, reality and uncertainty in distributed hydrological modelling. *Adv. Water Resour.* **16**, 41–51.

Beven, K. J. & Binley, A. M. (1992) The future of distributed models: model calibration and uncertainty prediction. *Hydrol. Processes* **6**(2), 279–298.

Beven, K. & Freer, J. (2001) Equifinality, data assimilation, and uncertainty estimation in mechanistic modelling of complex environmental systems using the GLUE methodology. *J. Hydrol.* **249**, 11–29.

Grayson, R. B, Moore, I. D. & McMahon, T. A. (1992) Physically based hydrologic modelling, 2, Is the concept realistic? *Water Resour. Res.* **28**, 2659–2666.

Guo, Shenglian, Li, Lan & Zeng, Guangming (1995) Uncertainty analysis of impact of climate change on hydrology and water resources. *Hydrology* **6**, 1–5.

Rui, Xiaofang, Liu, Fanggui & Xing Zhengxiang (2007) Advances in hydrology and some frontier prolems. *Adv. Sci. & Tech. Water Resour.* **27**(1), 75–79.

Sivapalan, M. K. K., Takeuchi, S. W. & Franks, S. W. (2003) IAHS decade of prediction in ungauged basins (PUB), 2003–2012: shaping an exiting future for the hydrological sciences. *Hydrol. Sci. J.* **48**(6), 857–879.

Xie, Hehai & Huang, Guoru (2006) Some calculation methods for topographic index. *J. Hohai University (Natural Sciences)* **34**(1), 46–50.

Xiong, Lihua & Guo, Shenglian (2004) Research on equifinality of the Xin'anjiang Model of three components. In: *Investigations and Advances in Complexity and Uncertainty of Water Problems* (ed. by Xia Jun). China Waterpower Press, Beijing, China.

Yang, Dawen, Xia, Jun & Zhang, Jianyun (2004) Investigations and advances on China PUB. In: *Investigations and Advances in Complexity and Uncertainty of Water Problems* (ed. by Xia Jun), 47–54. China Waterpower Press, Beijing, China.

Zhao, Renjun (1992) The Xin'anjiang model applied in China. *J. Hydrol.* **135**, 371–381.

Generation of triangulated irregular networks for distributed hydrological modelling

GUANGHENG NI, LEI WANG, FUQIANG TIAN & HEPING HU

Department of Hydraulic Engineering, State Key Laboratory of Hydroscience and Engineering, Tsinghua University, Beijing 100084, China

ghni@tsinghua.edu.cn

Abstract Taking slope, topographic and wetness index as the key criteria to select nodes, we generate triangulated irregular networks (TINs) that integrate multiple landscape properties. Also, comparison between two different interpolation methods is presented. The results indicate that for the same TIN, spatial variability of terrain is better depicted by using 5-degree polynomial interpolation than linear interpolation, and TIN can account for basin terrain and other landscape properties with fewer nodes while keeping the similar precision compared to a regular grid and hence can lead to the higher computation efficiency when applied to distributed hydrological modelling, especially for the large-scale watersheds.

Key words TIN; Grid; topographic index; slope; hydrological model

INTRODUCTION

Grid and TIN (triangulated irregular network) are two common ways used to reflect the spatial variability of topography, distribution of land use, etc. TINs are a series of triangles, which can represent terrain features as well as other properties of the surface, while grids are a series of points ranked in a matrix format, and the value of a point shows the surface information of the location. The data storage of TINs contains nodes of triangular mesh and their relationship, which is more complex than that of grids. TIN nodes can be distributed irregularly, bringing more flexibility of multiple resolutions.

In general, watershed boundary as well as special points and lines can be depicted more precisely by TINs than grids. With the same degree of precise presentation, the number of TIN meshes can be far smaller than that of grids. TINs are usually generated by selecting important nodes to capture the terrain variability. For example, a certain number of points on contour lines, generated by grid DEM, are selected as nodes of triangular mesh. Traditional TIN generation methods generally do not account for criteria other than the preservation of critical slopes, and TIN meshes have varying size and density. Vivoni *et al.* (2004) mentioned that traditional techniques ignore hydrographic and landscape features that are desirable within hydrological model domains and proposed an approach of the use of multiple resolution TIN meshes to represent surface terrain in a watershed model. Vazquez *et al.* (2007) analysed the effects of DEM gridding on the prediction of basin runoff. In this study, TIN and grid are compared as two ways to represent surface spatial characteristics. Terrain slope and topographic index are used as principal criteria to generate TINs, while spatial characteristics such as land use, soil type, hydrogeology, etc. are also considered. The resulting TINs can reflect both terrain variety and watershed hydrological features with fewer meshes, and are more applicable to hydrological modelling.

GENERATION OF TINS

Computation of slope and topographic index

Beven & Kirkby (1979) first proposed a topographic index in TOPMODEL. It is considered that runoff is mainly affected by terrain and landscape and it is concluded that areas with the same topographic index and soil features have hydrological response similarity. Distribution of the calculated topographic index over the whole watershed macroscopically indicates the change of runoff generation areas and the degree of saturation or wetness. Thus topographic index is usually used as an important parameter of physically based distributed hydrological models (Yang *et al.*,

2000). Therefore topographic index, as an index representing runoff generation features, is chosen as a significant factor in TIN generation. The topographic index is calculated as:

$$\lambda = \ln\left(\frac{\alpha}{\tan\beta}\right) \tag{1}$$

where α is pixel contributing area per unit width; $\tan\beta$ is local pixel slope.

Various methods to calculate topographic index have been proposed. In practice, the single flow direction algorithm (Kong *et al.*, 2003) and the D8 based multiple flow direction algorithm proposed by Quinn *et al.* (1991) are widely used. In this study, we used the single flow direction algorithm to compute topographic index with little amendment. In the calculation, the slope of one point is taken as the steepest slope along its neighboring eight directions, and it is assumed that effective length of contour line is equal to that of a grid cell or the diagonal one that is normal to the flow direction. The calculation is conducted with the aid of ArcInfo (Mike & Tom, 2005). Then the formula to calculate topographic index becomes:

$$\lambda = \ln\left(\frac{(A_{cc}+1)\cdot L^2}{L_e \tan\beta}\right) \tag{2}$$

where A_{cc} is the number of cells that flow into the point; L is the length of grid cell; and L_e is equal to L or $\sqrt{2}L$ depending on D8 flow direction.

The upper portion of the Qin River basin was chosen as the study area, which has an area of 1352 km^2, with the outlet of the basin located at the Kongjiapo hydrological station. The data used in this study are 100-m resolution grid DEM, which are based on SRTM 30-m DEM provided by NASA, and revised via GTOPO30 data and 1:250 000 contour lines of the Yellow River basin. If the slope of the point in the riverway is 0, we use the minimal slope of its neighbouring eight points instead to avoid the denominator being 0. Figure 1 illustrates the results of flow convergence calculation and the topographic index distribution.

TIN generation method

The nodes of TINs are generally selected from points of the grid DEM by a certain algorithm. For example, they are chosen from points on some elevation interval contours. This method has higher resolution along rugged hillslopes since the terrain variability is high, but flat regions, especially the river course where contours are sparse, are poorly resolved. In contrast, the analysis of topographic index indicates that topographic index isolines automatically remain dense within regions that frequently saturate. If only based on slope-preserving or topographic index, TINs will have excessively dense distribution in the hillslopes or plains, but attributes such as stream networks, land-use, etc. spatially couple with slope and topographic index. As a result, when we construct the TIN mesh, lots of information about basin landscape is synthetically considered. The main principles are as follows: (1) topographic index isolines are dominant near the valleys; (2) slope

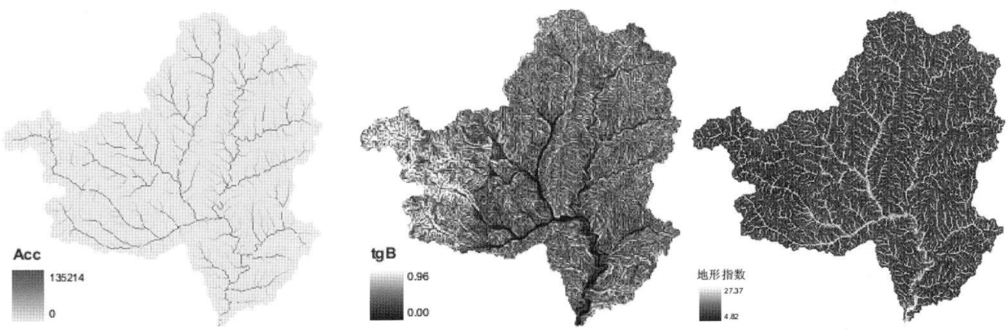

Fig. 1 Flow accumulation, slope and topographic index distribution.

isolines are dominant close to hillslopes and ridges; and (3) incorporate all the information on stream networks, sub-basins, land use, soil and Thiessen polygons formed by precipitation stations, and add TINs in some local areas where high spatial variability exists but meshes are sparse according to the above two criteria.

Depiction accuracy of topography based on different discretization forms of land surface

TIN meshes generated contain about 7000 triangles. TIN elevation error distribution compared with original high-resolution DEM (255 000 100-m grids, referred to as Grid in figures and tables) is shown in Fig. 2, which illustrates that the ratio of grids whose relative errors are within ±5% is over 95%, which means high precision, and if we use 5-degree polynomial interpolation, this ratio increases to 98%. In order to further study the depiction accuracy of topography by TINs, we utilized 500-m grids (referred to as Grid 500 in figures and tables) transformed from 100-m grids as a comparison, the number of which is about 10 000. The computed frequency distribution of the elevation error, slope and topographic index are shown in Fig. 2 and Table 1, Fig. 3 and Table 2, and Fig. 4 and Table 3, respectively. TIN1 stands for the TINs using linear interpolation, and TIN5 for the TINs using 5-degree polynomial interpolation, the original DEM (100-m) is indicated as grid and the coarsening DEM (500-m) as grid 500. The elevation error is based on the elevation values of 100-m grids. To account for the changes in size of grids, the number of units, the location of the centre of units, the distribution of upper stream catchment area of the unit (A_{cc}) and the pixel contribution area per unit width (α) are calculated and shown in Table 4 and Fig. 5.

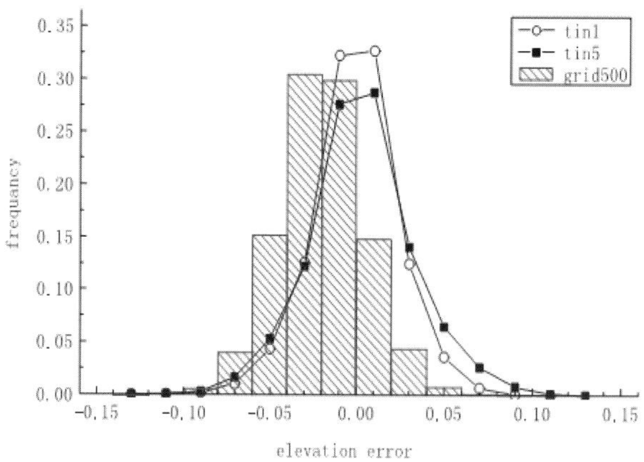

Fig. 2 Comparison of frequency distribution of elevation error.

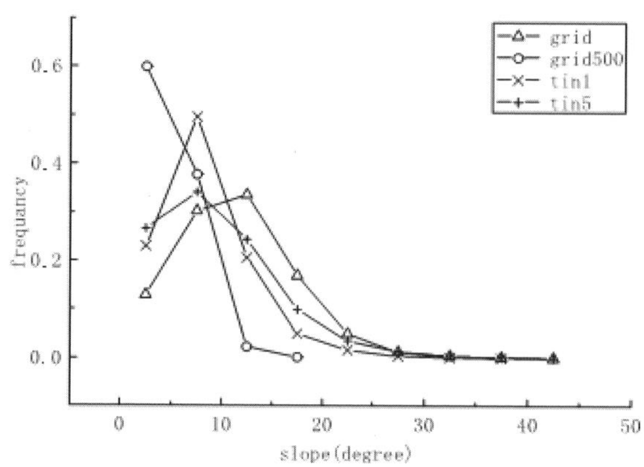

Fig. 3 Comparison of frequency distribution of slope.

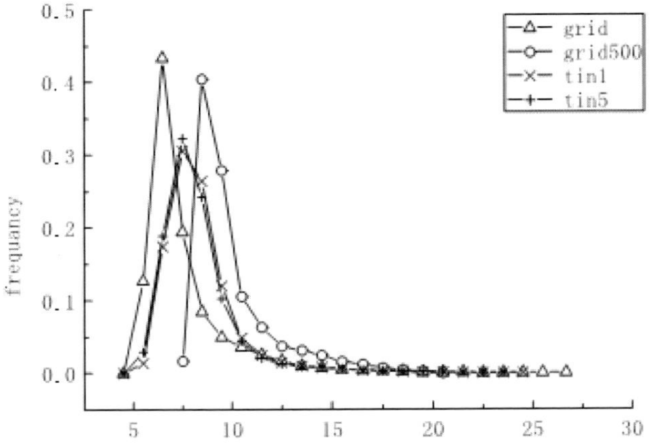

Fig. 4 Comparison of frequency distribution of topographic index.

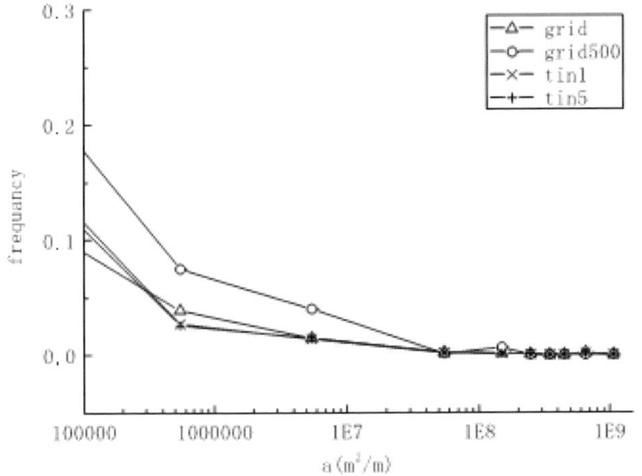

Fig. 5 Comparison of frequency distribution of α.

Table 1 Comparison of depiction accuracy of elevation.

Item	Grid	Grid500	TIN1	TIN5
Ratio of relative error within ±5%	–	0.95	0.98	0.94
Maximum (m)	2516	2490	2516	2516
Minimum (m)	999	999	999	999
Mean (m)	1445	1445	1441	1446
Standard deviation (m)	229	228	230	231

Table 2 Comparison of depiction accuracy of slope.

Item	Grid	Grid500	TIN1	TIN5
Maximum	43.92	17.78	37.56	42.39
Minimum	0	0.06	0.01	0.01
Mean	11.19	4.62	8.04	9.15
Standard deviation	5.56	2.27	4.44	5.85

Table 3 Comparison of depiction accuracy of topographic index.

Item	Grid	Grid500	TIN1	TIN5
Maximum	27.37	20.23	24.67	23.17
Minimum	4.82	7.44	4.93	4.77
Mean	7.53	9.91	8.35	8.24
Standard deviation	2.16	2.01	1.90	1.96

Table 4 Comparison of depiction accuracy of A_{cc}.

Item	Grid	Grid500	TIN1	TIN5
Maximum (hm^2)	135 214	134 075	134 813	134 750
Minimum (hm^2)	0	0	0	0
Mean (hm^2)	392	1586	338	356
Standard deviation(hm^2)	5474	10 607	5025	5214

The comparison results indicate that: (1) Both TINs and coarsening grids can depict the elevation well, but the former have a little higher precision. (2) Compared with original high-resolution grids, the TIN method depicts the steep regions well, yet smoothes the gentle slope. (3) Both the values and distribution of flow accumulation area per unit width are in good agreement for TINs using different interpolations and a 100-m grid model, while the results of 500-m grids differ from those of 100-m grids in the source areas far away from the river and are similar to those of TINs near the river where the flow accumulation area is large. (4) Regions near the watershed division become homogeneous when the mesh size is large. TINs have less impact than homogeneous 500-m grids. If we use 5-degree polynomial interpolation, the results improve a little, but ensemble distributions by the two interpolations differ very little.

CONCLUSIONS

- TIN mesh system, which is based on slope and topographic index and incorporates various kinds of hydrological information, has several advantages compared with the grid method of the spatial discretization of the land surface.
- TIN method, which uses continuous polylines to capture the curved borderlines, can handle watershed boundaries easily, while the grid method can only use serrated borders to approach actual ones.
- TIN method can accurately represent local details with fewer units when dealing with multi-scale data due to its inherent characteristic of variable resolution, while in grids, the number of nodes is in significant contradiction with accuracy for the single spatial resolution.
- TINs can incorporate multiple known landscape information in the process of constructing TIN mesh. Therefore, the mesh themselves contain a lot of basin hydrological information. While raster grids depend on the smallest scale of all the information and can only use points to depict information other than lines and surfaces, which leads to less accuracy.
- TINs are flexible to include additional discretization units in local areas to account for higher spatial variability, which makes it more effective than the grid method.

Acknowledgements This research was supported by the National Natural Science Foundation of China (project no. 50679031).

REFERENCES

Beven, K. J. & Kirkby, M. J. (1979) A physically-based variable contributing area model of basin hydrology. *Hydrol. Sci. Bull.* **24**, 43–69.

Kong, F. & Rui, X. (2003) Calculation method for the topographic index in TOPMODEL. *Adv. Water Sci.* **14**, 41–45.

Mike, K & Tom, F. (2005) Notes for ARCInfo users (EB/OL) ftp://e0srp01u,ecs.nasa.gov/srim/version2/Documentation/Notes for ARCInfo users.pdf.

Quinn, P., Beven, K., Chevallier, P. & Planchon, O. (1991) The prediction of hillslope flow paths for distributed hydrological modelling using digital terrain models. *Hydrol. Processes* **5**, 59–79.

Vazquez, R. F. & Feyen, J. (2007) Assessment of the effects of DEM gridding on the predictions of basin runoff using MIKE SHE and a modelling resolution of 600 m. *J. Hydrol.* **334**, 73–87.

Vivoni, E. R., Ivanov, V. Y., Bras, R. L. & Entekhabi, E. (2004) Generation of TIN networks based on hydrological similarity. *J. Hydrol. Engng* **9**, 288–302.

Yang, D., Herath, S. & Musiake, K. (2000) Comparison of different distributed hydrological models for characterization of catchment spatial variability. *Hydrol. Processes* **14**, 403–416.

Scale and resolution effects of topographic index by 2-D continuous wavelet transform

DINGBAO WANG & XIMING CAI
Ven Te Chow Hydrosystems Laboratory, Civil and Environmental Engineering, University of Illinois at Urbana-Champaign, Urbana 61801, USA
dwang3@uiuc.edu or xmcai@uiuc.edu

Abstract Topographic index (TI) is an important parameter in catchment hydrology. The resolution dependence of TI statistics such as mean and variance has been discussed extensively in literature, while the TI structure at different scales, a different issue from resolution, needs more attention. To deal with the directional characteristics of TI structure, this paper uses the two-dimensional continuous wavelet transform (CWT) to explore the TI structure at different scales under several resolutions. For the 90-m resolution, the TI structure can be retrieved up to the scale of 270 m; only the main features (such as main streams) can be retrieved with the scale of 360 m; and even the main structure cannot be retrieved with the 900 m scale. For the 180 m resolution, both high and low order streams can be retrieved at the 180 scale; but for 270 m resolution, only the main steam structure can be captured at the 270 scale. Such preliminary results can give some insight into a suitable DEM resolution for deriving TI and the scale characteristics of TI structure. For example, the 180 m resolution of TI can be used if the high resolution DEM, such as 30 m resolution, is not available; the 270 m resolution TI is not suitable for capturing the flow process in the second or higher order stream.

Key words topographic index; scale; direction; 2 dimensional continuous wavelet transform; Morlet wavelet

INTRODUCTION

Topographic index (TI) is an important parameter in catchment hydrology. For example, TI is a key concept in TOPMODEL (Beven & Kirkby, 1979), a catchment hydrological model that is widely used to predict spatial patterns of soil moisture and depth of the water table (Moore *et al.*, 1991). The prediction of TOPMODEL depends on the mean value and distribution of TI in a catchment. TI is derived from the digital elevation model (DEM); and then the TI statistics depend on the resolution of DEMs (Zhang & Montgomery, 1994). A DEM with low resolution may not capture the small scale features; whereas DEMs with high resolution may not be available.

The scale issue is closely related to the resolution issue, but they are two different concepts. With the decline of resolution (e.g. from 30 m to 90 m), small features, i.e. features at small scales, may disappear. Given a certain resolution, different physical phenomena or characteristics dominate at different scales. It is then worthwhile to explore the TI structure at different scales for a given resolution of TI. Cai & Wang (2006) explored the spatial autocorrelation of TI at different scales under the 30 m DEM; they found that positive spatial autocorrelation of TI exists over a wide range of spatial extents and the autocorrelation declines with the increase of spatial extents, but becomes relatively static beyond a threshold area of approx. 1 km^2. To further study the issue of resolution *vs* scale, this paper applies the two dimensional (2-D) continuous wavelet transform (CWT) to exploring the TI structure at different scales and resolutions. The wavelet transform technique is a useful tool to investigate the scale issue. By using the CWT, we test the following hypothesis – for a given scale, a certain DEM resolution may be necessary to derive the TI structure at the scale. On the other hand, for the same purpose, a fine DEM resolution may not be necessary because of data redundancy.

METHODOLOGY

The wavelet transform originated from seismic data analysis (Morlet, 1981, 1983). Grossmann *et al.* (1985) developed the geometrical formalism of the continuous wavelet transform based on invariance under the translation and dilation, which allowed the decomposition of a signal into contributions of both space and scale. The continuous wavelet transform is particularly suitable for analysing the local differentiability of a function and for detecting and characterizing possible singularities (Arnéodo *et al.*, 1988). The wavelet transform performs like a microscope that discriminates different scales in an *n*-dimensional field, and like a polarizer that separates the different angular contributions of the signal. The wavelet transform of a function $f(b)$ is defined as:

$$Wf(a,b) = \int_{-\infty}^{\infty} f(u)\overline{\psi}_{a,b}(u)du \quad a>0 \tag{1}$$

where

$$\psi_{a,b}(u) = \frac{1}{\sqrt{a}}\psi\left(\frac{u-b}{a}\right) \tag{2}$$

$\Psi_{a,b}(u)$ represents a family of functions, which is called wavelets; $\overline{\psi}_{a,b}(u)$ is the complex conjugate of $\Psi_{a,b}(u)$; and a is a scale parameter. If $a > 1$, it has a dilating effect on the wavelet, otherwise it has a contracting effect. b is a location parameter, and the analysis can be taken around different locations by changing the value of b. The wavelet transform defined by equation (1) is called the continuous wavelet transform (CWT) because the scale and location parameters have continuous values. Wavelet transforms implemented on discrete values are called discrete wavelet transforms (DWTs). DWTs have been used to segregate spatial rainfall into multi-scale local averages and multi-scale local fluctuations by Kumar & Foufoula-Georgiou (1993a,b); and technical details of DWTs are presented by Daubechies (1992). A comprehensive review of wavelet applications in geophysical studies can be found in Kumar & Foufoula-Georgiou (1997).

The 2-D CWT is used in this paper. For a 2-D image (such as a TI map) denoted as f, the wavelet transform corresponding to the analysing wavelet ψ is defined as (Antoine *et al.*, 1993):

$$Wf(a,\theta,\mathbf{b}) = a^{-1}\int d^2\mathbf{x}\overline{\psi(a^{-1}r_{-\theta}(\mathbf{x}-\mathbf{b}))}f(\mathbf{x}) = a\int d^2\mathbf{k}e^{i\mathbf{b}\cdot\mathbf{k}}\overline{\hat{\psi}(ar_{-\theta}\mathbf{k})}\hat{f}(\mathbf{k}) \tag{3}$$

where $\hat{f}(\mathbf{k})$ denotes a Fourier transform of $f(\mathbf{k})$, and the over bar is the conjugate complex. $Wf(a,\theta,\mathbf{b})$ is the projection of f on the wavelet Ψ, which is translated by $\mathbf{b}$, dilated by a and rotated by an angle θ. The 2-D CWT acts as a filter for four variables: location parameters $\mathbf{b} = (b_1,b_2)$, scale parameter a and direction parameter θ. Thus, there are two types of representation: (1) By fixing (a,θ), the transform is conducted as a function of $\mathbf{b}$ alone. This representation is useful in the detection of position, shape and contours of objects, form recognition and the identification of directional features. (2) With fixing $\mathbf{b}$, the CWT is a function of (a,θ), i.e. the scale-angle representation. In this representation, by fixing a and traversing along θ, we can extract directional information at a fixed scale a; while by fixing θ, scale information in a particular direction can be obtained (Kumar, 1995).

Among several available 2-D analysing wavelets such as the 2-D Mexican hat, difference wavelets and the optical wavelet (Antoine & Murenzi, 1996), the 2-D Morlet wavelet is chosen for this analysis because of its directional selectiveness capability of detecting oriented features and fine tuning to specific frequencies (Antoine *et al.*, 1993). The 2-D Morlet wavelet is defined as:

$$\psi(\mathbf{x}) = \exp(i\mathbf{k}_0\cdot\mathbf{x})\exp\left(-\frac{1}{2}|A\mathbf{x}|^2\right) \tag{4}$$

where $i=\sqrt{-1}$, $A = \mathrm{diag}\,[\varepsilon^{-1/2},1]$, $\varepsilon \geq 1$, is a 2×2 diagonal matrix that defines the anisotropy of the wavelet. $k_0 = (0,k_0)$ defines the frequency complex exponential $\exp(i\mathbf{k}_0\,\mathbf{x})$. The Morlet wavelet is characterized by the two parameters, ε and k_0.

In this paper, the TI structure is extracted at different scales. The TI structure, which reflects the stream network, is multidirectional. Thus, for each considered scale, the responses with maximum modulus over all possible orientations are extracted, i.e.:

$$M(\alpha,\mathbf{b}) = \max_{\theta}\left(|Wf(\alpha,\theta,\mathbf{b})|\right) \tag{5}$$

The Morlet wavelet transform is computed for θ, spanning from 0 up to π at an interval of $\pi/36$ (i.e. 5 degrees); and the maximum coefficients are taken. Because of the periodicity characteristic of Fourier transform, $|Wf(\alpha,\theta,\mathbf{b})|$ is equivalent to $|Wf(\alpha,\theta+\pi,\mathbf{b})|$ (Antoine & Murenzi, 1996).

To demonstrate the directional selectivity of Morlet wavelet, a 2-D random field is created and a horizontal line is added to the field (Fig. 1). θ indicates the direction in the Fourier domain. Thus large values along this direction in the Fourier domain correspond to large values in ($\pi/2 - \theta$) direction in the physical domain. Figure 2(a) shows the CWT modulus with $\theta = \pi/2$, the horizontal features can be retrieved, i.e. the line in Fig. 1. The orientation selectivity of Morlet wavelet depends on the two parameters: ε and $\mathbf{k}_0$, specifically the value of $\mathbf{k}_0\sqrt{\varepsilon}$ (Antoine *et al.*, 1993). Comparing Fig. 2(b) to Fig. 2(a), the values of ε and $\mathbf{k}_0$ are different and others are the same. Figure 2(b) has stronger selectivity in the horizontal direction.

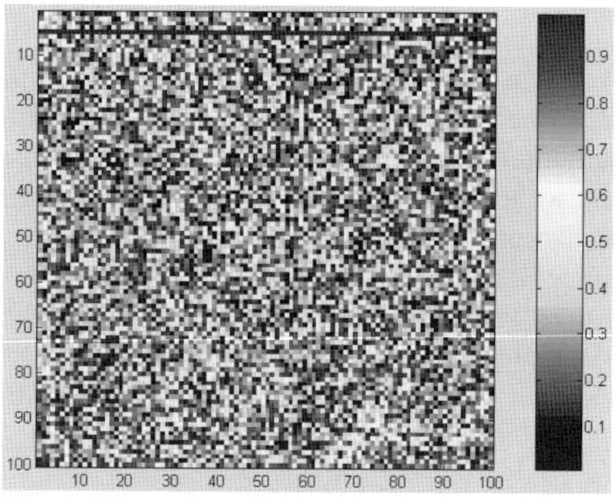

Fig. 1 2-D random field with a horizontal line.

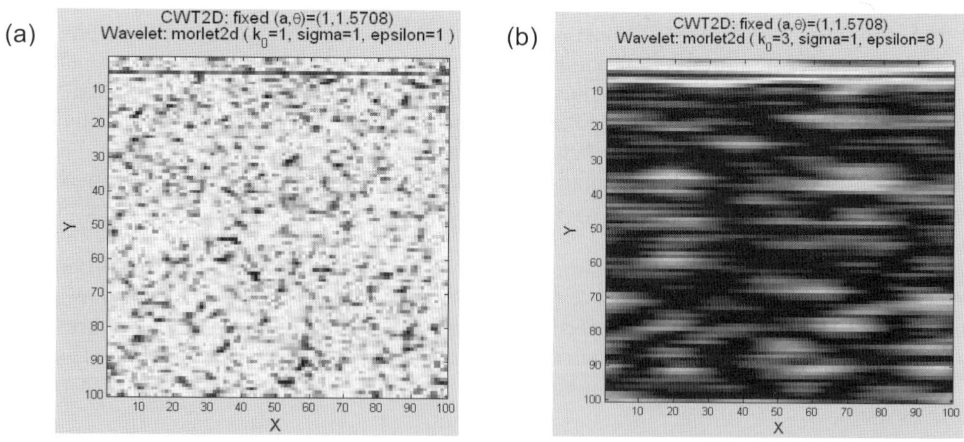

Fig. 2 CWT with the 2-D Morlet wavelet, (a) $a = 1$, $\theta = \pi/2$, $k_0 = 3$, $\varepsilon = 8$; (b) $a = 1$, $\theta = \pi/2$, $k_0 = 1$, $\varepsilon = 1$.

RESULT ANALYSIS

A TI map is analysed using the 2-D Morlet wavelet, and the analysis is implemented in ArcGIS and MatLab softwares. Figure 3(a) shows the TI map of a catchment, which includes the Mackinaw River basin located in Illinois. The original resolution of the DEM is about 90 m. Figure 3(b) shows the stream network derived from the DEM when the threshold for defining streams is set to 500 cells (i.e. 4.05 km^2 drainage area) using single flow direction algorithm. As shown in Fig. 3(a), the spatial distribution of TI matches the river network. Along the stream (the white lines) the TI values are large; the TI values gradually decrease away from the stream. The black line features are the ridges, which have small TI values. Thus, the directional characteristics are shown with the TI structure.

To extract the TI structure, the optimal values of parameters k_0 and ε are set to 3 and 2, respectively, by trial-and-error. Figure 4 shows the extracted features under a series of scale factors, $\alpha = 1, 2, 3, 4, 10$, which correspond to the scale of 90 m, 180 m, 270 m, 360 m and 900 m, respectively. From small to large scale, the retrieved features become larger and clearer, but the features at small scales cannot be retrieved (e.g. lower order streams). Under the scale of 360 m, only the main features (e.g. main streams) can be retrieved (Fig. 4(d)), but under the 900 m scale, the main structure cannot be retrieved, and the CWT results basically do not reflect the TI structure (Fig. 4(e)).

For the TI map at different resolutions, the $M(\alpha,\mathbf{b})$ can be retrieved at the same scale factor $\alpha = 1$. Figure 4(a) shows the retrieved features for TI at 90 m resolution, and Fig. 5(a) and (b) show the retrieved features at 180 m and 270 m resolution under a corresponding scale (i.e. 180 m or 270 m), respectively. As shown in the figures, under the 180 m resolution, both high and low order streams, but under 270 m resolution only the main steam structure can be captured at the corresponding scales. Comparing Fig. 5(a) to Fig. 4(b), the TI structures are retrieved under the 180 m scale based on TI derived from 180 m and 90 m DEM, respectively. The results from the

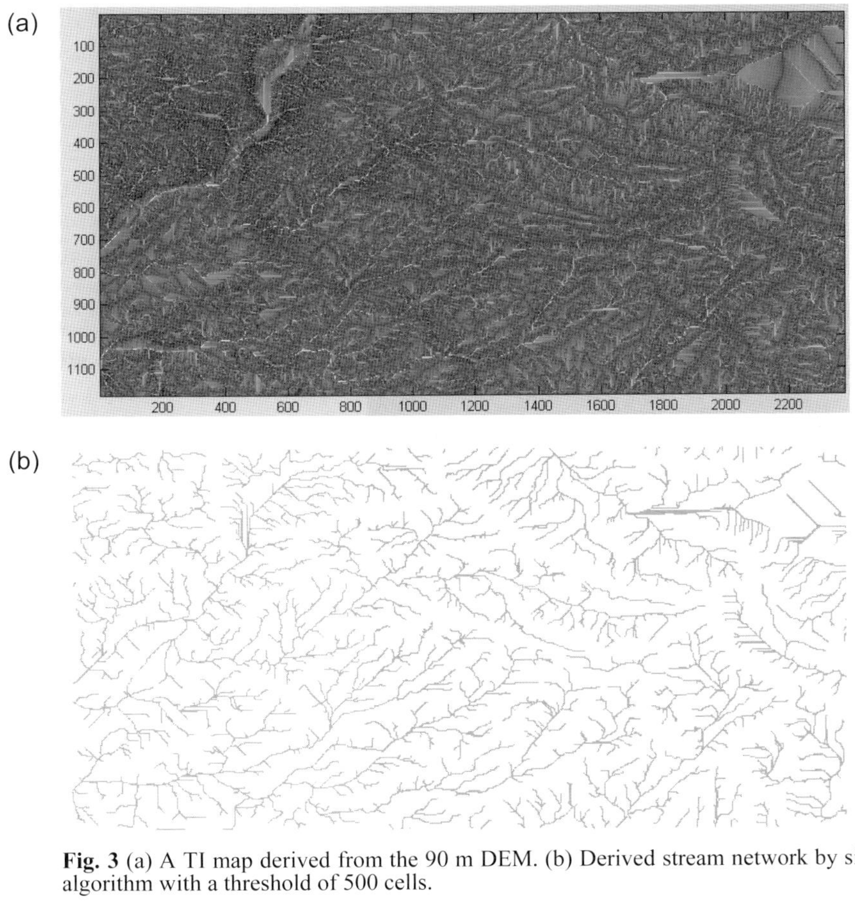

Fig. 3 (a) A TI map derived from the 90 m DEM. (b) Derived stream network by single flow direction algorithm with a threshold of 500 cells.

Antoine, J., Carrette, P., Murenzi, R. & Piette, B. (1993) Image analysis with two dimensional continuous wavelet transform. *Signal Processes* **31**, 241–272.

Arnéodo, A., Argoul, F., Elezgaray, J. & Grasseau, G. (1988) Wavelet transform analysis of gractals: applications to non-equilibrium phase transitions. In: *Proc. Conf. Nonlinear Dynamics* (ed. by G. Turchetti) (Bologna). World Scientific Singapore.

Beven, K. & Kirkby, M. J. (1979) A physically based, variable contributing area model of basin hydrology. *Hydrol. Sci. Bull.* **24**, 43–69.

Cai, X. & Wang, D. (2006) Spatial autocorrelation of topographic index in catchments. *J. Hydrol.* **328**, 581–591. doi: 10.1016/j.jhydrol.2006.01.009.

Daubechies, I. (1992) Ten lectures on wavelets. *Soc. For Ind. And Appl. Math.*, Philadelphia, Pennsylvania, USA.

Grossmann, A., Morlet, J. & Paul, T. (1985) Transforms associated to square integrable group representations I: general results. *J. Math. Phys.* **26**, 2473–2479.

Kumar, P. (1995) A wavelet-based methodology for scale-space anisotropic analysis. *Geophys. Res. Lett.* **22**(20), 2777–2780.

Kumar, P. & Foufoula-Georgiou, E. (1993a) A multicomponent decomposition of spatial rainfall fields. 1, Segregation of large- and small-scale features using wavelet transforms. *Water Resour. Res.* **29**(8), 2515–2532.

Kumar, P. & Foufoula-Georgiou, E. (1993b) A multicomponent decomposition of spatial rainfall fields. 2, Self-similarity in fluctuations. *Resour. Res.* **29**(8), 2533–2544.

Kumar, P. & Foufoula-Georgiou, E. (1997) Wavelet analysis for geophysical applications. *Rev. Geophys.* **35**(4), 385–412.

Moore, I. D., Grayson, R. B. & Ladson, A. R. (1991) Digital terrain modeling: a review of hydrological, geomorphologic and biological applications. *Hydrol. Processes* **5**, 3–30.

Morlet, J. (1981) Sampling theory and wave propagation. In: *Proc. 51st Annual Meeting Soc. Explor. Geophys.* Los Angeles, USA.

Morlet, J. (1983) Sampling theory and wave propagation. In: *NATO ASI, Vol. FI, Issues on Acoustic Signal/Image Processing and Recognition* (ed. by C. H. Chen). Springer, Berlin, Germany.

Zhang, W. & Montgomery, D. R. (1994) Digital elevation model grid size, landscape representation, and hydrologic simulations. *Water Resour. Res.* **30**, 1019–1028.

A long-term runoff forecast model based on association rules of data mining

FUQIANG WANG & SHIGUO XU

Institute of Environmental and Water Resources, Department of Civil Engineering, Dalian University of Technology, Dalian 116024, China

sgxu@dlut.edu.cn

Abstract Association rule is an important method for data mining. There are a lot of hydrological and forecasting data in the region area of long-term runoff forecasts. It is important to fully analyse and mine these data via various intelligent algorithms to formulate hydrological forecast for a precise forecast. Considering the characteristics of hydrological forecasting, the association rule mining method is applied to the long-term runoff forecast. The hydrological and meteorological data from 1956 to 2005 were selected to constitute the Jiangqiao station runoff forecast database at Nenjiang River. According to the min-support and min-confidence, the data was pre-treated before extracting association rules by the standards to find the strong association rules. In the practical example of Jiangqiao station, three strong association rules were mined and these rules reveal the effects of the North Pacific sea surface temperature (SST) on the flood season runoff at Jiangqiao hydrological station. The qualified rate of the model is 80%, and the results show that the model is highly effective for flood prediction of Jiangqiao station in the flood season. Furthermore, the association rule mining may be used as an effective tool for the long-term hydrological forecast.

Key words association rules; data mining; sea surface temperature (SST); long-term runoff forecast

INTRODUCTION

With the development of long-term hydrological forecasting and the increasing use of very large databases and data warehouses, the number of species and amount of hydrometeorological data has been increasing sharply. Revealing valuable knowledge hidden in massive data becomes more critical for decision making. When more data is collected and accumulated, extensive data analysis would be easier with effective and efficient data mining methods. How to store, manage and analyse these data has become an important problem in the research of long-term hydrological forecasts. How to extract the useful information and make long-term hydrological forecasts are more problems. The conventional long-term hydrological forecast pattern cannot deal with the numerous data. Furthermore, we expect that data is gathered and analysed by using the computer from the viewpoint of information management. Therefore, extraction of available implicit information to aid decision making has become a new and challenging task.

Data mining is a new research area that aims to extract implicit, previously unknown and potentially useful information from databases (Chen *et al.*, 1996; Abascal *et al.*, 2006; Karasozen *et al.*, 2006; Perner, 2006; Lavrac *et al.*, 2007). Many approaches have been proposed to extract data. Mining association rules are one of the most important. The problem of finding association rules was first introduced by Agrawal *et al.* (1993). This problem aims at discovering unknown relationships, providing results that can be the basis of forecast and decision making (Zaki, 2004). For example, supermarkets or catalogue companies collect sales data from sale orders. These orders usually consist of the sale date, the items in the transaction, and customer-ID. Through getting and analysing of association rules of these orders, a lot of valuable information such as the buying patterns of consumers can be inferred (Burdick *et al.*, 2005). Finding frequent item-sets is the most fundamental and essential problem in finding association rules, from which association rules can be generated directly (McGarry, 2005).

Considering the characteristics of hydrological forecast, the association rules mining method is applied to the long-term runoff forecast. A long-term runoff forecast model has been set up based on the association rules mining method, and it indicated that the method might be an effective tool for long-term hydrological forecasting by taking the flood season runoff forecast of Jiangqiao station as an example.

The rest of the paper is organized as follows: related works are reviewed and the contribution of this paper is discussed; the application of association rules techniques in long-term runoff forecast is described by taking the flood season runoff prediction of Jiangqiao station as an example; and conclusions.

RELATED WORKS

Conception of association rules

Association rules were introduced in Agrawal *et al.* (1993) as a method to find relationships among the attributes in a database. Using these techniques interesting qualitative information with which we can make later decisions, can be obtained. In general terms, an association rule is a relationship between attributes in the way $C_1 \Rightarrow C_2$, where C_1 and C_2 are pair conjunctions (attribute-value) in the way $A = v$ if it is a discrete attribute or $A \in [v_1, v_2]$ if the attribute is continuous or numeric. Generally, the antecedent is formed by a conjunction of pairs, while the consequence is usually a unique attribute-value pair.

There are a large number of rules of this kind in most databases, so it is essential to define some measures that allow us to only filter the most significant ones. The most used measures to define the interest of the rules were described in Agrawal *et al.* (1993):

- *Support.* It is a statistical measure that indicates the ratio of the population that satisfies both the antecedent and the consequent of the rule. A rule R: $C_1 \Rightarrow C_2$ has a support s, if s% of the records of the database contain C_1 and C_2.
- *Confidence.* This measure indicates the relative frequency of the rule, that is, the frequency with which the consequent is fulfilled when it also fulfilled the antecedent. A rule R: $C_1 \Rightarrow C_2$ has a confidence c, if the c% of the records of the database that contain C_1 also contain C_2.

The goal of the techniques that search for association rules is to extract only those that exceed *min-support* and *min-confidence* that are defined by the user. The basic data mining process may be summarized as follows: (a) find out all frequent itemsets and their support counts. A frequent itemset is a set of items which are contained in a sufficiently large number of transactions, with respect to a minimum support. (b) From the set of frequent itemsets found, find out all the association rules that have a confidence value exceeding a minimum confidence. Some of these algorithms can be seen (Agrawal *et al.*, 1994; Manila *et al.*, 1994; Park *et al.*, 1995; Savarese *et al.*, 1995).

Process of study

Our goal is to find association rules in a runoff forecast database, which have a big effect on the runoff. This work can be summarized in four phases: (a) The factors that exceed critical correlation coefficients are selected based on the correlative analysis between SST and flood season runoff of Jiangqiao station. (b) Data of "key months" SST and flood season runoff of Jiangqiao station is discretized to build the prediction databases. (c) The strong association rules which accord with the min-support and min-confidence can be found by use of Apriori algorithm (Aflori & Craus, 2007; Rodtook & Makhanov, 2007). (d) The long-term runoff forecast model is built by explaining the strong association rules.

PRACTICAL IMPLEMENTATION

Research area and data

Songhua River is one of the seven major Chinese rivers. Jiangqiao station on the Nenjiang River, which is a main tributary of Songhua River (Fig. 1). The annual mean precipitation is 400–600 mm in the Nenjiang River basin. Precipitation is concentrated in the summer (June to August) and

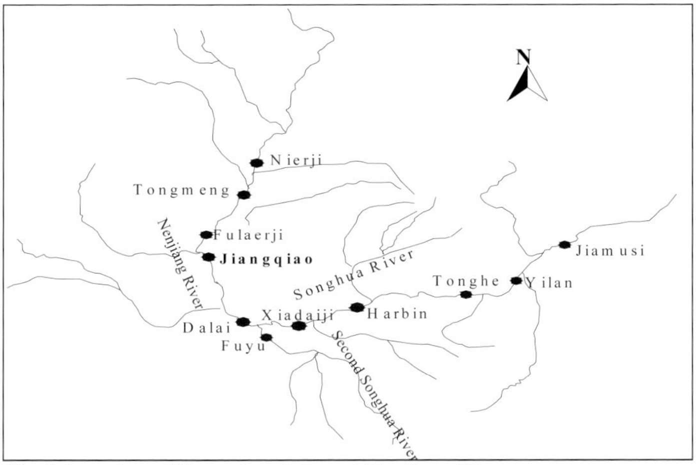

Fig. 1 Location of Jiangqiao station at Nenjiang River.

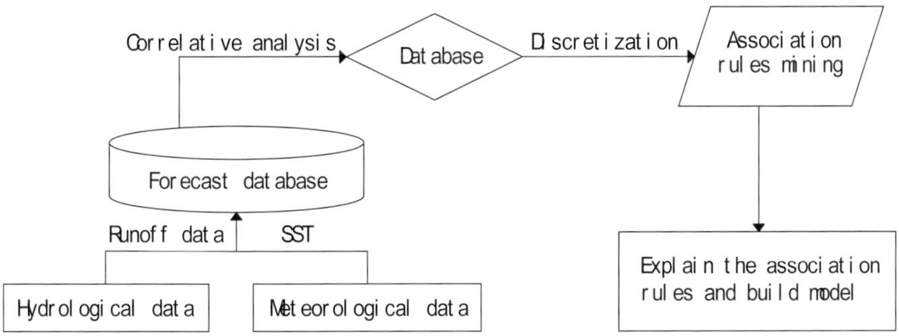

Fig. 2 The process of association rules mining

accounts for 80% of the precipitation in the whole year. In this situation it is important to prevent floods in the flood season.

Fifty years (1956–2005) of continuous monthly runoff data from Jiangqiao hydrological station are used in the study. Monthly SST data from the north pacific region, practically located in the area that spans from 10°S, 50°N and 120°E, 80°W (286 dots, 5° × 5°) are used. The data ranges from 1951 to 2005.

Process of association rules mining

Firstly, the physical factors whose correlation coefficients beyond critical correlation coefficients can be selected based on the correlative analysis between SST and flood season runoff of Jiangqiao station. After that, data of "key months" SST and flood season runoff of Jiangqiao station is discretized to build the prediction database. Thirdly, the strong association rules which accord with the min-support and min-confidence can be found by use of the *a priori* algorithm. Finally, the long-term runoff forecast model is built by explaining the strong association rules (Fig. 2).

Database

We should set up the long-term runoff forecast database first in order to extract association rules. Considering the delay effects of equatorial SST on the climate of research area (Liu *et al.*, 2001), the physical factors whose correlation coefficients are beyond critical correlation coefficients (0.4) can be selected for runoff prediction. We can find that SST of April, May, June and October have a strong effect on the flood season (June to September) runoff of Jiangqiao station via correlative analysis. April, May, June and October are seen as "key months" and their data is selected to set up the long-term runoff forecast database in the study.

Data pre-treatment

The data should be pretreated before extracting association rules. Runoff data for the period 1956–2000 can be classified based on the standard for hydrological information and hydrological forecasting (SL250-2000, 2000) (see Table 1).

Table 1 The classification standard for long-term qualitative flood forecast.

Classification	Low-runoff (1)	Lower-runoff (2)	Mean-runoff (3)	Higher-runoff (4)	High-runoff (5)
Factor Anomaly (FA) (%)	$FA < -20$	$-20 \le FA < -10$	$-10 \le FA \le 10$	$10 < FA \le 20$	$FA > 20$

The classification standard for SST data can be seen in Table 2. The data of the north pacific sea surface temperature is classified as three ranks.

Table 2 The classification standard for SST data.

Classification	Low-SST (–1)	Mean-SST (0)	High-SST (1)
Factor Anomaly(FA) (°C)	$FA < -0.5$	$-0.5 \le FA \le 0.5$	$FA > 0.5$

The database is set up based on the discrete data, as can be seen in Table 3.

Table 3 The database of long-term flood forecast.

Number	Year	Physical factors SST(April–June)	SST(October)	Jiangqiao Runoff
1	1956	–1	–1	5
2	1957	–1	–1	5
3	1958	1	1	3
4	1959	1	0	1
5	1960	0	0	5
—	—	—	—	—
45	2000	–1	–1	1

Table 4 Association rules.

Rules	Support	Confidence
$SST(4-6) = 1$ and $SST(10) = 1 \Rightarrow runoff = 5$	0.156	0.636
$SST(4-6) = -1$ and $SST(10) = -1 \Rightarrow runoff = 1$	0.178	0.727
$SST(4-6) = -1 \Rightarrow runoff = 1$	0.222	0.667

Association rules mining

The goal of the techniques that search for association rules is to extract only those that exceed *min-support* and *min-confidence* that are defined by the user. *A priori* algorithm is used to extract association rules in the long-term runoff forecast database in the study. The results that are discovered by using *a priori* algorithms are shown in Table 4.

Experimental results

- *Rule1: SST*(4 – 6) = *1 and SST*(10) = *1⇒runoff = 5 (15.6% support, 63.6% confidence)*
  This rule can be expressed as *SST*(4 – 6) = 1 *and SST*(10) = *1⇒runoff = 5 (15.6% support, 63.6% confidence)*, where 63.6% is the confidence level of the rule and 15.6% is the support level of the rule, indicating how frequently it appears simultaneously as both SSTA(4 – 6) > 0.5°C ∧ SSTA(10) > 0.5°C in the previous year and high runoff of Jiangqiao station in the next year.
- *Rule2: SST*(4 – 6) = *–1 and SST*(10) = *–1⇒runoff = 1 (17.8% support, 72.7% confidence)*
  Rule2 shows that sea surface temperature in April–June and October are all at a low level (<0.5°C) in the previous year, the flood season runoff of Jiangqiao station is also at a low level (runoff anomaly<20%) in the next year, where the support level is 17.8% and the confidence level is 72.7%.
- *Rule3: SST*(4 – 6) = *–1⇒runoff = 1 (22.2% support, 66.7% confidence)*

The rule reveals that the sea surface temperature of April to June is at a low level (<0.5°) in the previous year; the flood season runoff of Jiangqiao station is also at a low level (runoff anomaly >20%) in the next year, where 22.2% of support level and 66.7% of confidence level.

The results of the experiment indicate that when the sea surface temperature of April to June and October are all at a high level (>0.5°) or a low level (<0.5°) in the previous year, the flood season runoff of Jiangqiao station is also at a high-runoff level (runoff anomaly□20%) or a low-runoff level (runoff anomaly <20%) in the next year where there is higher support and confidence. At the same time, the changes of sea surface temperature of April to June have a more evident effect on the flood season runoff of Jiangqiao station in the next year. So more attention should be paid to the changes of SST of "key months" when we make a flood prediction. The other changes of SST do not have a significant effect on the flood season runoff of Jiangqiao station because the other rules cannot satisfy with *min-support* and *min-confidence*.

Model test

Five years (2001–2005) of continuous runoff data from Jiangqiao station are used to test the model in the study. The results can be seen in Table 5. From Table 5, we can see that the prediction results of 2001, 2002, 2003 and 2004 agree well with the measurement. The results of tests show that the qualified rate of the model comes to 80%, and the model is highly effective for the flood prediction of Jiangqiao station in the flood season. Furthermore, association rules mining may be an effective tool for long-term hydrological forecast.

Table 5 The results of the model test.

Year	Physical factors		Measurement	Prediction	Evaluation
	SST(April–June)	SST(October)			
2001	–1	–1	1	1	√
2002	–1	0	1	1	√
2003	1	1	5	5	√
2004	–1	0	1	1	√
2005	1	0	1	–	×

CONCLUSIONS AND FUTURE WORK

In this paper, the association rules mining method is used to discover the unknown, valuable and operable information in physical factors for long-term runoff prediction.

The study has two major contributions. The first is that some interesting rules, which have a strong effect on the changes of flood season runoff of Jiangqiao station, have been discovered.

These rules are useful for the flood prediction of Jiangqiao station in the flood season. The second contribution is that a new prediction method is introduced and identified for long-term hydrological forecasting. How to fully analyse and mine those data via various intelligent algorithms to formulate hydrological forecast and reservoir operation algorithm accordingly for precise forecasts and rational operation is revealed in this paper. Meanwhile, the paper demonstrates that association rules mining is an effective data mining analysis tool in long-term hydrological forecasts.

However, the paper has some possible extensions. How to express the rules in higher level concepts, etc. is very important for practical purposes. Without generalization, the generated rules may have too much detail and not be fit for decision makers. Therefore, by including the concept hierarchies we may produce rules that are more abstract and concrete. Another possible extension is to prune less interesting rules that are trivial or implied by other rules. Without removing the uninteresting rules, we may be overwhelmed by enormous rules. Besides, we may try to set up a database that includes many kinds of physical factors, such as SST, atmosphere circulation characteristics volumes, the height of 500 hpa and 100 hpa isobaric surface, etc. This will help us to make a more precise forecast and rational operation.

Acknowledgements The authors thank Mr Fan Chuiren (Changchun Natural Calamity Forecast and Research Service Center, Changchun, China) for offering SST and runoff data in this study. This study was partially supported by the National Natural Science Foundation of China (no. 50139020).

REFERENCES

Abascal, E., Garcia, L. I. & Mallor, F. (2006) Data mining in a bicriteria clustering problem. *European J. Operational Res.* **173**(3), 705–716.

Aflori, C. & Craus, M. (2007) Grid implementation of the *a priori* algorithm. *Adv. Engng Software* **38**(5), 295–300.

Agrawal, R., Imielinski, T. & Swami, A. (1993) Mining association rules between sets of items in large databases. In: *Proc. SIGMOD International Conference on Management of Data.* 207–216. Washington, DC, USA.

Agrawal, R. & Srikant, R. (1994) Fast algorithms for mining association rules. In: *Proc. Of the VLDB Conference.* 487–489. Santiago, Chile.

Burdick, D., Calimlim, M., Flannick, J., Gehrke, J. & Yiu, T. M. (2005) A maximal frequent itemset algorithm. *IEEE Trans Knowledge Data Engng* **17**(11), 1490–1504.

Chen, M.S., Han, J. & Yu, P. S. (1996) Data mining: an overview from a database perspective. *IEEE Trans Knowledge Data Engng* **8**, 866–883.

Karasozen, B., Rubinov, A. & Weber, G. (2006) Optimization in data mining. *European J. Operational Res.* **173**(3), 701–704.

Lavrac, N., Bohanec, M., Pur, A., Cestnik, B., Debeljak, M. & Kobler, A. (2007) Data mining and visualization for decision support and modeling of public health-care resources. *J. Biomedical Informatics* **40**(4), 438–447.

Liu, S. & Wang, N. (2001) The impacts of antecedent ENSO event on air temperature over northeast China in summer. *J. Tropical Meteorol.* **17**(3), 314–315.

Manila, H., Toivonen, H. & Verkamo, A. I. (1994) Efficient algorithms for discovering association rules. In: *KDD-94: AAAI Workshop on Knowledge Discovery in Databases*, 181–192. Seatle, Washington, USA.

McGarry, K. (2005) A survey of interestingness measures for knowledge discovery. *Knowledge Engng Rev.* **20**(1), 39–61.

Park, J. S., Chen, M. S. & Yu. P.S. (1995) An effective hash based algorithm for mining association rules. In: *Proc. ACM SIGMOD Int. Conf. on Management of Data.* 175–186. San Jose, California, USA.

Perner, P. (2006) Recent advances in data mining. *Engng Appl. Artificial Intelligence* **19**(4), 361–362.

Rodtook,A. & Makhanov, S. S. (2007) A filter bank method to construct rotationally invariant moments for pattern recognition. *Pattern Recognition Letters* **28**(12), 1492–1500.

Savarese, A., Omiecinski, E. & Navathe, S. (1995) An efficient algorithm for mining association rules in large databases. In: *Proc. of the VLDB Conference.* 432–444. Zurich, Switzerland.

SL250-2000, (2000) Standard for hydrological information and hydrological forecasting. Ministry of Water Resources, China.

Wu, F., Chiang, S. & Lin, J. (2007) A new approach to mine frequent patterns using item-transformation methods. *Information Systems* **32**(7), 1056–1072.

Zaki, M. J. (2004) Mining non-redundant association rules. *Data Mining & Knowledge Discovery* **9**(3), 223–248.

A macro-scale hydrological model based on a spatial averaging approach: model structure and an application

ZHIYONG YANG, HEPING HU & FUQIANG TIAN

Department of Hydraulic Engineering, State Key Laboratory of Hydroscience and Engineering, Tsinghua University, Beijing 100084, China

huhp@tsinghua.edu.cn

Abstract The inconsistency between the scale at which the hydrological process-governing equations are applicable and the scale at which the model application are practicable lies at the heart of many current hydrological challenges. The spatial averaging approach firstly proposed by Kavvas is one of the promising methods to address such scale issues. This paper presents a macro-scale hydrological model based on the spatial averaging approach, within which the watershed is divided into model computation units (MCUs) and the hydrological processes across MCUs are described by the macro-scale equations derived from point scale equation through spatial averaging approach. As a case study for long-term continuous simulation, the model is applied to the Qin River basin with a semi-arid climate. The continuous streamflow and individual flood events simulation shows that the model can predict streamflow pretty well at lower computation time consumption, which shows the prospects of the model and its potential capabilities for application to water resources and flood management at a large-scale watershed.

Key words heterogeneity; spatial averaging; hydrological model; the Qin River basin

INTRODUCTION

Watershed-scale hydrological response is greatly influenced by the heterogeneity of landscape properties, which are intensified by the nonlinearity of hydrological processes. Such a situation leads to the so-called "scale issue" widely recognized by the hydrological community (Kalma & Sivapalan, 1995), i.e. the mismatch between the scale at which the physical equations are applicable and the scale at which the model applications are practicable. To address the scale issue, Kavvas proposed the spatial averaging approach for upscaling the point scale equations through the analysis of landscape heterogeneity (1999) and nonlinearity of hydrological processes (2003). By adopting the macro-scale equations derived by the spatial averaging approach, the WEHY (Watershed Environmental Hydrology) model was developed by Kavvas *et al.* (2004). As a first attempt, however, the WEHY model suffered from the watershed discretization scheme, as well as the over-complex groundwater simulation scheme (Yang, 2007). Also, the WEHY model has not yet been applied to continuous long-term streamflow simulation.

In this paper, a new macro-scale hydrological model (MSH model) based on the spatial averaging approach, but with a different spatial discretization scheme and simplified groundwater component, is developed and applied to simulate the continuous long-term streamflow in the Qin River basin.

MODEL DESCRIPTION

Spatial discretization

According to its river network and catchment area threshold, a watershed can be divided into some sub-watersheds, and the sub-watershed can be approximately represented by a rectangle hillslope and a river reach. The rectangle hillslope is the model computation unit (MCU) for hillslope hydrological processes simulation and its length and width can be obtained according to the river network density and river length (Yang *et al.*, 2004). The landscape heterogeneity within the MCU is quantified by the statistical approach and directly incorporated into the equations of nonlinear hydrological processes through spatial averaging, which leads to the equations applicable directly at the macro-scale, i.e. MCU. The statistical characteristic values of point scale parameters, such as mean, variance, and covariance, are directly incorporated into macro-scale equations as parameters.

Model structure

The MCU is the computation unit for hillslope hydrological processes such as evapotranspiration, infiltration, subsurface flow, etc. and the generated runoff flows into river and routes to the watershed outlet.

Evapotranspiration Total evapotranspiration consists of four components: canopy interception evaporation (E_c), depression storage evaporation (E_d), soil surface evaporation (E_s), and vegetation transpiration (E_{tr}). They are estimated by atmosphere conditions (represented by potential evapotranspiration E_p), vegetation conditions, and available moisture (Yang *et al.*, 2002) as shown in equations (1), (2), (3) and (4):

$$E_c = K_c E_p \tag{1}$$

$$E_d = E_p \tag{2}$$

$$E_{tr} = K_c E_p f(\theta_{root}) LAI/LAI_m \tag{3}$$

$$E_s = E_p f(\theta_{surf}) \tag{4}$$

where K_c is the vegetation coefficient, LAI_m is the maximum LAI of vegetation, θ_{root} is the soil moisture in root zone, and θ_{surf} is the soil moisture at land surface.

Infiltration and unsaturated flow Infiltration and unsaturated flow determine infiltration excess runoff, soil moisture, and water recharge to saturated surface flow and groundwater. Soil characteristics vary dramatically in space, which results in complex impacts on infiltration and unsaturated flow. To account for such heterogeneity, a spatial averaging Green-Ampt model (Chen *et al.*, 1994a,b) is adopted in the MSH model. To confine the problem into a manageable level, only the horizontal heterogeneity of saturated hydraulic conductivity (K_s) is considered. The lognormal distribution is then used to quantify the probability distribution of K_s.

At a specific time t and rainfall density q_t, the wetting front depth and surface moisture are dependent with K_s. Therefore, vertical soil water flux ($q(K_s,z)$) is a function of K_s and depth z as shown in equation (5) (Kavvas *et al.*, 2004):

$$q(K_s,z) = \begin{cases} 0 & K_s < K_{sz} \\ K_s\left[1 + \dfrac{1}{Y\left(\dfrac{K_s}{q_t},\tau_v\right)}\right] & K_{sz} < K_s < K_{st} \\ q_t & K_s > K_{st} \quad or \quad K_s > K_{sz} > K_{st} \end{cases} \tag{5}$$

where K_{sz} and K_{st} are two specific K_s, and the wetting front just reaches z when $K_s = K_{sz}$ and the soil surface is just saturated when $K_s = K_{st}$; τ_v is a variable indicating soil water condition; and $Y(\cdot)$ is the function of q_t, K_s, and soil water condition (Kavvas *et al.*, 2004).

The ensemble average of soil water flux ($\langle q(z)\rangle$) at z can easily be obtained by integrating equation (5) over K_s, given by equation (6), and the infiltration excess runoff ($\langle q_r\rangle$) is given in equation (7). Considering the proportion of impermeable area (β_p), the average infiltration excess runoff ($\langle q_{ex}\rangle$) is given in equation (8).

$$\langle q(z)\rangle = \int_0^{+\infty} q(K_s,z) f(K_s) dK_s \tag{6}$$

$$\langle q_r\rangle = q_t - \langle q(z)\rangle\big|_{z=0} \tag{7}$$

$$\langle q_{ex}\rangle = \langle q_r\rangle \beta_p - q_t\left(1-\beta_p\right) \tag{8}$$

where $\langle\cdot\rangle$ is the spatial average operator and $f(K_s)$ is the probability distribution function (pdf) of K_s.

Saturated subsurface flow When unsaturated flow reaches the impermeable stratum, which is assumed parallel to land surface, soil water starts accumulating on the layer and moves downhill as saturated subsurface flow. When the water table rises to the bottom of the rills, saturated soil water will recharge the rills as return flow, and when the water table rises to the land surface, saturation excess runoff will occur.

To account for the heterogeneity along the hillslope transect, the ensemble averaged model based on Darcy's law (Dogrul *et al.*, 1998) is used to model saturated subsurface flow as shown in equation (9):

$$\frac{S_s}{R}\frac{\partial\langle h\rangle}{\partial t}=\left\{\langle K\rangle\frac{\partial}{\partial x}\left(\langle h\rangle\frac{\partial\langle h\rangle}{\partial x}\right)+\frac{\partial^2}{\partial x^2}\left[\langle h\rangle Cov(K,h)\right]\right\}\cos^2\theta$$
$$+\frac{1}{R}\left[\langle K\rangle\frac{\partial\langle h\rangle}{\partial x}+\frac{\partial Cov(K,h)}{\partial x}\right]\sin\theta\cos\theta+\frac{\langle P\rangle}{R^2} \tag{9}$$

where S_s is specific yield; K is saturated hydraulic conductivity; R is soil depth; h is the normalized saturated subsurface flow depth; θ is the slope of land surface; and P is the net recharge along the vertical direction.

Saturated subsurface flow (q_{sat}), return flow (q_{rf}), and saturation excess runoff (q_s) can be estimated by equations (10), (11), and (12), respectively:

$$q_{sat} = R\sin\theta\cos\theta(Kh_0 + \mathrm{cov}(h_0,K)) \tag{10}$$

$$q_{rf} = 2K\sin\theta_y\cos\theta_y R(d_r + h-1)N_{ro} \tag{11}$$

$$q_s = S_sR(h-1)/\Delta t \tag{12}$$

where h_0 is the normalized subsurface flow depth at the downstream transect; θ_y is the slope of the interrill area along the y-direction; d_r is the normalized rill depth; N_{ro} is the rill occurrence density; and Δt is time step.

Surface flow The land surface of hillslope contains irregular local topography, which is presented by a rill–interrill configuration. The influence of rill formation on soil erosion and flow dynamics is recognized in many experimental studies (Morris, 1979; Abrahams *et al.*, 1989; Govindaraju & Kavvas, 1992, 1994a,b; Parsons & Wainwright, 2006). To account for the effects of rill–interrill configuration, spatially averaged conservation equations for interacting rill–interrill area surface flow model was developed by Tayfur & Kavvas (1994) and improved by Yoon & Kavvas (2000). The Yoon's model is adopted in the MSH model with some improvements. The ensemble averaged equations for interrill areas and rills are as follows:

$$\frac{\partial\langle\bar{h}_{ir}\rangle}{\partial t}+c_1\frac{\partial}{\partial x}\left[\left(K_x(\langle\bar{r}\rangle)+\frac{1}{2}\sum_i\sum_j\mathrm{cov}(r_i,r_j)\frac{\partial^2K_x(\bar{r})}{\partial r_i\partial r_j}\right)\langle\bar{h}_{ir}\rangle^{5/3}\right]$$
$$=\langle q_{ex}\rangle+\langle q_s\rangle-c_2\rho_{ir}\left[K_y(\langle\bar{r}\rangle)+\frac{1}{2}\sum_i\sum_j\mathrm{cov}(r_i,r_j)\frac{\partial^2K_y(\bar{r})}{\partial r_i\partial r_j}\right]\langle\bar{h}_{ir}\rangle^{5/3} \tag{13}$$

$$\frac{\partial\langle\bar{h}_r\rangle}{\partial t}+\frac{\partial}{\partial x}\left\{K_r(\langle\bar{r}\rangle)\frac{\langle w_r\rangle^{1/2}\langle\bar{h}_r\rangle^{5/3}}{(\langle w_r\rangle+2\langle h_r\rangle)^{1/2}}+\frac{1}{2}\sum_i\sum_j\mathrm{cov}(r_i,r_j)\frac{\partial^2}{\partial r_i\partial r_j}\left[K_r(\bar{r})\frac{\langle w_r\rangle^{1/2}\langle\bar{h}_r\rangle^{5/3}}{(\langle w_r\rangle+2\langle h_r\rangle)^{1/2}}\right]\Bigg|_{\bar{r}=\langle\bar{r}\rangle}\right\}$$
$$=\langle q_{ex}\rangle+\langle q_s\rangle+\frac{\langle q_{rf}\rangle}{\lambda_{rf}}+2\rho_rc_2\left[K_y(\langle\bar{r}\rangle)+\frac{1}{2}\sum_i\sum_j\mathrm{cov}(r_i,r_j)\frac{\partial^2K_y(\bar{r})}{\partial r_i\partial r_j}\right]\langle\bar{h}_r\rangle^{5/3} \tag{14}$$

where $\bar{h}_{ir}$ and $\bar{h}_r$ are average flow depths of every interrill area and rill, $c_1 = 1.0944$, $c_2 = 2.1888$ when the flow depth curve within single interrill area are assumed to be a power function; λ_{ra} is the rill surface proportion, $\rho_r = N_{ro}/\lambda_{ra}$, $\rho_{ir} = N_{ro}/(1 - \lambda_{ra})$; w_r is the width of rills; l is the width of interrill area; $K_x(\langle\bar{r}\rangle) = \dfrac{\langle S_x\rangle^{1/2}}{\langle n_{ir}\rangle\left[1+\left(\dfrac{\langle S_x\rangle}{\langle S_y\rangle}\right)^2\right]^{1/4}}$, $K_y(\langle\bar{r}\rangle) = \dfrac{\langle S_y\rangle^{1/2}}{\langle n_{ir}\rangle\left[1+\left(\dfrac{\langle S_y\rangle}{\langle S_x\rangle}\right)^2\right]^{1/4}}$, $K_r(\langle\bar{r}\rangle) = \dfrac{1}{\langle n_r\rangle}$,

where S_x and S_y are bed slopes along x- and y-directions of interrill areas, respectively; S_r is bed slope of the rill; n_{ir} and n_r are the Manning roughness coefficients of interrill areas and rills respectively; and $\bar{r}$ is a vector of the hydraulic parameters mean value at hillslope scale, $\bar{r} = \{\bar{S}_x, \bar{S}_y, \bar{n}_{ir}, \bar{l}\}$ for interrill areas and $\bar{r} = \{\bar{S}_r, \bar{n}_r, \bar{w}_r\}$ for rills.

Groundwater Groundwater provides the baseflow in a stream. The water exchange between stream reaches and groundwater is estimated by equation (15):

$$Q_g = K_g(h_g^2 - h_r^2)/L \tag{15}$$

where K_g is hydraulic conductivity in an aquifer; h_g is water table of groundwater from the bed of stream reach; h_r is water depth of stream reach; and L is the width of hillslope.

River routing Stream reaches are fed by lateral inflow from neighbouring hillslopes, including surface flow, subsurface flow, and groundwater flow. River routing is simulated by 1-D kinematic wave equations (Chow, 1988).

MODEL APPLICATION

Study area and data

The Qin River is the upper stream of the Qin River basin, and it has a catchment area of 1355 km^2. The climate in the study area is typically semi-arid with annual pan-evaporation of 1700 mm and annual rainfall of 611 mm.

The input data of the model includes DEM, soil class, land use, leaf area index (LAI) data, and meteorological data. The watershed boundary and stream networks, as shown in Fig. 1, were generated from DEM with spatial resolution of 100 m. According to stream networks, the study area was divided into 35 MCUs. The geomorphologic parameters of each MCU were obtained through GIS analysis. The rainfall data of 12 raingauges (as shown in Fig. 1) were used to generate the model rainfall input. The daily stream discharge data at the outlet, i.e. Kongjiapo stream station, was used for model calibration and validation. Meteorological data such as temperature, sunshine hours, humidity, etc. were collected from a station located at the middle reach and used to estimate potential evapotranspiration. The land use of the study area was reclassified into six types: cultivated land, forest, grassland, water body, urban area, and other, as shown in Fig. 2. The cultivated land, forest, and grassland covered >99% of the study area. The soil class map was transferred from the FAO/UNESCO data set with spatial resolution of 10 km and the associated soil hydrological properties were taken from the same data set. The monthly *LAI* date is derived from the MODIS TERRA MOD15A2 product with spatial resolution of 1 km.

RESULT AND DISCUSSION

The model was applied to simulate continuous streamflow in the study area. Hourly rainfall data at 12 raingauge stations and daily discharge data at the Kongjiapo station are available from 1986 to 1989. The Thiessen method was used to assign representative areas for each station in order to determine the spatial distribution of rainfall. The period from 1986 to 1987 was selected for model warming up. The soil hydraulic parameters were estimated from the soil map and associated soil

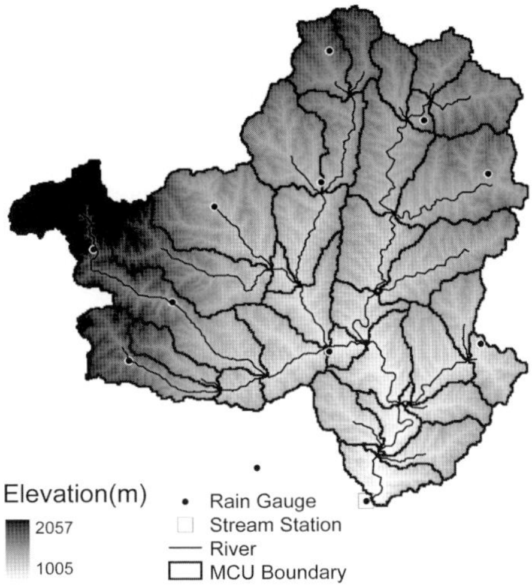

Fig. 1 DEM, MCUs, stream network and locations of rain and stream gauging station.

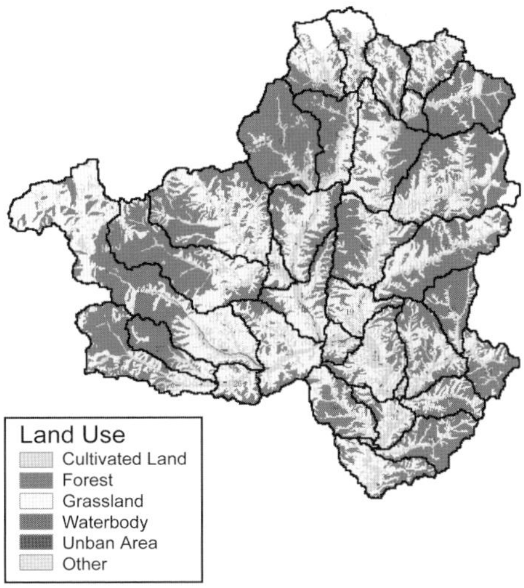

Fig. 2 Land use of the study area (overlaid with MCUs).

properties. The parameters of variance and covariance were determined with reference to former studies (Yoon & Kavvas, 2000).

The parameters such as soil depth (R), Manning's roughness of hillslope (n_s) and channel (n_r), soil anisotropy coefficient (C_a), and aquifer hydraulic conductivity (C_g), need to be estimated through model calibration. The period of 1988 was selected for model calibration. The pre-assigned range of the parameter and the calibrated results are listed in Table 1. The simulated hydrograph during the calibration period is shown in Fig. 3. The period of 1989 was selected for model validation. The hydrograph at the Kongjiapo Station during the validation period is shown in Fig. 4.

It can be seen from Figs 3 and 4 that the model-predicted hydrograph at the Kongjiapo station matched the observed data quite well. The Nash-Sutcliffe efficiency coefficient (E) (Nash & Sutcliffe, 1970) and coefficient of determination (R^2) were selected to quantify model performance (see Table 2). The results show that E in 1988 and 1989 reaches 0.88 and 0.94, respectively, and

R^2 reaches 0.89 and 0.97, respectively. Furthermore, the two coefficients during the validation period are similar to those during the calibration period (see Table. 2).

There are three flood events during calibration and validation. The model performance for a flood event is quantified simultaneously by relative error of flood peak discharge, the Nash-Sutcliffe efficiency coefficient (E), and coefficient of determination (R^2) during flood events and is listed in Table 3. The predicted flood peak and hydrograph matches the observed data very well.

Table 1 Parameter ranges and fitted values.

Parameter	R (m)	n_s	n_r	C_a	C_g
Range	0.5–5.0	0.1–1.0	0.01–0.1	1.0–10.0	0.01–1.0
Fitted Value	0.8,1.2,1.6[a]	0.5	0.05	5.0	0.05

[a] R = 0.8 ,1.2, and 1.6 for upstream, middle stream, and downstream of the study area.

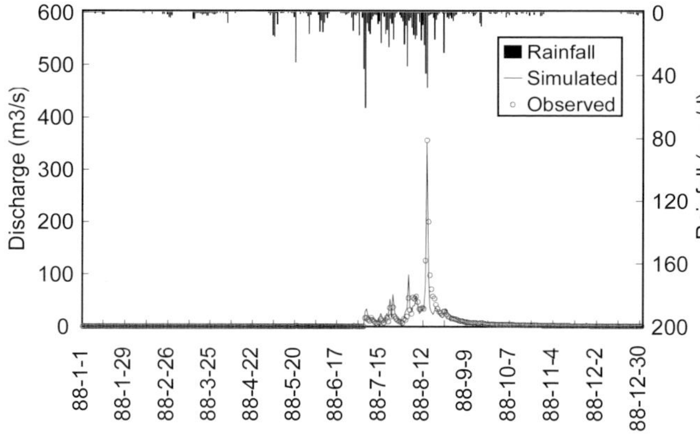

Fig. 3 Comparison of observed and simulated hydrographs during calibration period.

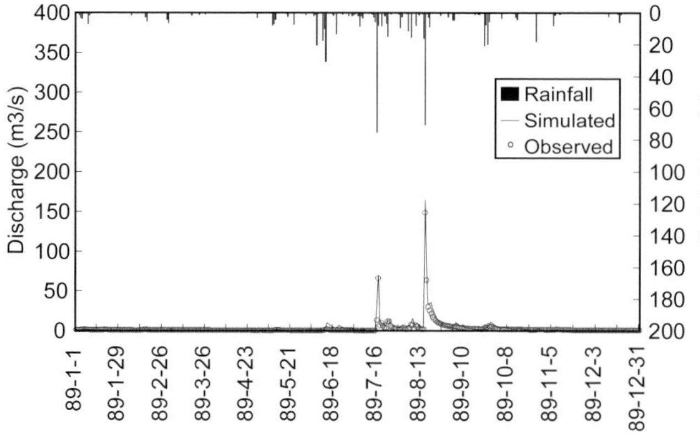

Fig. 4 Comparison of observed and simulated hydrographs during validation period.

Table 2 Model performance during calibration and validation periods.

Period	E	R^2
1988	0.88	0.89
1989	0.94	0.97

Table 3 Model performance during three flood events.

Flood events	Q_p (m^3 s^{-1}) Simulated	Observed	R_e (%)	E	R^2
7/05/1988–9/22/1988	350.63	356.00	−1.51	0.85	0.86
7/15/1989–7/22/1989	69.92	65.90	6.10	0.92	0.97
8/15/1988–9/21/1988	164.14	149.00	10.16	0.96	0.99

The high computation efficiency is obtained via coarse discretization of the watershed into MCUs, which does not lower the model performance. In the case study, 178 s was consumed for 35 064 h of hydrological modelling in a personal computer with an AMD Athlon 2500+ CPU and memory of 1GB.

SUMMARY

As a new method to address the scale issues in hydrological modelling, the spatial averaging approach is proposed to directly incorporate the micro-scale heterogeneity into the macro-scale process equations. A new macro-scale hydrological model is developed based on the spatial averaging approach. Compared to the former version of spatial averaging model, the WEHY model, our model modifies the spatial discretization scheme and simplifies the groundwater system. The model is intended to be able to simulate the long-term large scale hydrological processes at the watershed-scale with less computational burden. Our case study in the Qin River basin (1355 km^2) shows that the model can obtain the higher computation efficiency, as well as the sound performance of runoff simulation, which show the potential capabilities of the model for water resources and flood management application in semiarid regions.

Acknowledgements This research was supported by the National Natural Science Foundation of China (50721140161 and 50579031).

REFERENCES

Abrahams, A. D., Parsons, A. J. & Luk, S. H. (1989) Distribution of depth of overland flow on desert hillslopes and its implications for modelling soil erosion. *J. Hydrol.* **106**(1–2), 177–184.

Chen, Z. Q., Govindaraju, R. S. & Kavvas, M. L. (1994a) Spatial averaging of unsaturated flow equations under infiltration conditions over areally heterogeneous fields. 1. Development of models. *Water Resour. Res.* **30**(2), 523–533.

Chen, Z. Q., Govindaraju, R. S. & Kavvas, M. L. (1994b) Spatial averaging of unsaturated flow equations under infiltration conditions over areally heterogeneous fields. 2. Numerical simulations. *Water Resour. Res.* **30**(2), 535–548.

Chow, V. T. (1988) *Applied Hydrology.* McGraw-Hill, New York, USA.

Dogrul, E. C., Kavvas, M. L. & Chen, Z. Q. (1998) Prediction of subsurface stormflow in heterogeneous sloping aquifers. *J. Hydrol. Engng ASCE.* **3**(4), 258–267.

Govindaraju, R. S. & Kavvas, M. L. (1992) Characterization of the rill geometry over straight hillslopes through spatial scales. *J. Hydrol.* **130**(1–4), 339–365.

Govindaraju, R. S. & Kavvas, M. L. (1994a) A spectral approach for analyzing the rill structure over hillslopes. Part 1. Development of stochastic theory. *J. Hydrol.* **158**(3–4), 333–347.

Govindaraju, R. S. & Kavvas, M. L. (1994b) A spectral approach for analyzing the rill structure over hillslopes. Part 2. Application. *J. Hydrol.* **158**(3–4), 349–362.

Kalma, J. D. & Sivapalan, M. (eds) (1995) *Scale Issues in Hydrological Modeling.* John Wiley & Sons Inc., London, UK.

Kavvas, M. L. (1999) On the coarse-graining of hydrologic processes with increasing scales. *J. Hydrol.* **217**(3–4), 191–202.

Kavvas, M. L. (2003) Nonlinear hydrologic processes: Conservation equations for determining their means and probability distributions. *J. Hydrol. Engng ASCE* **8**(2), 44–53.

Kavvas, M. L., Chen, Z. Q., Dogrul, C., Yoon, J. Y., Ohara, N., Liang, L., Aksoy, H., Anderson, M. L., Yoshitani, J., Fukami, K. & Matsuurall, T. (2004) Watershed Environmental Hydrology (WEHY) model based on upscaled conservation equations: hydrologic module. *J. Hydrol. Engng ASCE* **9**(6), 450–464.

Morris, E. M. (1979) The effect of the small-slope approximation and lower boundary conditions on solutions of the Saint-Venant equations. *J. Hydrol.* **40**(1–2), 31–47.

Nash, J. & Sutcliffe, J. (1970) River flow forecasting through conceptual models-PT1. *J. Hydrol.* **10**(3), 282–290.

Parsons, A. J. & Wainwright, J. (2006) Depth distribution of interrill overland flow and the formation of rills. *Hydrol. Processes* **20**(7), 1511–1523.

Tayfur, G. & Kavvas, M. L. (1994) Spatially averaged conservation equations for interacting rill–interrill area overland flows. *J. Hydraul. Engng* **120**(12), 1426–1448.

Yang, D., Herath, S. & Musiake, K. (2002) A hillslope-based hydrological model using catchment area and width functions. *Hydrol. Sci. J.* **47**(1), 49–65.

Yang, D., Li, C., Ni, G. & Hu, H. (2004) Application of a distributed hydrological model to the Yellow River Basin. *Acta Geographica Sinica* **59**(1), 143–154.

Yang, Z. (2007) Study on spatial averaging watershed hydrological model at macro-scale based on probability approach. PhD Thesis, Tsinghua University, Beijing, China.

Yoon, J. & Kavvas, M. L. (2000) Spatial averaging of overland flow equations for the hillslope with rill–interrill configuration. In: *Proc. ICHE-2000, Published on CD, Seoul, Korea, 26–29 September.*

Assessment of hydrological model structure based on parameter identifiability

HONGLI ZHAO, XIAOLIU YANG & YUNZHONG JIANG
Department of Water Resources, Institute of Water Resources and Hydropower Research, 100044, China
zhaohl@iwhr.com

Abstract The uncertainty in structures of hydrological models results in uncertainties in estimation of parameter values, and simulation or prediction results of models. One of the most direct ways to reduce model uncertainty is to improve model structure. This improvement has traditionally been done, dependant on modeller's intuitions and experiences, not quantitative indexes. Parameters are always set to correspond to model components. The identifiability of parameters is intimately related to the rationality of model structure. Based on the relationship between them, this paper presents an index of parameter identifiability as a measurement of the rationality of model structure. The measurement can be used as the guiding information for adjusting model structure. The GRJ model is selected here as the case study. Parameters identifiabilities of different versions of GRJ models are calculated as the measurements of evaluating model structures. Finally, some suggestions for improving model structures are presented based on the analysis result.

Key words hydrological model; model assessment; uncertainty; parameter identifiability; GRJ model

INTRODUCTION

Hydrological models are the basic tools for research into water resources and the water environment. Hydrologists have devoted most of their time to improving the precision of models in the past two decades. Their research means that model structures are getting more and more complicated. Meanwhile, in order to solve the problems due to correlativity among model parameters in multidimensional space, more and more emphasis is put on the research of efficiency and precision of optimum algorithms of model parameter, but the efforts were not satisfied in many applications. Compared to the uncertainty of data observed and model parameters, the model structure achieved from the existing knowledge system is the biggest source of uncertainty of hydrological model (Beck, 1987; Liu *et al.*, 2002). The structure identification is a very difficult process. Sometimes models with different structures may yield equally good results. This ambiguity has serious impacts on parameter calibration and predictive accuracy, and therefore limits the application of hydrological models, e.g. for the simulation of land use or climate change scenarios, or regionalization studies.

To address the credit crisis of model calculation results due to uncertainty, two reactions were found in hydrological literature. The first is the increased use of parsimonious model structures (e.g. Jakeman & Hornberger, 1993; Young *et al.*, 1996; Wagener *et al.*, 2002), i.e. structures only containing those parameters, and therefore model components, that can be identified from observed output. Some hydrologists even considered that the general hydrological observation can only satisfy the identification of 3–5 model parameters (Beven, 1989; Jakeman & Horberger, 1993). However, the increase in identifiability of model parameter is at the cost of the decrease in the number of separate hydrological processes that can be described by the model. There is therefore a danger of building a model structure that is too simplistic for the anticipated purpose. The second reaction is the search for better calibration methods to mine more information hidden in time series of observed discharge or groundwater level, etc. Some researchers suggested that multi-objective calibration methods can help to retrieve more information (DYNIA, Wagener *et al.*, 2003).

The most primary solution for model uncertainty is the adjustment of model structure, since the parameters' uncertainty largely results from model structure. Traditionally, the adjustment of model structure is implemented based on intuition and experience of the modeller, which is not an objective criterion. Reducing the uncertainty of the calculation result should be considered as an important aim of structure adjustment. Therefore, it is necessary to quantify the structural causes of model uncertainty and supply hints for model structure adjustment.

This paper makes use of the Regional Sensitivity Analysis (RSA, Spear & Hornberger, 1980; Hornberger & Spear, 1981), the Generalized Likelihood Uncertainty Estimation framework (GLUE, Beven & Binley, 1992), the Dynamic Identifiability Analysis (DYNIA, Wagener *et al.*, 2003) and constructs the assessment criterion of structure rationality and reliability using the measure of information content to help improve the model structure. For the goal that orientates model development in the right direction, this paper focuses on structure assessment for different versions of one model, instead of structure evaluation for different models.

MODEL STRUCTURE ASSESSMENT AND PARAMETER IDENTIFICATION

Hydrologists construct models to describe the behaviour of the real world based on their perception. Due to the limit of human perceiving capability, simplification and generalization are adopted to deal with the parts beyond perception or difficult to describe. It causes the difference between the result and real world behaviour. The enhancement of abilities of perceiving and describing model structure could definitely be improved. The review of hydrological model development indicates that the update period of the latter is always shorter than the former. Therefore, there are some invalid modifications in the "improvement" process. To keep the model development in a healthy and sustainable way, these invalid modifications should be identified.

The ultimate objective of model building is to supply information and remove doubt in the hydrological process step by step. From the view of uncertainty, the process of identification and improvement of model structure is just about the process of reducing uncertainty. So model structure should be modified following the guidance of the assessment of whether or not the structure adjustment reduces the uncertainty of simulation or forecast.

Most uncertainty analysis methods belong to the statistics category, which is a kind of data, and information analysis based on the theory of probability and statistics. Informatics deems that statistics methods can make up the deficiency of mathematics and physics methods. Most research on hydrological uncertainty focus on parameter uncertainty and study to find out the most appropriate parameter set corresponding to some structure, such as the popular GLUE. However, compared to model parameters, the imperfect model structure is more likely a primary cause for model uncertainty. Therefore, it will be more efficient to reduce model uncertainty through structure adjustment.

A model is the combination of structure and parameters. Usually there is an intimate relationship between parameters and structure components. Generally, the unidentifiability of parameters resulted from over-simplification or too much detail to be supported by practical data condition. The parameter uncertainty is the reflection of structure illogicality. So the faultiness of a model structure can be found out using the analysis methods of parameter uncertainty.

PARAMETER INDENTIFIABILITY

The parameter indentifiability can be defined as the confirmation degree of a parameter based on the information which is available and can be validated in practice. The basic steps in the calculation procedure are as follows:

- Set the initial range and prior distribution of parameters on the basis of physical argument or experience. A methodology for sampling the parameter space is required to get enough parameter sets. In most of the analysis methods of parameter uncertainty this has been done by Monte Carlo simulation, using uniform random sampling across the specified parameter range.
- Define the likelihood measure and criteria for acceptance or rejection of parameter sets. The measure describes the coincident degree between simulation and observation with a special parameter set. The measure value will go up monotonously following the increase of coincident degree. The Nash and Sutcliffe efficiency criterion has been chosen as the basic likelihood measure by many studies. Its form is as follows:

$$L(\underline{\theta_i} \mid \underline{Y}) = 1 - \frac{\sigma_i^2}{\sigma_o^2} \tag{1}$$

where $L(\underline{\theta_i} \mid \underline{Y})$ is the likelihood measure for the *i*th parameter set conditional on the observation, σ_i^2 is the associated error variance for the *i*th parameter set, σ_o^2 is the observed variance for the period under consideration.

Calculate the likelihood measure for each parameter set obtained from step 1; reject the parameter sets whose likelihood measures are less than the criterion set in step 2 by setting their likelihood measure as 0.

Following rejection, the likelihood measures associated with the retained parameter sets can be rescaled to give a cumulative sum of 1. Sort by the likelihood measure for each parameter and get their 90% confidence limit.

The identifiability = 1 – the width of 90% confidence limits

The lower parameter identifiability means that the corresponding structure component is difficult to express accurately, can not work actively and should be considered to adjust.

GRJ MODELS

The model selected for this study is the GRJ model series. The parameter identifiabilities of different versions of GRJ models are analysed and the value of structure adjustment in their development progress is assessed.

The GRJ model is a lumped daily rainfall–runoff model developed by Edijatno *et al.* (1999). The original intention of the model development is to build a representation for the process of rainfall–runoff in a manner as parsimonious as possible. The model should have the ability of self-adopting to guarantee the accuracy in different applications. The trace of model development passes through several stages, which have 1, 2, 3, and 4 parameters, accordingly named GR1J, GR2J, GR3J, and GR4J models. The only discussion about this model is the value of structure adjustment from GR3J to GR4J with the measure of parameter identifiability.

The GRJ model series belong to the family of soil moisture accounting models. The presentation of GR3J and GR4J are provided in Figs 1 and 2.

$$Ps = \frac{P_n\left(1-\left(\frac{S}{A}\right)^2\right)}{1+\frac{P_n}{A}\left(\frac{S}{A}\right)} \qquad Es = \frac{E_n\frac{S}{A}\left(2-\frac{S}{A}\right)}{1+\frac{E_n}{A}\left(2-\frac{S}{A}\right)}$$

$$0 < t < X3, \qquad SH1(t) = \left(\frac{t}{X3}\right)^3$$

$$t \geq X3, \qquad SH1(t) = 1$$

$$0 < t \leq X3, \qquad SH2(t) = \frac{1}{2}\left(\frac{t}{X3}\right)^3$$

$$X3 < t < 2.X3, \qquad SH2(t) = 1 - \frac{1}{2}\left(2-\frac{t}{X3}\right)^3$$

$$t \geq 2.X3, \qquad SH2(t) = 1$$

$$F = X1.\left(\frac{R}{X2}\right)^4 \qquad R = \max\,(0\,;\,R + Q9 + F)$$

$$Qr = R.\left\{1-\left[1+\left(\frac{R}{X2}\right)^4\right]^{-\frac{1}{4}}\right\}$$

$$R = R - Qr \qquad Q = Qr + Qd$$

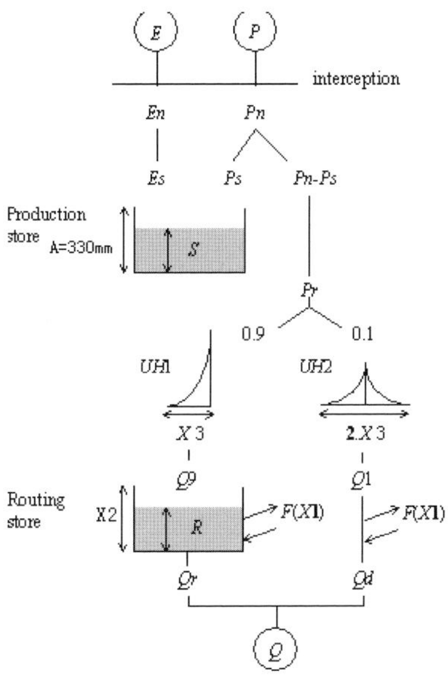

Fig. 1 Scheme of the GR3J rainfall–flow model.

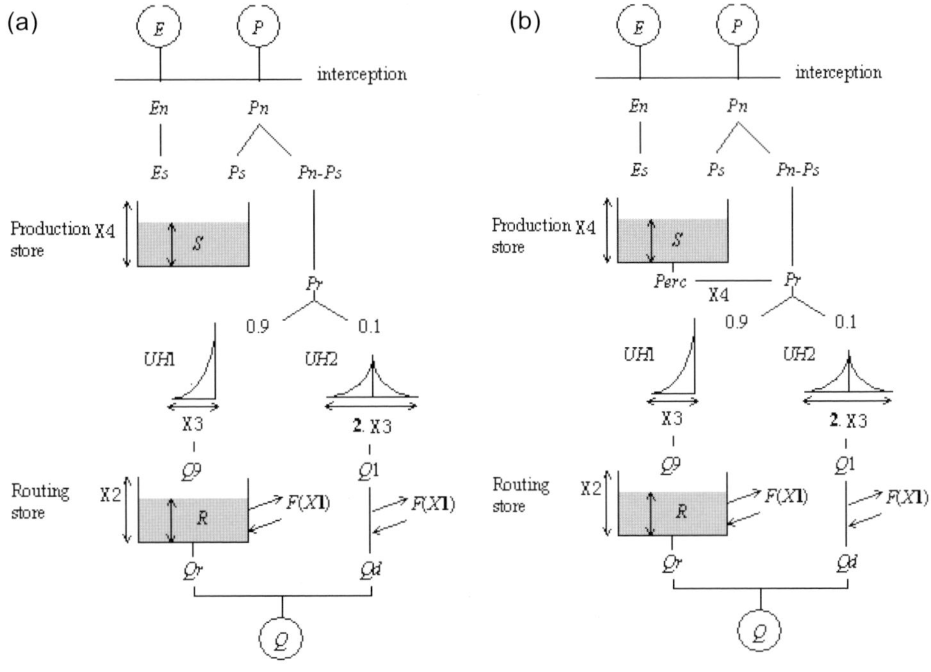

Fig. 2 Scheme of the GR4J rainfall-flow model. (a) GR4J-a model; (b) GR4J-b model.

A summary description of the GR3J model is shown in Fig. 1, as proposed by Edijatno *et al.* (1999). The three parameters in GR3J represent groundwater exchange coefficient (X1), one day ahead maximum capacity of the routing store (X2), and time base of unit hydrograph UH1(X3), respectively. *P*, the rainfall depth and *E*, the potential evapotranspiration, are inputs to the model. *Pn* is net rainfall, *En* is net evapotranspiration capacity and *Ps* is the part of *Pn* that fills the production store. *S* is the level of the production store, whose maximum capacity *A* is a fixed parameter with a value of 330 mm. The total quantity *Pr* of water that reaches the routing functions is divided into two flow components according to a fixed split: 90% of *Pr* is routed by a unit hydrograph *UH*1 and then a nonlinear routing store, and the remaining 10% of *Pr* is routed by a single unit hydrograph *UH*2. *SH*1 and *SH*2 are denoted as the ordinates of both unit hydrographs *UH*1 and *UH*2, respectively. *F* is a groundwater exchange that acts on both flow components. *R* is the level in the routing store and *Qr* is the outflow of the reservoir. *Q* is the total streamflow.

Compared with GR3J, the fixed maximum capacity *A* of the production store is replaced by parameter X4 in GR4J-a, as shown in Fig. 2(a), a percolation leakage *Perc* from the production store is introduced into GR4J-b in addition, as shown in Fig. 2(b). The change in computation equations from GR3J to GR4J is as follows:

$$Ps = \frac{X4.\left(1-\left(\frac{S}{X4}\right)^2\right).\tanh\left(\frac{Pn}{X4}\right)}{1+\frac{S}{X4}.\tanh\left(\frac{Pn}{X4}\right)} \tag{2}$$

$$Es = \frac{S.\left(2-\frac{S}{X4}\right).\tanh\left(\frac{En}{X4}\right)}{1+\left(1-\frac{S}{X4}\right).\tanh\left(\frac{En}{X4}\right)} \tag{3}$$

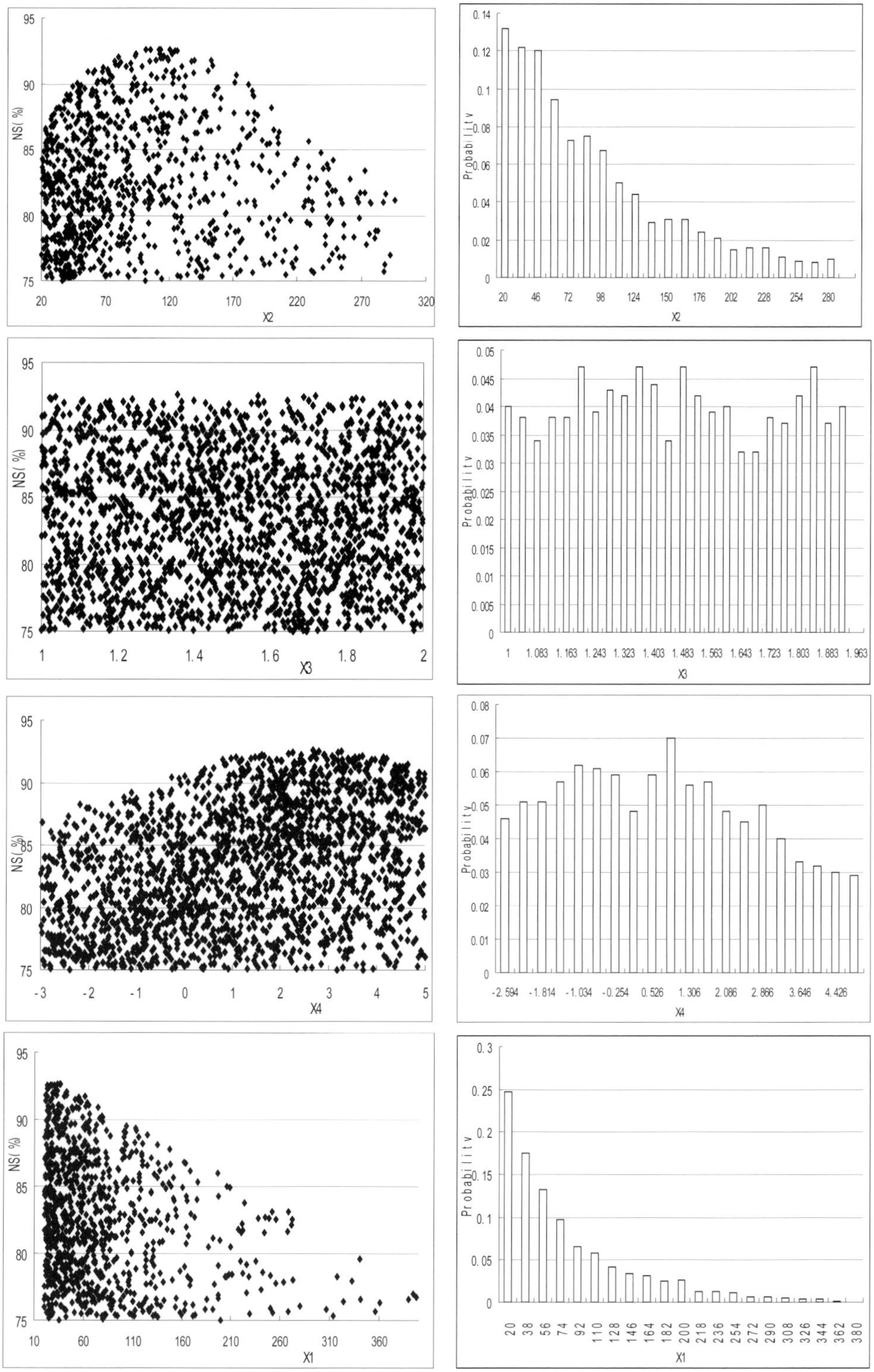

Fig. 3 Parameter distribution and model efficiency of GR4J-b.

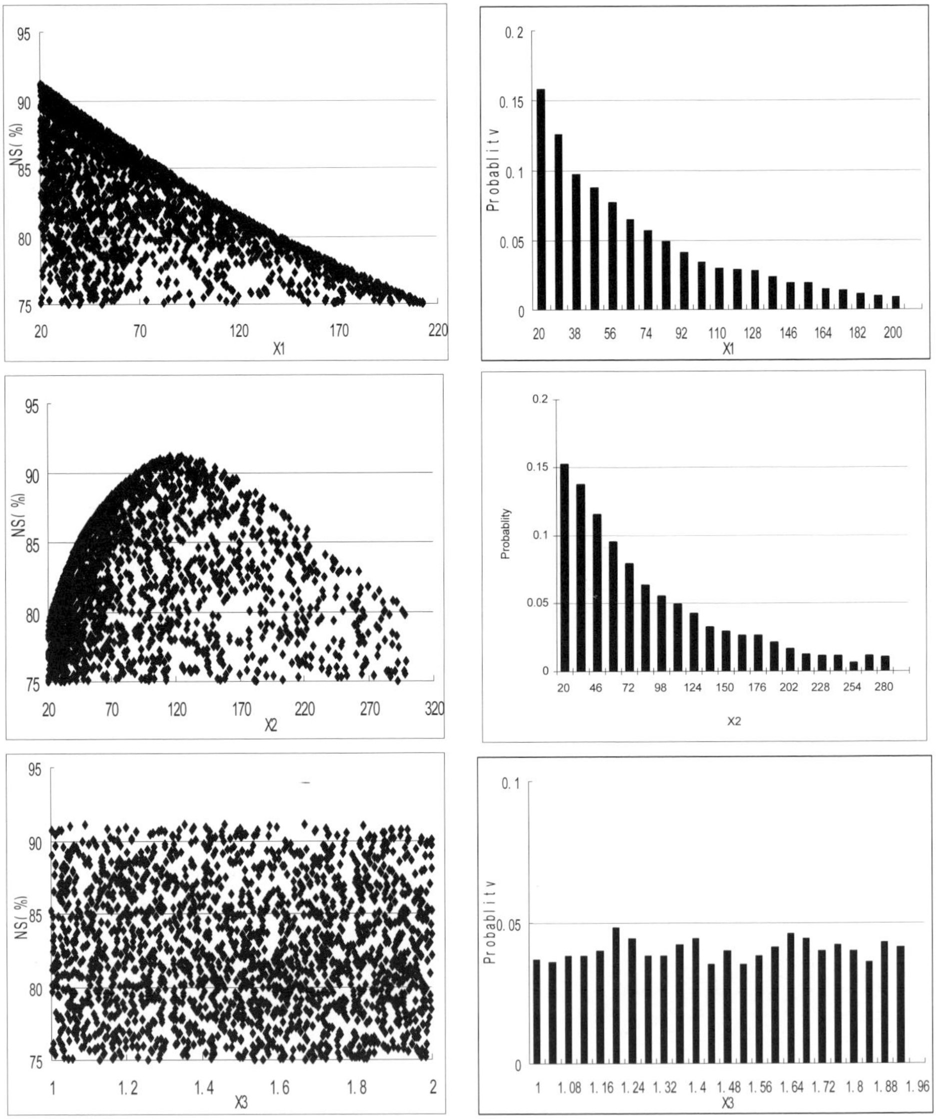

Fig. 4 Parameter distribution and model efficiency of GR3J.

$$Perc = S.\left\{1-\left[1+\left(\frac{4}{9}\frac{S}{X4}\right)^{4}\right]^{-\frac{1}{4}}\right\} \tag{4}$$

PARAMETERS INDENTIFIABILITY OF GRJ MODELS

With a daily time step, a case study was carried out in Nandian catchment, with an area of 765 km^2, located in the Taizihe River watershed, tributary of Liaohe River, northeast China. There are seven raingauges in this catchment. The daily observed time series of rainfall, evaporation and discharge from 1979 to 1987 were selected for analysis of parameter identifiability of GRJ models. The average annual precipitation in the period reaches 949 mm.

According to experience in parameters setting of GR3J and GR4J in the literature, the initial ranges of parameters were set as follows:

*X*1: –3~5
*X*2: 20~320
*X*3: 1~2
*X*4: 20~120

The Monte Carlo simulation approach using uniform random sampling is applied to the analysis of this case. The Nash and Sutcliffe efficiency (NSE) criterion is chosen as the likelihood measure. The criterion for acceptance or rejection is 75% of NSE value. The parameter sets whose NSE are higher than 75% are called behavioural. The distribution of behavioural parameters and their corresponding NSE, the probability distribution histogram of behavioural parameters of GR3J and GR4J-b, are shown in Figs 3 and 4.

The parameters identifiabilities of GR3J and GR4J models calculated using the method mentioned above are listed in Table 1.

As showed in Table 1, compared with GR3J, the parameters identifiabilities of GR4J have been generally improved. Furthermore, with the additional description of percolation leakage from the production store, the parameter identifiability of *X*4, which corresponds to the component of production reservoir in the model structure, has evidently increased from 0.381 to 0.55. It means that the part of structure change is reasonable, since it results in a lower uncertainty of its parameter. However, it is found that the identifiability of parameter *X*3 in GR4J-b with a more complicated structure is lower than the identifiability of it in GR4J-a.

Table 1 Parameters identifiabilities of three versions of GRJ.

Model	Max NS	*X*1	*X*2	*X*3	*X*4
GR3J	0.92	0.107	0.382	0.077	
GR4J-a	0.92	0.155	0.428	0.125	0.381
GR4J-b	0.93	0.158	0.428	0.083	0.55

In all versions of GRJ models, the identifiability of parameter *X*3 is always low. The structure it describes is the routing progress using a unit hydrograph. This is the most extraordinary design in the model: two unit hydrographs parsimoniously depend on the same time parameter *X*3 expressed in days. However, the measure of parameter identifiability indicates it is difficult to get a satisfactory value for this parameter as its uncertainty is huge. This means that this part of the structure should be adjusted in the future.

CONCLUSIONS AND DISCUSSIONS

For watershed hydrological simulation or prediction, there is not a most optimized model structure or parameter set, which results in the uncertainty of calculation results. The uncertainty is intimately related with model structure. Neither parsimonious nor complicated approaches can be treated as the assessment criteria of structure improvement. The parameter identifiability reveals the rationality of a model structure under the given data condition and it is instructive information for structure adjustment. A measure index of parameter identifiability is presented in this paper and used for assessment of different versions of GRJ model structures. The improvement degree in performance of structure adjustment is demonstrated with this measure and the existing problems in the present version are suggested.

The identification of model structure failure is relatively straightforward and objective. However, the analysis of why a failure occurred and how the model structure can be improved very much depends on the experience and creativity of the modeller himself. As Beck (1985) pointed out: "there is no systematic "algorithm" for changing an inadequate structure that is equivalent to increasing a polynomial order from n, say, to $(n + 1)$, as would be possible for a class III (data-based) model structure". The modeller's task is to draw an inference from the type of failure that has occurred with respect to the hypothesis underlying the specific model component in order to develop an improved version. Obviously, the adjustment of model structure only based on the measure of parameter identifiability is not enough. Many efforts on selecting more information about structure failure have to be done in the next steps.

Acknowledgements The authors would like to thank the data support from Prof. X. M. Xie of IWHR, China. Thanks are also due to the reviewers for their valuable comments.

REFERENCES

Beck, M. B. (1985) Structure, failure, inferences, and prediction. In: *Identification and System Parameter Estimation* (ed. by M. A. Baker & P. C. Young) , 1443–1448. (Proc. 7th Symposium Vol. 2, July 1985), IFAC/IFORS, York, UK.

Beck, M. B. (1987) Water quality modeling: a review of the analysis of uncertainty. *Water Resour. Res.* **23**(8), 1393–1442.

Beven, K. J. (1989) Change ideas in hydrology – the case of physically-based models. *J. Hydrol.* **23**, 1393–1442.

Beven, K. J. & Binley, A. M. (1992) The future of distributed models: model calibration and uncertainty in prediction. *Hydrol. Processes* **6**, 279–298.

Edijatno, N. O., Yang, X. L., Makhlouf, Z. & Michel, C. (1999) GR3J: a daily watershed model with three free parameters. *Hydrol. Sci. J.* **44**(2).

Hornberger, G. M, & Spear, R. C. (1981) An approach to the preliminary analysis of environmental system. *J. Environ. Manage.* **12**, 7–18.

Jakeman, A. J. & Hornberger, G. M. (1993) How complexity is warranted in a rainfall–runoff model? *Water Resour. Res.* **29**, 2637–2649.

Liu, Yi, Chen, Jining & Du, Pengfei, (2002) Parameters identification and uncertainty analysis for environment model. *Environ. Sci.* **23**(6), 6–10 (in Chinese).

Mo, Xingguo & Liu, Suxia (2004) The application of GLUE on hydrological uncertainty analysis, In: *Advances in Research of Complexity and Uncertainty of Water Problem*. China Water publishing Press, China (in Chinese).

Perrin, C., Michel, C. & Andréassian, V. (2003) Improvement of a parsimonious model for streamflow simulation. *J. Hydrol.* **279**, 275–289.

Spear, R. C. & Hornberger, G. M. (1980) Eutrophication in Peel Inlet, II, identification of critical via generalised sensitivity analysis. *Water Resour. Res.* **14**, 43–49.

Wagener, T. Camacho, L A. Lees, M. J. & Wheater, H. S. (2001) Dynamic parameter identifiability of a solute transport model. In: *Advances in Design Sciences and Technology* (ed. by R. Beheshti), 251–264. (Proc. Europa'8 Delft, April).

Wagener, T., Mclntyre, N., Lees, M. J., Wheater, H. S. & Gupter, H. V. (2003) Towards reduced uncertainty in conceptual rainfall–runoff modeling: dynamic identifiability analysis. *Hydrol. Processes* **17**, 455–476.

Yang, X. L. & Michel, C. (2000) Flood forecast with a watershed model: a new method of parameter updating. *Hydrol. Sci. J.* **45**(4).

Young, P. C., Parkinson, S. & Lees, M. J. (1996) Simplicity out of complexity in environmental modeling: Occham's razor revisited. *J. App. Statistics* **23**, 165–210.

Zhang, Xuewen (1981) *The Information Analysis of Weather Forecast*. Science Publishing Press, China (in Chinese).

Uncertainty of the box-counting method for estimating the fractal dimension of river networks

MIAO ZHOU, FUQIANG TIAN & HEPING HU
Department of Hydraulic Engineering, State Key Laboratory of Hydroscience and Engineering, Tsinghua University, Beijing 100084, China
tianfq@tsinghua.edu.cn

Abstract The fractal dimension of river networks has become a significant geomorphologic parameter for identifying hydrological patterns. Several definitions for fractal dimension are used in hydrological literature, among which the box-counting dimension via Digital Elevation Model (DEM) is widely applied. To explore the uncertainty of the box-counting method along with the mathematical implementing processing, three factors, i.e. threshold area for river networks extracted from DEM, range and increment ratio of box sizes, were selected to identify their influence on the value of the box-dimension. One real watershed, the Chabagou watershed, and one ideal watershed generated by an iterative algorithm, were selected for numerical experiments. The results suggest the following properties of the box-dimension of river networks: it shows an asymptotic power law behaviour for larger threshold areas; it has a logarithmic dependence for smaller threshold areas; and it is monotonic to the range of calculated box sizes while insensitive to the increment ratio of calculated box sizes. These relationships reveal the complexity of the box-dimension of fractal river networks.

Key words fractal river network; box-counting method; digital elevation model (DEM); uncertainty

INTRODUCTION

Natural channel networks show a fascinating variety of shapes and forms, which present a deep sense of regularity in the relationship among the parts. This regularity is presented as self-similarity by Horton (1945). Mandelbrot (1983) proved that natural channel networks have a fractal character. La Barbera & Rosso (1989) promoted a fractal dimension in terms of Horton's laws. Thereafter, several approaches to calculate the fractal dimension of networks have been applied. Rodriguez-Iturbe & Rinaldo (1997) mentioned the Hausdorff dimension, box-counting dimension, similar dimension, correlation dimension, mass dimension, information dimension, etc. Using three different approaches, Veltri *et al.* (1996) counted the fractal dimensions of 17 Calbrian drainage basins.

Over the last two decades, with the help of Digital Elevation Models (DEMs), the fractal dimension of some river basins has been counted. La Barbera & Rosso (1989) claimed that different networks, or channel networks in different physiographic locations, may display different values of fractal dimension. As fractal dimensions of many river basins were counted, the underlying connection between the fractal dimension and hydrological character were also studied. For example, Kusumayudha *et al.* (2000) showed that the box-counting dimension of river networks is related to the erosion of rock, as well as to the fractal dimension of underground river pattern.

Consequently, for further investigation on the relationship between the fractal character of rivers and their hydrological properties, it is of great importance to determine the fractal dimension accurately and subtly. However, it has been pointed out that the fractal dimension calculated by different approaches will probably be different. Therefore, to compare the applicability of those approaches for estimating the fractal dimension of river networks, further investigations are indispensable. Moreover, even for a particular method, there are some other issues that can bring uncertainty to the results, e.g. Helmlinger *et al.* (1993) found that the range of box sizes affects the fitting of box-counting dimensions. Particularly, Ichoku *et al.* (1996) showed that those four kinds of fractal dimensions decrease as the threshold of confluence area increases, albeit with only five threshold areas. Radziejewski *et al.* (1997) counted the box-counting dimension for excursion of flow process at a threshold level, and observed the same trend between dimension and threshold areas; however, the study lacks adequate data for more detailed discussion. Yang *et al.* (1997, 2001) proposed that the multifractal spectra of width function and area function were also sensitive to the threshold area. Huang (1996) studied the factors affecting the reliability of box-counting dimension from a mathematical aspect. Up until now, the sensitivity of the box-counting dimension of river networks to the threshold area has not yet been discussed comprehensively, nor has the uncertainty coming from the factors of chosen method; further study of the relationship needs more and adequate data.

In this paper, we investigate the uncertainty of fractal characteristics in river networks along with the mathematical dealing processes, with the box-counting dimension as an example due to its simplicity, and thus popularity in the hydrological literature. Previous work has discussed this method itself and the effects of threshold area and box size in its calculation. However, there is still a lot of work that needs to be done in this area. The paper first presents a comparative study using a theoretical fractal to validate the result. Then it takes the Chabagou watershed as an example, and focuses on the investigation of the factors determining the box-counting dimension.

STUDY AREA AND NUMERICAL ANALYSIS METHOD

The analysis is based on the case study in the Chabagou watershed, a branch of the Yellow River basin. The Chabagou watershed is located at longitude 109°47′E and latitude 37°31′N, with a total area of 187 km^2. Figure 1 shows the DEM map of the study area with the resolution of 50m.

Fig. 1 DEM of the Chabagou River basin.

To calculate the box-counting dimension for a fractal F, we first assume this fractal lies on an evenly-spaced grid, and count the number of boxes in the grid required to cover the set. The box-counting dimension is then calculated by measuring how this number changes as we make the grid finer. Let r be a positive number and F a non empty and bounded subset of a metric space (X, d). Suppose that $N(r)$ is the number of boxes with side length r required to cover the set F. As $r \to 0$, if $\log N(r)/\log(1/r)$ converges to a finite value, then the box-counting dimension is defined as:

$$D_{box} = \lim_{r \to 0} \frac{\log N(r)}{\log(1/r)} \tag{1}$$

The box-counting dimension is used in application for its compatibility in a numerical experiment by the definition. A sketch of such a numerical algorithm, described by Lovejoy *et al.* (1987), is as follows. First, put set F into a grid, which is divided into boxes of size r. Second, count the number of boxes $N(r)$ that actually cover F. In practice, r is chosen from a finite range in a geometric series, and $N(r)$ is obtained correspondingly. Finally, make a $\log N \sim \log 1/r$ plot and the slope is the box-counting dimension.

According to this method, we set up the following steps to approach the box-counting dimension of the Chabagou River. (1) Extract the river networks from the DEM map of the Chabagou watershed, with different thresholds of confluence areas; the threshold of confluence area is a critical factor in defining a river network. The smaller the threshold area is, the denser the river network will be. Consequently it strongly affects the characteristics of river networks, including the fractal dimension. (2) Generate boxes with different sizes, and use them to cover the river networks. Count the number of boxes $N(r)$ which contain the river segments. The choice of sizes, e.g. the range and the increment ratio of r, will impact the final result. In principle, an infinitesimal increment ratio of r and an infinitely large range are necessary to get the true fractal dimension. In order to reduce errors resulting from imperfectness of the range and increment ratio of r, we used small increment ratio and large ranges. Meanwhile, we also investigate the effects of their variation on the fractal dimension. (3) Generate a logarithmic plot, with the slope showing the box-counting dimension of river networks.

Before the detailed calculation of the box-dimension of the Chabagou River, it is necessary to verify that our numerical analysis method is technically correct. Hence as a control example, a treelike model, which is an ideal fractal and an approximate representation of natural river networks (mentioned by Nilora, 1993), is taken to estimate the box-counting dimension using this method. This ideal fractal network is generated using a simple iterative algorithm (cf. Fig. 2). According to the iterative algorithm, it shows that the real box-counting dimension is $\log 5/\log 3 = 1.46$. Then we calculate its box-counting dimension by the same method, which will be applied in the Chabagou River networks.

In Fig. 3, we find a nearly perfect straight linear fit of data points on the $\log N \sim \log (1/r)$ plot, with slope 1.43 being consistent with the above theoretical prediction. Even though the coefficient of determination is up to 0.99, the box-counting dimension varies with different ranges of box sizes. This can be explained as follows: when the box size is smaller than the smallest bifurcation of this ideal fractal simulation, its shape is not fractal anymore because its bifurcation ends here. In this case the fractal dimension measured by those boxes will become one. Therefore the slope is flatter when the box size is smaller. For the same reason, the box-counting dimension we estimate is slightly less than the real one.

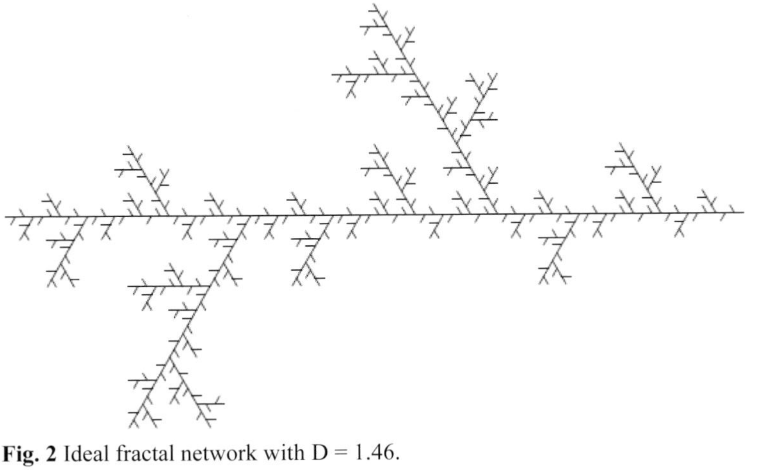

Fig. 2 Ideal fractal network with D = 1.46.

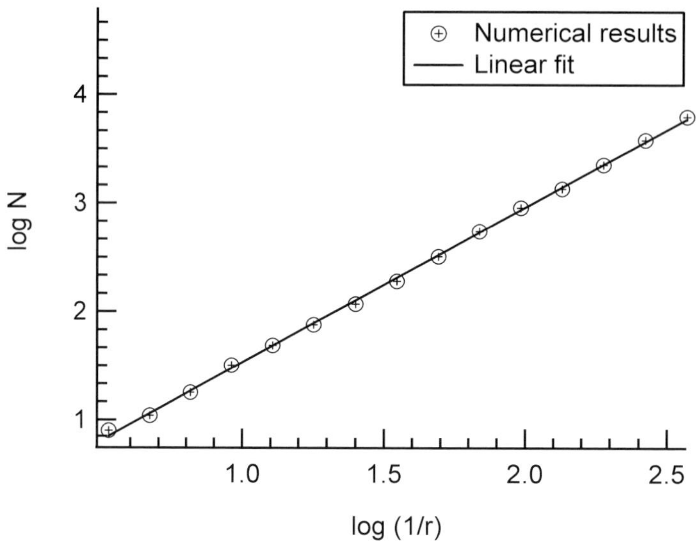

Fig. 3 Box-counting dimension of the ideal fractal with six iterations. The fitted line is log N = 1.43 log(1/r) + 0.10 with R^2 = 0.999.

RESULTS AND DISCUSSIONS

In this method, there are three critical factors in calculating box-dimension of river network. The first one is the threshold of the confluence area, which is determined while generating the river networks. The other two are the increment ratio and range of the boxes sizes. Our work focuses on how each factor influences the calculated box-counting dimension of river networks, and furthermore figuring out how to mitigate the uncertainty imposed by the three factors.

Relationship between the increment ratio of box sizes and the box-counting dimension

According to the definition of box-counting dimension, as the box size r converges to zero, the limitation of the formula requires r to decrease continually, and the increment ratio of box sizes should be as small as possible. However, in numerical analysis, the increment ratio can not be too small. Hence, to study the relationship between the increment ratio and the box-counting dimension, we should try to find out the appropriate increment ratio and exclude the influence of this factor.

From Table 1, it is shown that in a certain range the ratio of box sizes impacts little on the result of the ideal fractal. The result values fluctuate when the increment ratio of box sizes is over 2, but is identical for any ratio smaller than 2. When the ratio is smaller than 1.4, the error is less than 1%. Therefore, as long as the ratio is less than two, the effect of discreteness of box lengths can be neglected.

Relationship between the range of the boxes sizes and the box-counting dimension

Take the network with threshold area equal to 150 as an example. The range of box sizes is from 40 to 4000. Figure 4 shows that the slope changes with the variation of the range of boxes. For

Table 1 Relationship between the increment ratio of box sizes and the box-counting dimension.

Increment ratio of box sizes	1.10	1.21	1.46	2.74
Box-counting dimension	1.24	1.25	1.27	1.28
One standard deviation	0.01	0.01	0.02	0.04

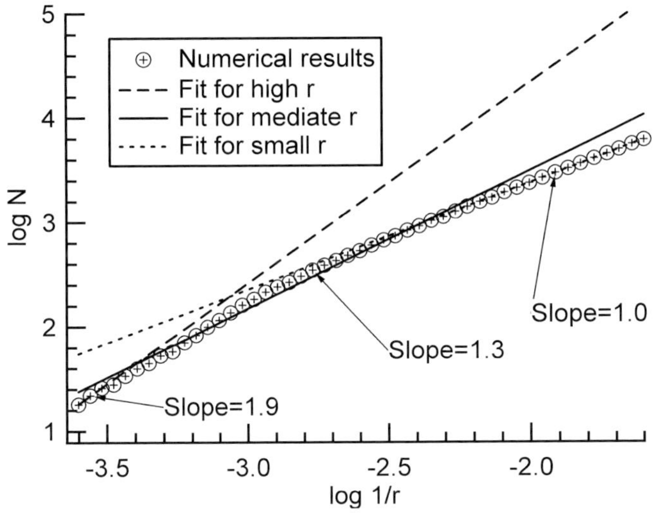

Fig. 4 Relationship between the range of box sizes and the box-counting dimension.

the smallest box size the slope of the curve approaches to 1, indicating D = 1, which is reasonable for the individual stream links consisting of straight lines. For the largest box sizes the slope will approach slope = 2 because all the points fall in only a few boxes. It can be concluded that the slope of the curve depends on the range of the boxes size, and it changes monotonically and slowly from 1 to 2, which indicates the box-counting dimension is monotonic to the range of calculated box size. This phenomenon also occurs when calculating the box-counting dimension of river networks with other threshold areas.

Theoretically, only when the size of boxes converges to 0, can one get the true value of the box-counting dimension. However, in numerical analysis, the box size can not really converge to 0 in mathematical terms. Our study suggests that in numerical analysis, smaller boxes can not necessarily bring more accurate results; in contrast, they sometimes make the results less accurate or even incorrect. Thus, the size of boxes does not necessarily need to converge to 0 in this numerical analysis.

It can be illustrated that the slope of the curve depends on the range of the box size, and it changes slowly from 1 to 2. The analysis shows that there is a theoretical causation of the unreliability of the box-counting method: the box length could only be limited discrete values and could not be less than the pixel length. This superficially dissatisfies the requirement of the definition of the box-counting dimension: the box-counting dimension is only reached when the box length is limited to zero. In contrast, at a scale smaller than the pixel, the generated river network is no longer fractal, which is natural for simulation at the pixel scale. Although the above preliminary conclusions are addressed, we cannot conclude from the analysis which range should be used to obtain the true fractal dimension unless further studies are carried out.

Relationship between the confluence area and the box-counting dimension

In this part, we calculate the box-counting dimension of Chabagou River networks, which is derived with threshold area from 22 to 7000. Results show they decrease as the threshold area increases.

From Fig. 5(a) and (b), it is found that the values of the box-counting dimensions vary according to the support area thresholds. Here the question arises of "identifying the exact value of the box-counting dimension of the real river channel". For the increasing application of fractal dimension as a parameter in hydrological responses and catchments geomorphology, etc. it is of consequence to answer this question.

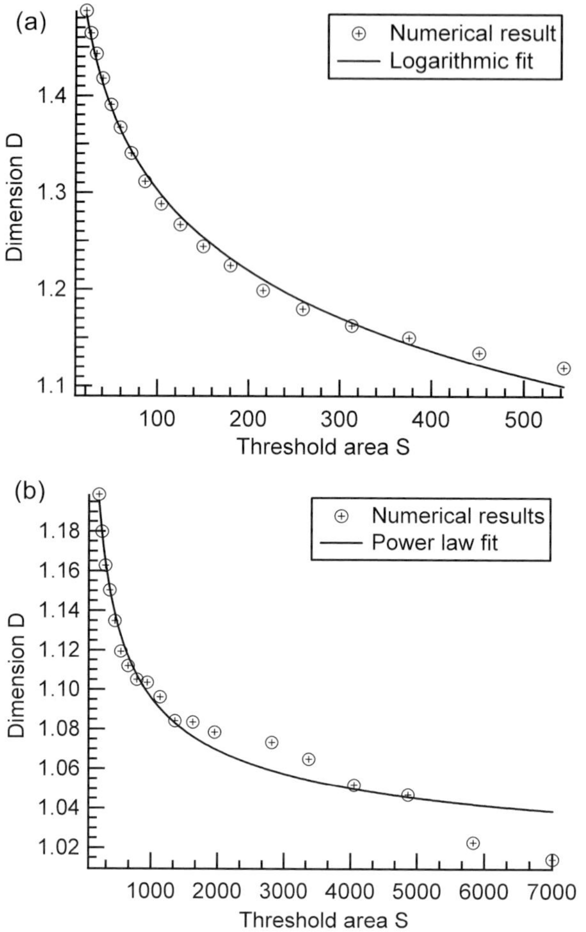

Fig. 5 Relationship between the threshold area and the box-counting dimension: (a) when threshold area is small, the fitting is $D = -0.12\ln S + 1.85$ with R^2=0.994; (b) when threshold area is large, the fitting is $D = 2.9S^{-0.50}$ with $R^2 = 0.87$.

For different threshold areas, the behaviour of the change is different, and therefore should be discussed separately. When the threshold area is small (e.g. from 22 to 546), there is a logarithmic fit (cf. Fig. 5(a)) between the threshold area and box-counting dimension. The relationship can be expressed as $D = -0.12\ln S + 1.85$, with the coefficient of determination equal to 0.99. However, from the figure it is noticed that when S becomes large, the fitting line falls below the true data value. It is probably because the logarithmic function diverges when S becomes large, while the box-counting dimension converges to 1. Hence, when the threshold area is large (e.g. from 546 to 7000), another behaviour is presented, which can be fit into a power law (cf. Fig. 5(b)) relation $D = 2.9S^{-0.50}$. The results are consistent with the observation which indicates that when the threshold area increases, the box-counting dimension decreases to be closer to 1.

If the fittings are appropriate, we can obtain the box-counting dimension D of a certain network by S (the threshold area which defined the network). Also, based on our fittings, we carry on theoretical analysis in the future to provide the possibility of investigating the real fractal river network according to the relationship between S and D, which could eliminate the effect of threshold area. Figure 5(a) shows the logarithmic relationship between the values of box-counting dimension and the support area thresholds, which may be demonstrated by the multifractal theory.

Our present work is only an exploratory approach, and further investigation remains to be done. Additional data from other natural rivers are needed in future investigations.

CONCLUSION

In the present work we have applied the digital elevation model (DEM) to analyse the box-counting dimension of the Chabagou River channel. The results obtained in this work support the observation of the complicated nature of fractal river networks and the sensitivity of the character of river to the threshold of confluence area. (1) The box-counting dimension is monotonic to the range of box sizes while insensitive to the increment ratio of box sizes. (2) The box-counting dimension has a logarithmic dependence on S; $D = -0.12\ln S + 1.85$ for smaller threshold areas; it has a power law asymptotic behaviour $D = 2.9S^{-0.50}$ for larger threshold areas.

In this paper an approach is proposed to evaluate the sensitivity of threshold area in box-counting dimension calculations, and it is expected that real box-counting dimensions can be identified from the calculation. However, dimensions derived from this method are not accurate enough because of its dependence on the discreteness of box sizes. Meanwhile, considerable uncertainty coming from the fitting and calculation are added together and make the final result more inaccurate than we can accept. Moreover, these uncertainties are intrinsic and cannot be eliminated by technical advances such as use of a finer grid.

On the other hand, our research is based on the assumption that the river is mono-fractal. Nevertheless, there is all likelihood that the nature river networks are multifractal. No matter how mono-fractal or multifractal the river network is, the problem of scaling always exists.

The results presented in this paper only touch upon the geometrical properties of the river network, which may serve as a foundation for the analysis of scaling properties of hydrophysical and hydrological processes.

Acknowledgements The authors would like to thank Professor Dawen Yang for his valuable comments. This research was supported by the National Natural Science Foundation of China (project no. 50779022).

REFERENCES

Helmlinger, K. R., Kumar, P. & Foufoula-Geogriou, E. (1993) On the use of digital elevation model data for Hortonian and fractal analyses of channel networks. *Water Resour. Res.* **29**(8), 2599–2613.

Horton, R. E. (1945) Erosional development of streams and their drainage basins: Hydrophysical approach to quantitative morphology. *Geol. Soc. Am. Bull.* **56**, 275–370.

Huang, Y. (1996) How to get a reliable fractal dimension by box-counting. *Communications in Nonlinear Science* **2**, 22–27.

Ichoku, C., Karnieli, A. & Verchovsky, I. (1996) Application of fractal techniques to the comparative evaluation of two methods of extracting channel networks form digital elevation models. *Water Resour. Res.* **32**(2), 389–399.

Kusumayudha, S. B., Zen, M. T., Notosiswoyo, S. & Gautama, R. S. (2000) Fractal analysis of the Oyo River, cave systems, and topography of the Gunungsewu karst area, central Java, Indonesia. *Hydrogeology J.* **8**, 271–278.

La Barbera, P. & Rosso, R. (1989) On the fractal dimension of stream networks. *Water Resour. Res.* **25**(4), 735–741.

Lovejoy, S., Schertzer, D. & Tsonis, A. A. (1987) Functional box counting and multiple elliptical dimensions in rain. *Science* **235**, 1036–1038.

Mandelbrot, B. B. (1983) *Fractal Geometry of Nature*. W. H. Freeman, New York, USA.

Nikora, V. I. & Sapozhnikov, V. B. (1993) River network fractal geometry and its computer simulation. *Water Resour. Res.* **29**(10), 3569–3575.

Radziejewski, M. & Kundzewicz, Z. W. (1997) Fractal analysis of flow of the River Warta. *J. Hydrol.* **200**, 280–294.

Rodríguez-Iturbe, I. & Rinaldo, A. (1997) *Fractal River Basins: Chance and Self-Organization.* Cambridge University Press, New York, USA.

Veltri, M., Veltri, P. & Maiolo, M. (1996) On the fractal description of natural channel networks. *J. Hydrol.* **187**, 137–144.

Yang, D., Herath, S. & Musiake, K. (1997). Analysis of stability of river network geomorphologic parameters using DEM data of Japanese catchments. In: *Proc. 27th Congress of the Intl. Association for Hydraulic Research*, vol. 2, 973–978.

Yang, D., Herath, S. & Musiake, K. (2001). Spatial resolution sensitivity of catchment geomorphologic properties and the effect on hydrological simulation. *Hydro. Processes* **15**, 2085–2099.

3 Applications

A variable fuzzy-sets assessment model and its application to regional water resources assessment

SHOUYU CHEN[1], CHUNLING CHAI[1,2] & MIN LI[1]

1 *School of Civil and Hydraulic Engineering, Dalian University of Technology, Dalian, Liaoning 116024, China*
chaiccl@yahoo.com.cn

2 *College of Urban and Rural Construction, Agricultural University of Hebei, Baoding, Hebei 071000, China*

Abstract Regional water resources assessments aim at estimating the current status and trends of water resources, providing a scientific basis for the development, utilization and protection of water resources, and optimizing a sustainable utilization plan. Therefore, it is essential to conduct a holistic water resources assessment. Most of the current assessment models for regional water resources only focus on water quantity and water quality. The assessment of hydroenergy resources is usually ignored. Furthermore the assessment result is only compared with that of other models without analysing the stability itself. To target these issues, a variable fuzzy-sets model for regional water resources assessment was developed. Water quantity, water quality, and hydroenergy are considered in this model, and its stability is also analysed with varying parameters. The given case study focusing on water quality shows that the variable fuzzy-sets model is useful to guide regional water resources assessment.

Key words hydroenergy; variable fuzzy-sets assessment model; water quality; water quantity; water resources

INTRODUCTION

Water is an important natural resource for human living, production, and societal development. River basins with abundant water of high quality, together with the convenience for utilization, are historically preferred places for human habitation and development. Today, unwise use of water resources has led to many problems, such as eco-environmental damage, water pollution and abnormal hydrological cycles, etc. Thus, water resources utilization should be well planned. This calls for a need to assess water resources reasonably, scientifically and holistically.

The earliest national water resources assessment was established in the USA in 1965. Later, other countries carried out similar projects, e.g. Japan (1975), India (1975), and China (1985) (Wang, *et al.*, 2006). Many models or methods were established with the development of water resources assessment, e.g. the DEA method (Charnes *et al.*, 1978), the method of fuzzy comprehensive evaluation (He, 1983), the matter-element model (Cai, 1990), the grey system theory (Deng, 1991) and the fuzzy pattern recognition theory (Chen, 1997), etc. The models were applied widely. In this paper, the variable fuzzy-sets assessment (VFSA) model (Chen, 2005) for regional water resources assessments is described. This model also considers hydroenergy besides water quantity and water quality. Stability analysis is conducted with varying parameters. A case study for the model, focusing on water quality, is applied to the water resources assessment of the Three Gorges reach. This study is anticipated to help guide future regional water resources assessments.

Principle of VFSA

The law of contradiction points out that opposite and unity relations are concurrent between contradictions, which are the power of events movement and development.

Definition of variable fuzzy sets (VFS)

Supposing that U is an opposite fuzzy concept (alternative or phenomenon), $\underset{\sim}{A}$ and $\underset{\sim}{A}^c$ denote the attractability and repellency, to any element u ($u \in U$), which falls into the intervals of $[1,0]$ ($\underset{\sim}{A}$) and $[0,1]$ ($\underset{\sim}{A}^c$), their relative membership degrees (RMD) of attraction and repellence are denoted by $\mu_{\underset{\sim}{A}}(u)$ and $\mu_{\underset{\sim}{A}^c}(u)$, respectively (Chen, 2005). Let:

$$\mu_{\underset{\sim}{A}}(u)+\mu_{\underset{\sim}{A}^c}(u)=1 \quad (1)$$

The boundary conditions are:

$$0\le \mu_{\underset{\sim}{A}}(u)\le 1,\ 0\le \mu_{\underset{\sim}{A}^c}(u)\le 1 \quad (2)$$

Let:

$$\underset{\sim}{V}=\{(u,\mu)\,|\,u\in U,\mu_{\underset{\sim}{A}}(u)+\mu_{\underset{\sim}{A}^c}(u)=1,\mu\in(0,1]\} \quad (3)$$

Here $\underset{\sim}{V}$ is the VFS of U. The attraction sets, repellence sets and balance boundary or qualitative change boundary are:

$$A_+=\{u\,|\,u\in U,\mu_{\underset{\sim}{A}}(u)>\mu_{\underset{\sim}{A}^c}(u)\} \quad (4)$$

$$A_-=\{u\,|\,u\in U,\mu_{\underset{\sim}{A}}(u)<\mu_{\underset{\sim}{A}^c}(u)\} \quad (5)$$

$$A_0=\{u\,|\,u\in U,\mu_{\underset{\sim}{A}}(u)=\mu_{\underset{\sim}{A}^c}(u)\} \quad (6)$$

Suppose that C is the variable factor sets of $\underset{\sim}{V}$:

$$C=\{C_A,C_B,C_C\} \quad (7)$$

where C_A are the model variable sets; C_B are the model parameters variable sets (e.g. weight); C_c are a variable other factors sets, except model and its parameters.

The qualitative change sets of VFS $\underset{\sim}{V}$ to variable elements sets C are:

$$A^+=C(A_-)=\{u\,|\,u\in U,\mu_{\underset{\sim}{A}}(u)<\mu_{\underset{\sim}{A}^c}(u),\mu_{\underset{\sim}{A}}(c(u))>\mu_{\underset{\sim}{A}^c}(c(u))\} \quad (8)$$

$$A^-=C(A_+)=\{u\,|\,u\in U,\mu_{\underset{\sim}{A}}(u)>\mu_{\underset{\sim}{A}^c}(u),\mu_{\underset{\sim}{A}}(c(u))<\mu_{\underset{\sim}{A}^c}(c(u))\} \quad (9)$$

The quantitative change sets of VFS $\underset{\sim}{V}$ to variable elements sets C are:

$$A_{(+)}=C(A_+)=\{u\,|\,u\in U,\mu_{\underset{\sim}{A}}(u)>\mu_{\underset{\sim}{A}^c}(u),\mu_{\underset{\sim}{A}}(c(u))>\mu_{\underset{\sim}{A}^c}(c(u))\} \quad (10)$$

$$A_{(-)}=C(A_-)=\{u\,|\,u\in U,\mu_{\underset{\sim}{A}}(u)<\mu_{\underset{\sim}{A}^c}(u),\mu_{\underset{\sim}{A}}(c(u))<\mu_{\underset{\sim}{A}^c}(c(u))\} \quad (11)$$

Model of relative membership function (RMF)

The model of RMF considering intervals can be expressed as follows (Chen, 2005). We suppose that $X_0=[a,b]$ is the attraction sets of VFS $\underset{\sim}{V}$ on the real axis, i.e. interval of $\mu_{\underset{\sim}{A}}(u)>\mu_{\underset{\sim}{A}^c}(u)$, and $X=[c,d]$ is a certain interval containing X_0 ($X_0\subset X$). Relationships are shown in Fig. 1.

According to the definition of VFS, intervals $[c,a]$ and $[b,d]$ are repelling sets of VFS, i.e. interval of $\mu_{\underset{\sim}{A}}(u)<\mu_{\underset{\sim}{A}^c}(u)$. Suppose that M is the point of $\mu_{\underset{\sim}{A}}(u)=1$ in attraction set $[a,b]$, according to actual problems M may not be the median value of interval $[a,b]$.

It is assumed that x is a random point in interval X, and if x locates on the left side of M, the model of RMF is as follows:

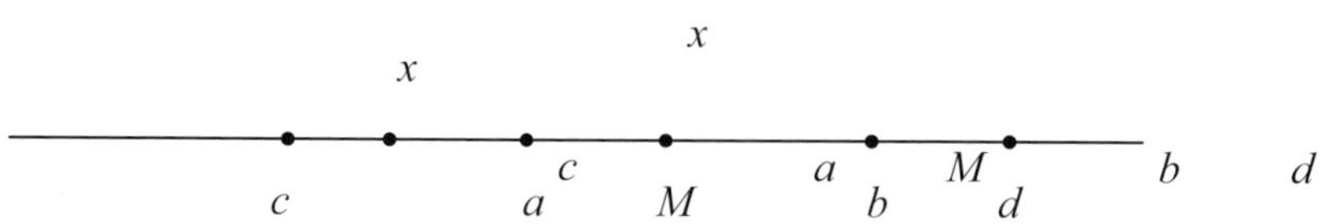

Fig. 1 Relationships between points x , M and intervals X_0, X.

$$\mu_{\underset{\sim}{A}}(u)=\begin{cases}0.5+(x-a)/2(M-a) & x\in[a,M]\\ 0.5-(x-a)/2(c-a) & x\in[c,a]\end{cases} \tag{12}$$

If x locates at right side of M, the model of RMF is as follows:

$$\mu_{\underset{\sim}{A}}(u)=\begin{cases}0.5+(x-b)/2(M-b) & x\in[M,b]\\ 0.5-(x-b)/2(d-b) & x\in[b,d]\end{cases} \tag{13}$$

If x locates out of interval X, the model of RMF is:

$$\mu_{\underset{\sim}{A}}(u)=0 \quad x\notin[c,d] \tag{14}$$

The model of VFSA

The model VFSA is expressed as follows:

$$v_{\underset{\sim}{A}}(u)_h=\left[1+\left(d_{gh}/d_{bh}\right)^{\alpha}\right]^{-1} \tag{15}$$

where α is the rule parameter, h is grade number and h = 1, 2, …, c. Through calculating $v_{\underset{\sim}{A}}(u)_h$ of each grade, we get synthetic RMD vector $v_{\underset{\sim}{A}}(u)$.

$$d_{gh}=\left\{\sum_{i=1}^{m}\left[w_i\left(1-\mu_{\underset{\sim}{A}}(u)_{ih}\right)\right]^p\right\}^{1/p} \tag{16}$$

$$d_{bh}=\left\{\sum_{i=1}^{m}\left(w_i\mu_{\underset{\sim}{A}}(u)_{ih}\right)^p\right\}^{1/p} \tag{17}$$

where i is an index number and I = 1, 2, …, m, p is distance parameter.

Generally, α = 1, 2 and p = 1, 2. α = 1 is least single model and α = 2 is least square model; p = 1 is hamming distance and p = 2 is Euclidean distance. Therefore, α and p can be arranged in four pairs: (a), $\alpha = 1, p = 1$; (b), $\alpha = 1, p = 2$; (c), $\alpha = 2, p = 1$; (d), $\alpha = 2, p = 2$.

Through normalizing $v_{\underset{\sim}{A}}(u)$, RMD $v^{\circ}_{\underset{\sim}{A}}(u)$ can be obtained, and then the rank feature values (RFV) can be calculated by the RFV equation.

$$H=(1, 2, \dots c)\ v^{\circ}_{\underset{\sim}{A}}(u)^T \tag{18}$$

In the VFSA model, model variability and model parameters variability can be used to analyse the assessment results stability.

Case study

Three sub-systems i.e. water quality, water quantity and hydroenergy, should be synthetically considered in regional water resources assessments (Fig. 2).

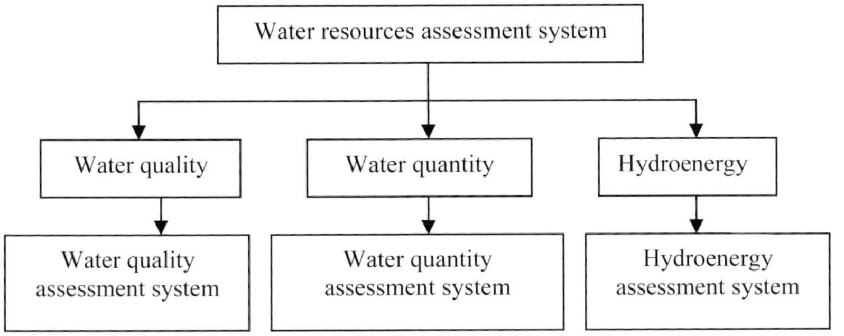

Fig. 2 Water resources assessing system.

Table 1 Index values for water resources assessment (mg/L).

No.	Cross-section	COD_{mn}	BOD_5	NH_3-N
1	Qingxichang (flood period)	8.26	1.80	1.79
2	Tuokou (flood period)	7.54	1.12	1.98
3	Saiwangba (flood period)	8.00	2.20	2.10
4	Fengjie (flood period)	8.13	2.18	2.05
5	Qingxichang (dry period)	2.71	1.79	1.98
6	Tuokou (dry period)	2.49	1.67	2.08
7	Saiwangba (dry period)	2.65	1.54	2.12
8	Fengjie (dry period)	2.54	2.01	2.19

Table 2 The grades intervals of assessment indices.

Index	Grade				
	I	II	III	IV	V
COD_{mn}	0~2	2~4	4~6	6~10	10~15
BOD_5	0~3	3~3	3~4	4~6	6~10
NH_3-N	0~0.15	0.15~0.5	0.5~1.0	1~1.5	1.5~2

The VFSA model was applied for water resources assessment to the Three Gorges reach. Wang (2004) monitored and collected to the data for the four cross-sections of the Three Gorges reach, viz. Qingxichang, Tuokou, Saiwangba and Fengjie. Owing to the limitation of referring data and criteria, only water quality assessment is analysed in this paper, i.e. $w_2 = w_3 = 0$. According to referring data, three indices are used (Table 1).

Based on the State Criterion of Surface Water Environmental Quality (GB3838-2002), assessment indices and their grades standard intervals are shown in Table 2.

According to Table 2 and Chen (2005), the matrixes $\mathbf{I}_{[a,b]}$, $\mathbf{I}_{[c,d]}$ and $\mathbf{I}_M$ are set up.

$$\mathbf{I}_{[a,b]} = \begin{bmatrix} [0,2] & [2,4] & [4,6] & [6,10] & [10,15] \\ [0,3] & [3,3] & [3,4] & [4,6] & [6,10] \\ [0,0.15] & [0.15,0.5] & [0.5,1] & [1,1.5] & [1.5,2] \end{bmatrix}$$

$$\mathbf{I}_{[c,d]} = \begin{bmatrix} [0,4] & [0,6] & [2,10] & [4,15] & [6,15] \\ [0,3] & [0,4] & [3,6] & [3,10] & [4,10] \\ [0,0.5] & [0,1] & [0.15,1.5] & [0.5,2] & [1,2] \end{bmatrix}$$

$$\mathbf{I}_M = \begin{bmatrix} 0 & 2 & 5 & 10 & 15 \\ 0 & 3 & 3.5 & 6 & 10 \\ 0 & 0.15 & 0.75 & 1.5 & 2 \end{bmatrix}$$

Using both equations (12) and (13), we can obtain RMDs $\mu_{\underset{\sim}{A}}(u)_{ih}$, here i is assessment index number and $i = 1,2,3$; h is grade number and $h = 1,2,3,4,5$. For example, RMD $\mu_{\underset{\sim}{A}}(u)_{ih}$ of sample 1 is:

$$\mu_{\underset{\sim}{A}}(u)_{ih(3\times5)} = \begin{bmatrix} 0.00 & 0.00 & 0.22 & 0.78 & 0.28 \\ 0.70 & 0.30 & 0.00 & 0.00 & 0.00 \\ 0.00 & 0.00 & 0.00 & 0.21 & 0.79 \end{bmatrix}$$

The weight vector (Wang *et al.*, 2004) is:

$$\vec{w} = (1/3, \quad 1/3, \quad 1/3)$$

Then we can use the VFSA model (equation (15)) to obtain synthetic RMDs to five grades. Here synthetic RMDs of sample 1 are calculated under two conditions: $\alpha = 1, p = 1$ and $\alpha = 1, p = 2$.

When $\alpha = 1, p = 1$, synthetic RMD vector $v_A(u)$ can be calculated:

$$v_A(u) = (0.233 \quad 0.100 \quad 0.073 \quad 0.331 \quad 0.358)$$

After normalization, $v_A^{\circ}(u)$ is obtained.

$$v_A^{\circ}(u)_k = (0.213 \quad 0.091 \quad 0.066 \quad 0.302 \quad 0.327)$$

Using RFV equation (18), we calculate RFV of water quality:

$$H = (1, 2, \ldots c) \; v_A^{\circ}(u)^T$$

$$= (1 \quad 2 \quad 3 \quad 4 \quad 5)\square(0.213 \quad 0.091 \quad 0.066 \quad 0.302 \quad 0.327)^T = 3.438$$

So, the RFV of sample 1 is 3.438. Similarly, we can generate RFV of sample 1 under $\alpha = 1, p = 2$ (see Table 3). With the same process, the assessment results of other samples can be calculated (see Table 3).

Table 3 Assessment results of water quality with $\vec{w}$.

	1	2	3	4	5	6	7	8
$\alpha = 1, p = 1$	3.438	3.390	3.497	3.509	2.642	2.599	2.624	2.626
$\alpha = 1, p = 2$	3.271	3.267	3.317	3.324	2.734	2.693	2.710	2.718
mean	3.354	3.328	3.407	3.416	2.688	2.646	2.667	2.672

H intervals corresponding to water quality grades are:

(a) $H \in [1.0, \quad 1.5]$: I grade;
(b) $H \in (1.5, \quad 2.5]$: II grade;
(c) $H \in (2.5, \quad 3.5]$: III grade;
(d) $H \in (3.5, \quad 4.5]$: IV grade;
(e) $H \in (4.5, \quad 5.0]$: V grade.

Hence, from Table 3, we observe that the four cross-sections' water quality (samples 1–4) is between grade III and grade IV, and part of water quality is grade IV during flood period; while during dry periods, four cross-sections water quality (samples 5–8) are generally between grade II and grade III with partial to grade III. We also can see that the water quality of the Tuokou cross-section is the best in the four cross-sections. The assessment using VFSA model is similar to Wang's (2004), but, water quality assessment with grades is more accurate.

Generally, the weights vector defined by the theories and experience exhibits the importance of indices. Undoubtedly, it contains the subjectivity of decision-makers. Aiming at the above problem, this paper obtains the weight matrix, $\mu_A(u)$. Based on $\mu_A(u)$ and the principle of larger value $\mu_A(u)$ obtain larger weights, the variable weight matrix w' of three assessment indices of eight samples is obtained:

$$w' = \begin{bmatrix} 0.399 & 0.388 & 0.386 & 0.387 & 0.231 & 0.224 & 0.230 & 0.224 \\ 0.128 & 0.118 & 0.132 & 0.131 & 0.159 & 0.158 & 0.155 & 0.163 \\ 0.472 & 0.495 & 0.482 & 0.481 & 0.610 & 0.618 & 0.615 & 0.612 \end{bmatrix}$$

Table 4 Assessment results of water quality with w'.

No.	1	2	3	4	5	6	7	8
$\alpha = 1, p = 1$	4.048	4.101	4.123	4.134	3.556	3.551	3.569	3.550
$\alpha = 1, p = 2$	3.936	3.968	3.972	3.983	3.672	3.671	3.673	3.673
mean	3.992	4.034	4.048	4.058	3.614	3.611	3.621	3.611

Following the above, the calculating process i.e. the assessment results, are shown in (Table 4).

We found that four cross-sections water quality (samples 1–4) belongs to grade IV during a flood period; while during a dry period, four cross-sections of water quality (sample 5–8) generally fall between grade III and grade IV, with partial to grade IV. Here, COD_{mn} and NH_3-N concentrations are high, over grade V. The weights of these 2 indices are enhanced. Therefore comparing to assessment with $\vec{w}$, this assessment is reliable.

CONCLUSIONS

It is essential to integrate water quantity, water quality, and hydroenergy into water resource assessment. The VFSA model for regional water resources assessment is introduced and a tidy water resources assessment system is established in this paper. The four cross-sections of the Three Gorges reach are selected for a case study. Because regional water resources assessment is complex, nonlinear and fuzzy, the VFSA model, a developed fuzzy tool, is valuable, compared to mathematical models. The stability of the model is also analysed by variable parameters. The case study focusing on water quality shows that the VFSA model is useful for regional water resources assessment.

Acknowledgements This research was supported by grant no. 50779005 from the National Natural Science Foundation of China and no. TRS-28-3.2 from the Water Resources Department of Hebei Province of China.

REFERENCES

Cai, W. (1990) The extension set and non-compatible problems. In: *Advances Mathematics and Mechanics in China* (ed. by W. Ch). International Academic Publishers, China (in Chinese).

Charnes, A., Cooper, W. W. & Phodes, E. (1978) Measuring the efficiency of decision making units. *European J. Operational Res.* **3**(2), 429–444.

Chen, S. (1997) *The Fuzzy Sets Theory and Practice for Engineering Hydrology and Water Resources System.* Dalian University of Technology Press, China (in Chinese).

Chen, S. (2005) *Theories and Methods of Variable Fuzzy Sets in Water Resources and Flood Control System.* Dalian University of Technology Press, China (in Chinese).

Deng, J. (1991) The comparability of representation in grey system theory. *J. Grey System* **2**, 107–128 (in Chinese).

He, Z. (1983) *Fuzzy Mathematics and Application.* Tianjin Science and Technology Press, China (in Chinese).

Wang, H., Wang, J., Jia, Y., Qin, D., Qiu, Y. & Wang, L. (2006) A study on the method of water resources assessment in river basin under the present environment. *J. China Hydrol.* **26**(3), 18–22 (in Chinese).

Wang, M., Fu, H. & Lu, P. (2004) Synchronous observation and evaluation for the water quality in Three Gorges reach. *J. Chongqing Jiaotong Univ.* **23**(5), 122–125 (in Chinese).

Implementation of subgrid-scale spatial variability of parameters in a regional climate–hydrology coupled model

XING CHEN[1,2], ZHONGBO YU[1,2], YONG HUANG[3], GUANGBAI CUI[1,2], LILIANG REN[1,2] & XI CHEN[2]

1 *Key Laboratory of Hydrology-Water Resources and Hydraulic Engineering, Hohai University, no. 1 Xikang Road, Nanjing 210098, China*
chenxing@hhu.edu.cn

2 *College of Hydrology and Water Resources, Hohai University, no. 1 Xikang Road, Nanjing 210098, China*

3 *College of Civil Engineering, Hohai University, no. 1 Xikang Road, Nanjing 210098, China*

Abstract Climate variability has an important impact on various hydrological processes. In order to represent small-scale hydrological processes on the climate and hydrology interactions, it is necessary to couple the regional climate model with a fine-scale hydrological model. In this paper, the parameterization of subgrid spatial variability is implemented in a coupled climate–hydrology model. The subgrid-scale spatial variability in hydraulic conductivity and precipitation is considered in the simulation to represent small-scale hydrological processes in the regional model. The results show that the simulated streamflow hydrograph at the basin outlet is consistent with the observed one. The incorporation of subgrid-scale spatial variability in hydraulic conductivity and precipitation could greatly improve the simulation. It is necessary to incorporate these subgrid hydrological processes to obtain realistic and accurate simulations of hydrological responses.

Key words climate–hydrology coupled model; subgrid; spatial variability; runoff generation mechanism

INTRODUCTION

Hydrological processes such as soil water movement, evaporation and runoff generation are directly influenced by the spatiotemporal distribution of rainfall, the variation of air temperature, humidity and wind velocity, as well as of solar radiation intensity induced by the climate change. In the form of evaporation, precipitation and runoff, the hydrological cycle is the major carrier for the energy and substance exchange in climate and hydrological systems (Liu, 1997). There is a close interrelationship between climate and hydrological systems. The integrated simulation of climatic and hydrological responses and parameterization of hydrological processes have become focal points of recent research (Xue *et al.*, 2005).

However, coupling hydrological and climate models poses new challenges for researchers (Liu, 2003). The first problem is the complex mechanism of rainfall–runoff generation. It is known that the infiltration excess (Horton) runoff and saturation excess (Dunne) runoff are two main mechanisms for rainfall–runoff generation. Generally, Horton runoff is the control mechanism in dry areas, while Dunne runoff occurs in wet areas. However, the Horton runoff can happen in wet areas during a high intensity rainstorm. Precipitation with a long duration can also produce saturation excess runoff in dry areas. The rainfall–runoff generation mechanism is controlled by the characteristics of landform, soil, geology and climate. There is still a lack of an effective method to describe hydrological processes such as soil water movement and evapotranspiration due to the high heterogeneity of land cover and soil texture in hydrological models. Traditional conceptual models, with features of strong reliability on data and indirect relationship with geographical features, are not suitable for coupling climate–hydrology models (Zeng *et al.*, 2003). Distributed hydrological models have become an important tool for the climatic and hydrological study of various processes (Beven & Kirkby, 1979; Paniconi & Wood, 1993; Ren & Liu, 2000; Walko *et al.*, 2000), but most distributed models still have deficiencies in characterizing non-homogeneity of rainfall–runoff generation and considering the flow concentration process in a grid or between grids (Yu & Schwartz, 1998). Secondly, the scale discrepancy between climate models of large scale and hydrological models of relatively smaller scale is still an issue for coupling two systems (Wood *et al.*, 1990). Hydrological processes such as rainfall, runoff, evapotranspiration and infiltration all show subgrid variability. So it is difficult to develop a parameterization scheme which can describe actual physical processes in large spatial grids. The previous studies show that streamflow would be underestimated without considering such a subgrid variability (Yu *et al.*, 2001). Further, research on the data decomposition in smaller time and space intervals is also an effective way to solve the problem of scale discrepancy between climate and hydrological models.

In this study, the Hydrologic Model System (HMS) (Yu *et al.*, 2002), which is physically based and linked with a regional climate model (RCM), is used for the simulation of various hydrological processes (Fig. 1). The method to describe the spatial variation of precipitation and hydraulic parameters is also implemented in the model (Genuchten, 1980; Thomas & Henderson, 1991).

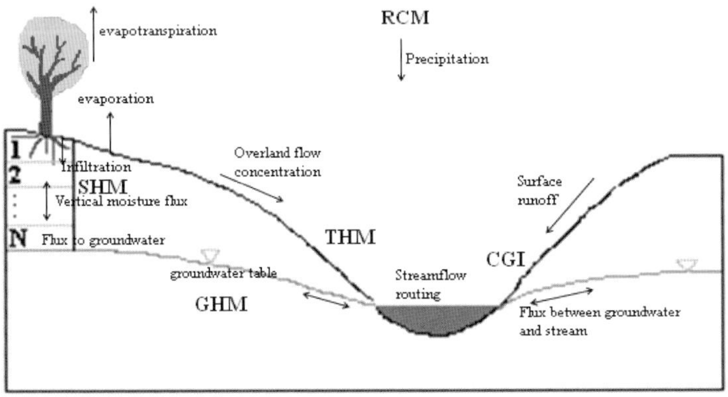

Fig. 1 Schematic map of climate and hydrology models.

METHODS

Climate and hydrology models

Regional climate change is greatly influenced by local characteristics, which are difficult to represent in the large-scale grid of global climate models. The regional climate model (RCM) used in the study has a finer resolution, which is adapted from MM5. RCM has the multi-level Blackadar-type planetary boundary layer parameterization and Grell cumulus parameterization (Grell, 1993). RCM models 20 layers distributed unevenly in a vertical direction from the surface to a height of 100 mb. The distance between two layers is smaller near the surface (Yu *et al.*, 1999, 2000). In this study, RCM is used in the non-hydrostatic mode and simulate precipitation per hour in the resolution of 4 km.

The Hydrologic Model System (HMS) is a physically-based, parameter distributed model system. It includes four submodels, a Soil Hydrologic Model (SHM), a Terrestrial Hydrologic Model (THM), a Ground-Water Hydrologic Model (GHM), and a Channel Ground-Water Interaction Model (CGI) (Yu & Schwartz, 1998). In SHM, based on the water storage including precipitation and flow from neighboring grids, the Richards equation is used to simulate the vertical soil water flow in the unsaturated zone, which is solved with the Crank-Nicholson method numerically. The evapotranspiration is calculated using the Penman-Monteith method. The Green-Ampt method is implemented to partition rainfall into the infiltration and surface runoff. The spatial distribution of soil moisture content simulated with SHM is used for the calculation of other hydrological components in every time step. In THM, the kinematic wave algorithm is used to model the overland flow based on the flow direction derived from DEM. The Muskingum-Cunge procedure is used to perform the channel routing from grid cell to grid cell. Through these flow simulations, the flow through each grid is routed to the basin outlet. The second order partial differential equation is applied to describe groundwater flow in the saturated zone, which is solved with a finite difference method in GHM. Darcy's law is used in CGI to calculate the interactive flux between stream and groundwater based on the hydraulic gradient between groundwater level and channel water level in the stream cell.

HMS uses the remotely sensed data and digital data as the parameters in the model system and they have physical meaning. The four submodels in HMS are interrelated to describe actual physical processes. For example, the interaction of groundwater and streamflow are considered and the loss and acquisition of water in a grid cell or between grid cells can also be described in the model.

Characterizing the subgrid scale spatial variability

As stated in the previous section, it is important to apply subgrid-scale spatial variability in climate and distributed hydrological models, not only for better coupling two systems, but also for improving the forecasting precision of hydrological processes (i.e. soil moisture, runoff). The hydrological responses strongly rely on the spatial variation of precipitation and hydraulic characteristics of soil (Pitman *et al.*, 1990). The precipitation simulated with the RCM is the spatially-averaged value in a large-scale grid cell, and this averaging may lower the actual rainfall intensity. In the Horton method, the rainfall intensity plays an important role in runoff generation and leads to the underestimation of runoff. In this study, the probability density distribution is implemented to represent the first-order approximation of the spatial variation of precipitation in each RCM grid cell. Based on the observed data, the exponential distribution is used to downscale the simulated precipitation in each RCM grid cell (at a resolution of 4 km) into hydrological grid cells (at a resolution of 1 km), which fall into each RCM grid cell (Yu *et al.*, 2000):

$$f(p_i) = \frac{c}{P}\exp(-\frac{cp_i}{P}), \quad \int_0^\infty f(p_i)dp_i = 1 \tag{1}$$

where f is the area fraction of a RCM grid cell with precipitation p_i, P is the precipitation simulated by RCM, p_i is the downscaled precipitation, c is the conversion coefficient which range from 0.3 for convective rainfall to 1.0 for large scale rainfall (Thomas & Henderson, 1991). The downscaled precipitation for a RCM grid cell is assigned to every corresponding hydrological grid cell randomly.

With the same method, the exponential distribution is also used to downscale the hydraulic conductivity into subdivisions within each hydrological grid cell (Yu *et al.*, 2000):

$$f(k_i) = \frac{1}{K}\exp(-\frac{k_i}{K}), \quad \int_0^\infty f(k_i)dk_i = 1 \tag{2}$$

where f is the area fraction of a HMS grid cell with hydraulic conductivity k_i, K is the hydraulic conductivity in a hydrological grid cell, k_i is the downscaled hydraulic conductivity.

With this method, the precipitation in each RCM grid cell is disaggregated to hydrological grid cells and the hydraulic conductivity in hydrological grid cells is downscaled into more detailed hydrological subdivisions. The hydrological simulation is conducted in each subdivision. The runoff and infiltration in subdivisions within a hydrological grid cell are summed for the calculation of water flow and water balance.

APPLICATION

Study area

The upper West Branch watershed with a drainage area of 14 710 km^2 is a subbasin of the Susquehanna River basin in Pennsylvania, USA (Fig. 2). Most of the upper West Branch watershed lies within the Appalachian Plateau physiographic province, which is of broad upland plain separated by steep-walled, narrow valleys. The southern portion of the upper West Branch watershed lies within the Valley Ridge province of the Appalachian, which has long, low, even-crested ridges with elevation of approx. 400–500 m a.m.s.l (Fig. 3). The land cover of the upper West Branch watershed is predominately forest, especially in the northern and western of the basin. Agricultural and urban lands are primarily distributed in the eastern and southern parts of the watershed.

Fig. 2 Location map of the study area.

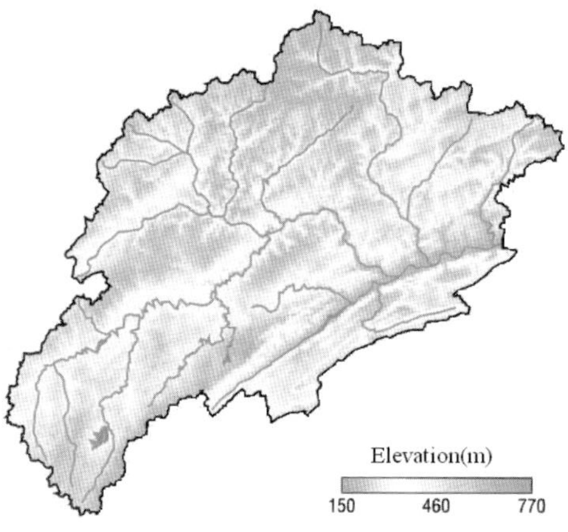

Fig. 3 DEM of the upper West Branch watershed.

Data processing

As a distributed hydrological model system, HMS uses remote sensed and other data sets as input data source. These image data have to be processed into digital ones which can be recognized by the model. After the correction of atmospheric attenuation, radiance and temperature values can be extracted from AVHRRR and Landsat TM images according to the grid resolution. The fractional vegetation cover, impervious surface area, evapotranspiration and surface soil moisture content can also be derived through processing these images. The land cover information is drawn based on basic land cover types. The spatial distribution of hydraulic parameters is obtained based on the soil data from the State Soil Geographic (STATSGO) and land cover information (Yu *et al.*, 2001). Assuming that hydraulic parameters are unchanged in a hydrological grid cell, values of two hydraulic parameters used in SHM, hydraulic conductivity and average capillary suction, are assigned into each hydrological grid cell with a resolution of 1 km.

A digital elevation model (DEM) is the basis for the distributed hydrological simulation and coupling of climate and hydrological systems. The DEM is generated from USGS 3-arc second data in the study. In order to match other data in HMS, the DEM is reprocessed into 1-km^2. The DEM is used to derive basin features such as grid slope, drainage area and flow direction, and to delineate the basin boundary and channel network in HMS.

Model calibration

Trial-and-error is still the most commonly used way to calibrate the model. In general, the calibration can then be classified into direct algorithm and indirect algorithm. The former uses the inverse operator while the latter adopts the iterative method. The calibration method for a distributed watershed model (Yu & Schwartz, 1999) was implemented in this study, which belongs to the indirect algorithm. Values of parameters in their own ranges are adjusted to optimize the model performance based on the multi-objectives such as soil moisture, groundwater level and discharge in the basin outlet. The model calibration has been conducted in previous studies (Yu *et al.*, 1999) for the upper West Branch watershed. The calibrated model can be applied to simulate hydrological responses in this watershed and describe the mechanisms and interrelationship of various hydrological processes.

RESULTS AND ANALYSIS

Influenced by the cyclone on 14 April 1986, the storm event began on 15 April and precipitation continued until 18 April. RCM is used to simulate the precipitation for this event. Using observed and simulated precipitation, HMS is run from 14 to 22 April with a time step of 10 minutes. The results are shown in Figs 4, 5 and 6. S1 and S2 represent the streamflow simulation with or without considering subgrid-scale variability of hydraulic conductivity with observed precipitation. S3 represents the simulation considering subgrid-scale variability of hydraulic conductivity with simulated precipitation. S4 represents the simulation considering subgrid-scale variability of hydraulic conductivity and precipitation with simulated precipitation.

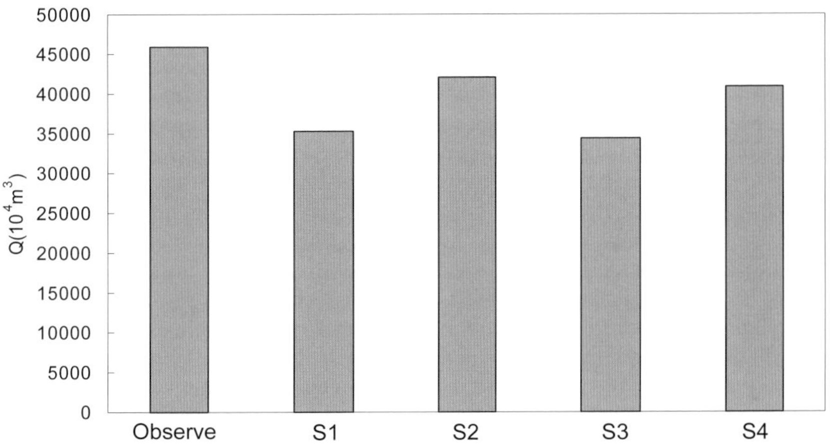

Fig. 4 Observed and simulated flood water volume.

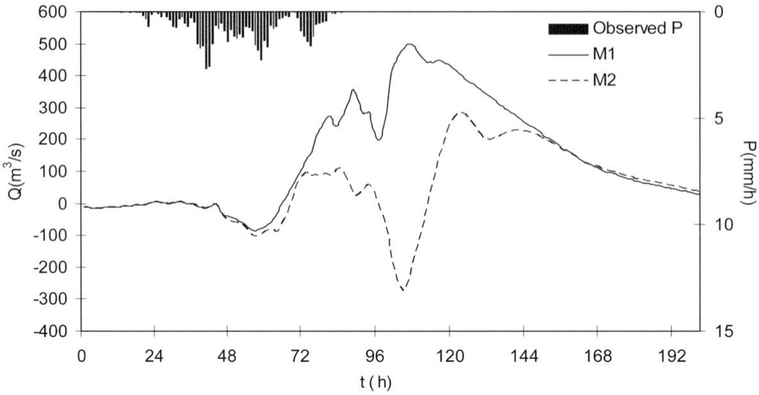

Fig. 5 Hydrographs of discharge differences between observed and simulated streamflow with observed precipitation.

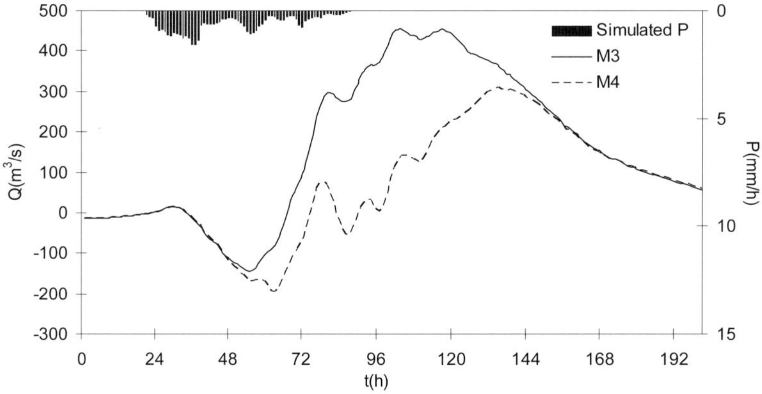

Fig. 6 Hydrographs of discharge differences between observed and simulated streamflow with simulated precipitation.

Comparing Figs 5 and 6, the precipitation simulated with RCM begins at the same time, with the observed one and the histogram shape for the simulated and observed precipitation being basically identical. However, the histogram of simulated rainfall is flatter than the observed, and the highest rainfall intensity generated with RCM is smaller than the observed. The reason for this phenomenon is that the grid cell size of RCM is so large that the simulated precipitation is the averaged value over grid cell and the spatial variability of rainfall is neglected. The spatial variation of rainfall could be large and should not be ignored, especially for the convective rainfall event. Therefore, this spatial averaging also causes the underestimation of precipitation.

In Fig. 4 it is obvious that the flood water volumes of S3 and S4 are less than those of S1 and S2, respectively, which is mainly caused by the underestimation of precipitation. With the observed precipitation and considering subgrid-scale variability of hydraulic conductivity, the discharge amount of S2 is still less than that observed. This is because this rainfall event is of long duration and low intensity so that the saturation excess runoff is the dominant runoff generation mechanism. According to previous studies (Dunne & Black, 1970), the expansion of the near stream area is the dominating mechanism of runoff generation in the study area. Through considering subgrid-scale variability of hydraulic conductivity, the HMS can describe the runoff generated with the expansion of saturated area implicitly, but because there is no explicit spatial description of this process in a grid cell, the streamflow is still underestimated.

For a better comparison, the hydrographs in Figs 5 and 6 show the differences between observed and simulated discharges. M1, M2, M3 and M4 represent the differences between observed discharge and discharge of S1, S2, S3 and S4, respectively. It can be seen that M1 shows poor performance because the spatial variability of hydraulic conductivity is not considered. With this variability, the performance of M2 improves a lot. Comparing hydrographs M3 and M4, it can be concluded that considering the subgrid-scale variability of precipitation can also enhance the performance for the streamflow simulations.

In Figs 5 and 6, there are relatively large differences between observed and simulated discharges in some sections of hydrographs, which may be primarily due to the DEM effect. DEM in a resolution of 1 km may ignore many terrain details, which causes the underestimation of discharge. It is also worth noticing that part of simulated values during the small and large discharges may be larger than observed ones, although most simulated discharge is less than the observed. On one hand, it is caused by the smoothing effect of flow calculations. On the other hand, not considering human activities (i.e. reservoirs) may also contribute to this problem.

CONCLUSION

Climate variability is the driving force for changes on various hydrological processes. Integrated study of climate and hydrological processes is necessary to further understand the interrelationship among climate, hydrology, ecology and human activities so that we can improve water resources planning and water quality management.

This paper describes how to downscale the precipitation generated with RCM and hydraulic parameters, which is not only necessary for better coupling of climate models with large-scale grids and hydrological models with small-scale grids, but also improves the description of runoff generation processes. Results show that the simulated precipitation is smaller than the observed in general due to the averaging effect of the large-scale grid in RCM. The spatial variability of precipitation and hydraulic parameters is important for the runoff generation process. When this spatial variability is described in the model, the simulated discharge hydrographs are improved.

Results also show that such small-scale processes as expansion of saturation area still can not be explicitly described in the model, although the discharge at the outlet has been greatly improved. So how to represent such processes in regional models while keeping their natural temporal and spatial scales is the next research point. Human activities should also be included in the model for more realistic and comprehensive simulations.

Acknowledgements This work has been supported in part by the US National Science Foundation under collaborative grant ATM 9972956 and ATM 0002637, Natural Science Foundation of China (no. 50239030 and no. 50679018) and the program for Changjiang Scholars and Innovative Research Team in University (IRT0717).

REFERENCES

Beven, K. & Kirkby, M. J. (1979) A physically based, variable contributing area model of basin hydrology. *Hydrol Sci Bull* **24**, 43–69.

Dunne, T. & Black, R. D. (1970) Partial area contributions to storm runoff in a small New England watershed. *Water Resour Res*, **6**, 1296–1311.

Genuchten, M. T. (1980) A closed-form equation for predicting the hydraulic conductivity of unsaturated soils. *Soil Sci. Soc. Am. J.* **44**, 892–898.

Grell, G. A. (1993) Prognostic evaluation of assumptions used by cumulus parameterizations. *Mon. Weath. Rev.* **121**, 764–787.

Liu, C. Z. (2003) The review of Impact study of Climate variability and change on the water cycle. *J. Hydrol.* **23**(4), 1–7.

Liu, G. W. (1997) *Atmosphere Processes for Hydrologic Circulation.* Science Press, Beijing, China.

Paniconi, C. & Wood, E. F. (1993) A detailed model for simulation of catchment scale subsurface hydrologic processes. *Water Resour. Res.* **29**, 1601–1620.

Pitman, A. J., Henderson-Sellers, A. & Yang, Z. L. (1990) Sensitivity of regional climates to localised precipitation in global models. *Nature* **346**, 734–737.

Ren, L. L. & Liu, X. R. (2000) Hydrological processes modeling based on digital basin. *J. Natural Disasters* **9**(4), 45–52.

Thomas, G. & Henderson, S. A. (1991) An evaluation of proposed representations of subgrid hydrologic processes in climate models. *J. Climate* **4**, 898–910.

Walko, R. L., Band, L. E. & Baron, J. (2000) Coupled atmosphere-biophysics hydrology models for environmental modeling. *J. Appl. Met.* **39**, 931–944

Wood, E. F., Sivapalan, M. & Beven, K. (1990) Similarity and scale in catchment storm response. *Rev. Geophys.* **28**, 1–18.

Xue, G. Y., Zhou, S. Q., Chen, G. Y., Yang, Y. C. & Chen, H. M. (2005) Advances in research on land surface progress. *Bull. Sci. Technol.* **21**(4), 378–486.

Yu, Z. & Schwartz, F. W. (1998) Application of integrated Basin-Scale Hydrologic Model to simulate surface water and ground-water interactions in Big Darby Creek Watershed, Ohio. *J. Am. Water Resour. Assoc.* **34**, 409-425.

Yu, Z. & Schwartz, F. W. (1999) Automated calibration applied to constrained groundwater flow modeling. *Hydrol. Processes* **13**, 191–209.

Yu, Z., Lakhtakia, M. N., Yarnal, B., Johnson, D. L., White, R. A., Miller, D. A., Barron, E. J., Duffy, C. & Schwartz, F. W. (1999) Simulation of the hydrologic response to atmospheric forcing in large river basins: linking a mesoscale meteorological model and a hydrologic model system. *J. Hydrol.* **218**, 72–91.

Yu, Z., Barron, E. J. & Schwartz, F. W. (2000) Retrospective simulation of a storm event: a first step in coupled climate/hydrologic modeling. *Geophy. Res. Lett.* **27**, 2561–2565.

Yu, Z., Carlson, T. N., Barron, E. J. & Schwartz, F. W. (2001) On evaluating the spatial-temporal variation of soil moisture in the Susquehanna River basin. *Water Resour. Res.* **37**(5), 1313–1326.

Yu, Z., Barron, E. J., Yarnal, B., Lakhtakia, M. N., White, R A., Pollard, D. & Miller, D. A. (2002) Evaluation of basin scale hydrologic response to a multi storm simulation. *J. Hydrol.* **257**, 212–225.

Zeng, X. M., Zhao, M., Su, B. K., Tang, J. P., Zheng, Y. Q., Gui, Q. J. & Zhou, Z. G. (2003) Simulations of a regional climate model as coupled with a hydrological model. *J. Nanjing University (Natural Sciences)* **39**(1), 91–96.

Impacts of climate change on hydrological processes and water resources in the headwater area of the Yellow River

HUI GAO, YANGWEN JIA, CUNWEN NIU, YAQIN QIU & HAO WANG

Department of Water resources, Institute of Water resources & Hydropower Research (IWHR), 20 Che-Gong-Zhuang West Road, Beijing 100044, China

jiayw@iwhr.com

Abstract The headwater area upstream of the Tang-Nai-Hai hydrological gauge station is one of the main runoff generation areas of the Yellow River Basin. Climate change is one of the main reasons for water resources decrease and ecological deterioration in the headwater area in recent years, and analysing of future further change impact is important for water resources planning and management in the basin. In this article, we simulated the changes of both annual and monthly runoffs in the headstream of the Yellow River under air temperature and precipitation change by using the WEP model. WEP is a physically-based distributed model and can reflect the impact of air temperature change on water resources through evapotranspiration change, snow storage and melting change and infiltration capability change in the frozen soil layer. The model was validated using the observed daily discharge data from 1956 to 2000 at the Tang-Nai-Hai station. After validating the model, we assumed eight different schemes with temperature change of ±1, ±2°C and precipitation change by ±10%, ±20% on the basis of historical observed meteorological data. The results indicated that air temperature change had different impacts on annual and monthly runoffs. The temperature increase causes an annual runoff decrease, with an obvious decrease of monthly runoff from May to October because of the evapotranspiration increase, and an increase from November to the following April because of snow storage and melting, and frozen soil infiltration capability changes. The maximum increase is 63.7% in March, 1989, when assuming the air temperature increase of 2°C. Precipitation increase or decrease causes runoff increase or decrease to different extents, and the runoff has a larger change rate than precipitation.

Key words headwater area of the Yellow River; climate change; distributed model; Tang-Nai-Hai

INTRODUCTION

Due to a sharp increment of the world's population and quick economic development, a large quantity of greenhouse gases was released, and some climate conditions such as precipitation and temperature were changed, resulting in a change in the water resources situation and water resources and environmental problems. Although the United States National Research Association (USNA) held a conference to discuss the relationship between climate, climate change and water supply early in 1977, the research about how climate change influences water resources did not draw much attention from the international hydrology group until the mid-1980s. In China, special research has been carried out in the Seventh, Eighth, Ninth "Five-Years" programmes and the GAME programme since the 1980s (Jiang *et al.*, 2003).

There are several uncertainties in the research on climate change, which are caused by the amount of greenhouse gas discharged, the response of the climate system and the levity of nature (Metoffice, 2002). To study the impacts of climate change on water resources, researchers present different hydrology models, including statistical models, concept models, and distributed models. By analysing the climate factors (precipitation and temperature) and the hydrology factors (runoff and flood frequency), the statistical models were used to set up the statistical relationship between each variable with which we could analyse the influence of climate change on water resources. Stockton (1979) set up the experiential function among precipitation, air temperature and runoff, and analysed runoff change in main American rivers under different assumed scenarios. Lan *et al.* (2001) studied the response of runoff to climate change on the basis of the precipitation, air temperature, and runoff data in the Qilian Mountainous Area and the Hexi corridor area. Concept models were used to simulate the runoff under the assumed scenarios, study the influence of climate change on runoff, evaporation, flood peak and its occurrence time, and the spatial distribution of hydrological factors. Lettenmaier & Gan (1990) analysed the water process response to the climate warming using the Sacramento Model; Liu (1997) studied the slope runoff using the improved monthly runoff model (Schaake & Liu, 1989); Wang & Wang (2000) studied the impacts on the annual runoff at the upstream of the Yellow River. Distributed models could be used to study the influence of climate change on the runoff at a macroscopic scale by coupling the climate model and the distributed hydrology model, or making the output of the climate model the input of the hydrology model. Yuan *et al.* (2005) studied the trend of water resource change under the climate change by coupling the Variable Infiltration Capacity model and the Providing Regional Climate for Impacts Studies model. Chen & Gao (2003) discussed the impact of climate change on the runoff of the middle reach of the Yangtze River using a Two-parameter Monthly Water Balance Model under the climate scenarios provided by two global climate models. The coupling of distributed models and general circulation models (GCMs) is a new study direction of the impacts of climate change on the water cycle.

The headwater area of the Yellow River is the main water source of the basin. The environmental problems of glacier degradation, frozen soil degeneration, and lake drying up in the headwater area interact and aggravate each other, and have weakened the carrying-capacity of the water resources. The problem of water resources change will influence the socio-economies of the Yellow River Basin for a long time. It is necessary to simulate and analyse the runoff response to climate change in the headwater area of the Yellow River Basin, which is helpful for doing research and forecasting the change of water resources.

There is lots of research on the problem of water resources change in the headwater area and some conclusions have been obtained, e.g. the response to climate change in the headwater area is sensitive, if temperature falls and precipitation increases, the runoff will increase by different degrees. But these conclusions were discussed in the time scale of one year (Wang & Wang, 2000; Li *et al.*, 2004; Yang *et al.*, 2004), not including the seasonal and monthly responses. In this article, we analysed not only the change of annual runoff, but also the change of monthly runoff under climate change by assuming eight scenarios of precipitation, considering the contribution of glacier and thaw to the runoff and using the WEP model.

MODEL INTRODUCTION

WEP-L (Water and Energy transfer Processes in Large river basins) (Jia *et al.*, 2005, 2006), a distributed model, was developed in a National 973 Program of China entitled "Evolutionary Laws and Renewable Mechanism of Water Resources in the Yellow River Basin".

The horizontal structure of WEP model is shown in Fig. 1. Runoff routing on slopes is carried out based on elevation, gradient and Manning roughness by applying a one-dimensional (1-D) kinematic wave approach or dynamic wave approach to trace the overland flow from upstream to downstream. Numerical simulation is made for groundwater flows in mountain and plain areas separately, with consideration of the groundwater exchange with surface water, soil moisture and streamflow.

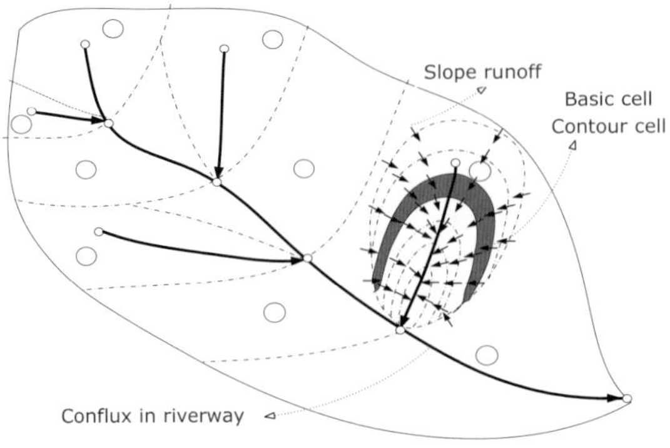

Fig. 1 Horizontal structure of the WEP-L model.

The vertical structure of the WEP model is shown in Fig. 2. From top to bottom, the vertical structure includes an interception layer of vegetation or buildings, a land surface depression layer, unsaturated soil layers, a transition zone, and unconfined and confined groundwater aquifers. State variables include interception by vegetation, storage in land surface depressions, temperature of land surface, soil moisture content, storage in transition zone, groundwater level and streamwater level. The main parameters are maximum interception, soil permeability, and characteristic curve parameters of soil moisture and suction, groundwater transmission coefficient, groundwater specific yield and storage coefficient, river bed permeability and Manning roughness. The "mosaic" method is applied to take unevenness of land use in the calculation unit into consideration by classifying the land into groups of bare land-vegetation, irrigated farmland, non-irrigated farmland, waters and impermeable areas for the calculation of water-heat flux through the land surface. The bare land-vegetation is further subdivided into bare land, grassland and woodland, and impermeable land into urban land and urban building. In addition, in order to reflect change in surface soil moisture content with depth and describe evaporation from soil and water uptake by root systems of grass, crops and trees, the top soils of permeable areas are divided into three layers.

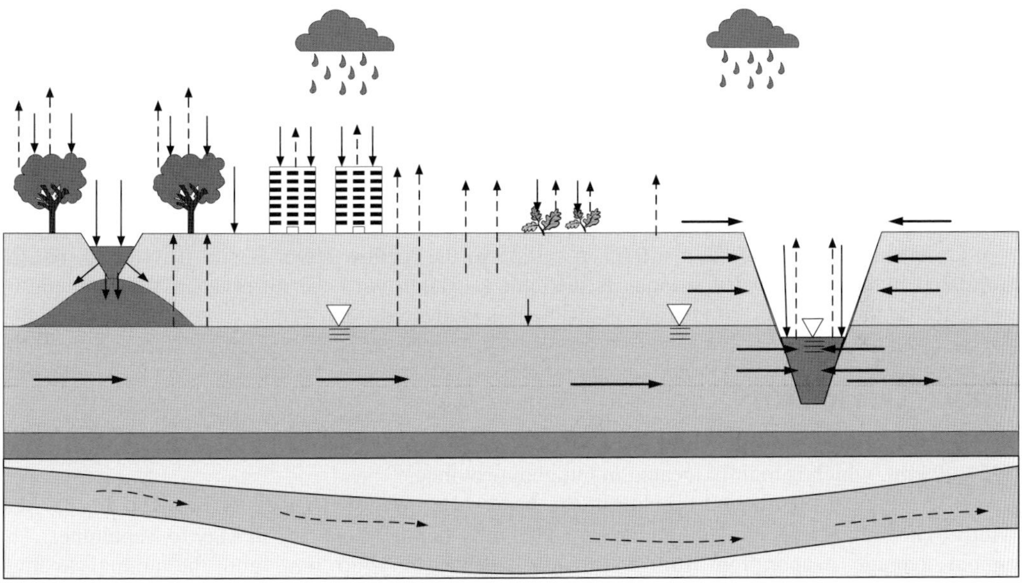

Fig. 2 Vertical structure of the WEP-L model.

The main characteristics of the WEP-L model are as follows:

(a) It integrates the merits of a distributed hydrological model and the SVATS model, couples the water cycle and energy processes, and calculates the vegetation canopy transpiration, surface evaporation, vegetation, soil and bare soil evaporation, water surface evaporation, and vegetation transpiration in each land-use area particularly.

(b) It generates the river system and divides sub-basins containing spatial topological information based on DEM and observed river vector graph, adopts "contour belt within sub-basin" as calculation unit, and considers diversity of land cover within the unit by means of a "mosaic" method, to avoid distortions of water balancing and runoff concentration route resulting from "coarse grids of large area" and to describe spatial variation of hydrological variables rationally.

(c) It performs simulations by adopting "varied time intervals" (such as 1 hour duration for the infiltration–runoff yield process of intensive rainfall, 6 hours duration for runoff concentration process on slopes and in the river course, and 1 day duration for groundwater flow) for each element of the hydrological cycle to ensure rational description of dynamic mechanisms, and efficient calculation.
(d) It incarnates the theory of variable source area (VSA) in the calculation, which can simulate the runoff yield under saturated storage, runoff yield under excess infiltration, and spring flow out, and calculate surface water, groundwater, and soil water dynamically.
(e) It has mutual-feedbacks with ROWAS (a lumped water allocation model) to realize close coupling of the natural hydrological cycle with the artificial water system.
(f) It has a quick calculation speed (11 minutes for simulation of the Yellow River Basin for one year duration), and functions of both flood forecasting and continuous calculation for the long-term.

The detailed content is referred to in Jia *et al.* (2005, 2006).

It should be mentioned that the temperature and precipitation changes, along with the topography, the snow and thaw simulation, and the moisture movement in frozen soil in the Yellow River Basin, are considered in the WEP-L model. The temperature and precipitation change along with the topography are interpolated based on the observations at gauge stations, DEM and the lapse rates to altitude. The snow and the thaw are computed by the degree-day index method, and the temperature influence to the hydraulic conductivity in frozen soil is computed as:

$$k_f = \begin{cases} k_0 & T_a \geq T_c \\ k_0 e^{a(T_a+b)} & T_a < T_c \end{cases} \tag{1}$$

where k_f is the hydraulic conduction coefficient; k_0 is the hydraulic conduction coefficient when frozen soils melt; T_a is the average daily temperature; T_c is the critical temperature of melting; and a, b are constants: $T_c = -5°C$, $a = 0.05$, $b = 0.25$ in the Yellow River Basin. It is found that the simulated hydrograph in dry seasons is difficult to match with the observed if the temperature influence on the hydraulic conduction coefficient is neglected.

MODEL APPLICATION

The headwater area is defined as the region upstream of the Tang-Nai-Hai station, with an area of 122 000 km^2, which is mainly located in Qinghai province.

It is a typical continental climate in the headwater area of the Yellow River: cold, dry and windy. The annual average temperature is –4°C, only being above 0°C in May to September. The annual average evaporation from the water surface is 1400 mm, and the annual average precipitation is 300 mm. It freezes from November to April every year, >160 days.

We selected the daily precipitation data of 14 rainfall stations, the daily weather data of 10 weather stations, and daily runoff data at 7 hydrology stations. The Earth surface data consists of 1:100 000 land use data in three periods of time (1986, 1996, 2000), 1:250 000 DEM terrain data, and soil data. Figure 3 shows the area upstream of Tang-Nai-Hai, Level-3 sub-basins and the distribution of weather and rainfall stations.

The calculation units of this simulation are contour belts of sub-basins divided on the base of DEM of the Yellow River Basin. According to the direction of flow, the whole basin is divided into several sub-basins, and then according to the elevation, each sub-basin is divided into several contour belts. Spatially, the precipitation data interpolated into sub-basins is enough. The detailed calculation steps are as follows: calculate the related coefficient of data between every two stations and fix a threshold of these coefficients. If the related coefficient of two stations is larger than the threshold, the two stations can be supposed to be related and can be reference stations for each other. Otherwise, if the related coefficients of a station to all the other stations are smaller than the threshold, interpolation with the Thiessen polygon method is needed.

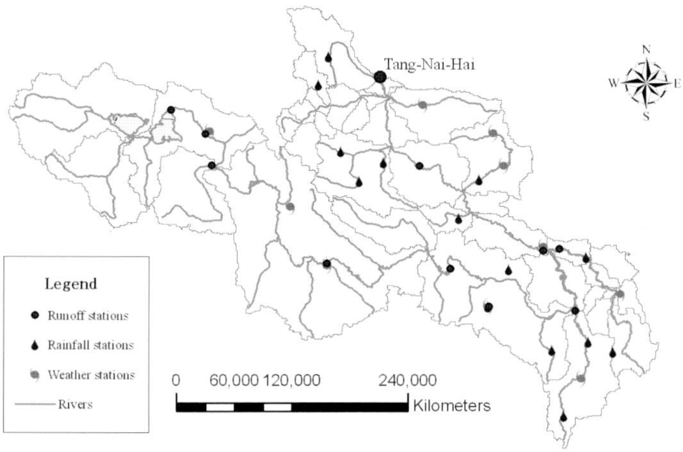

Fig. 3 Headwater area of the Yellow River and distribution of weather and rainfall stations.

Model calibration

In this article, we calibrated the model by the runoff in 1956–1979 with a time step of day. Then we validated the parameters by the runoff in 1980–2000 and got a good result, the relative error is not more than 4%, and the Nash coefficient is larger than 0.8 (see Table 1), which proved that the model fitted for the headwater area. Figure 4 shows the comparison between the simulated runoff and the observed ones.

Table 1 Comparison of the observed runoffs and calculated ones.

year	Observed runoff ($10^8 m^3$)	Calculated runoff ($10^8 m^3$)	Relatively error (%)	Nash coefficient
1956–2000	203.963	198.037	–2.90	0.821
1956–1979	202.145	197.609	–2.20	0.829
1980–2000	206.042	198.526	–3.60	0.814

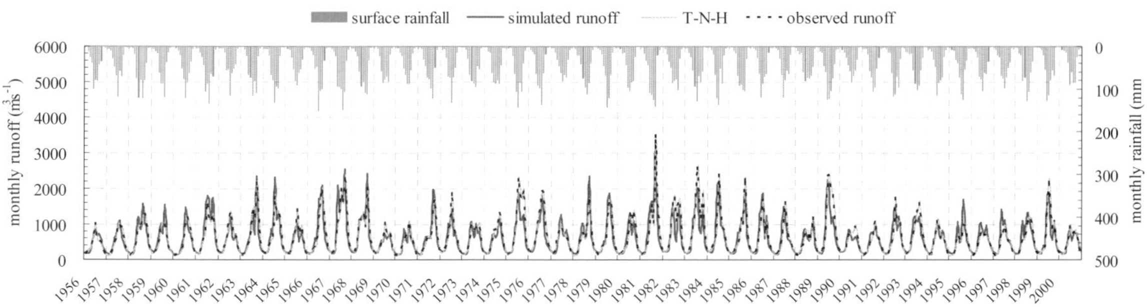

Fig. 4 Comparison of the simulated runoff and the observed ones.

Scenarios enactment

Considering the further trend of climate change and to analyse the runoff response clearly, we supposed that precipitation and temperature changed independently. In this research, we assumed eight different schemes with temperature change by ±1, ±2°C and precipitation change by ±10%,

±20% on the basis of historical observed meteorological data. Table 2 shows the results of these eight schemes' simulation.

Table 2 Simulation results of eight scenarios.

	Temperature change (°C)				Proportion of precipitation change (%)			
Climate change	–2	–1	+1	+2	–20	–10	+10	+20
Proportion of yearly runoff change (%)	+22.59	+10.43	–8.68	–15.77	–30.43	–15.96	+17.26	+35.64

Table 3 Comparison of simulation results with temperature changes.

Year		Actual case		Scenario case	
	Precipitation difference (mm)	Temperature difference (°C)	Runoff change (%)	Temperature difference (°C)	Runoff change (%)
1965 *vs* 1956	4.2	–1.94	20.30	–2.0	22.59
1977 *vs* 1969	–6.0	–1.22	11.30	–1.0	10.43
1989 *vs* 1967	–12.38	0.68	–15.74	1.0	–8.68

Table 4 Comparison of simulation results with precipitation changes.

year		Actual case		Scenario case	
	Temperature change (°C)	Precipitation difference (%)	Runoff change (%)	Precipitation difference (%)	Runoff change (%)
1983 *vs* 1965	0.01	–22.0	–31.4	–20.0	–30.43
1962 *vs* 1963	0.00	22.0	28.0	20.0	35.64
1966 *vs* 1980	0.01	–11.0	–20.8	–10.0	–15.96

Table 3 and Table 4 are comparisons of simulation results with temperature and precipitation changes. It is obvious that annual runoff changes with different rates when temperature and precipitation change. Annual runoff increases along with temperature decreasing. Precipitation change also affects the annual runoff. When precipitation increases, the annual runoff increases at a greater rate than precipitation does. The response of runoff to precipitation is more obvious than that to temperature. The maximal change rate is 35.64% when the precipitation increases by 20%.

Analysis of rationality of scenarios simulation results

In order to qualitatively validate the rationality of the simulation results, we analysed the response to temperature and precipitation separately. We chose the two years in which annual average precipitation was close and the average temperature discrepancy was about 1 or 2°C, from the data series, and then compared the annual runoff. For convenience, it was called "actual case" and the assumed scenario was called "scenario case".

In the actual case, the runoff change rates in every set of two years in which the annual average precipitation was very close and average temperature discrepancy was about 1°C or 2°C, were consistent with the runoff change rates of the scenario case. It proved that the scenario analysis results of temperature change were rational.

Similarly, we chose the two years in which annual average temperatures were close and average precipitation discrepancy was about 10% or 20% from the data series, and compared the annual runoff.

In the actual case, runoff in the two-years in which annual average temperatures were close and the average precipitation discrepancy was about 10% or 20% was uniform in the simulation result. It proved that the simulation result of precipitation change was rational.

Scenarios analysis

(a) Analysis of annual runoff change The proportions of runoff change in 45-year series were simulated based on the scenarios supposition. The results are shown in Figs 5 and 6, and the black line is the state without runoff change.

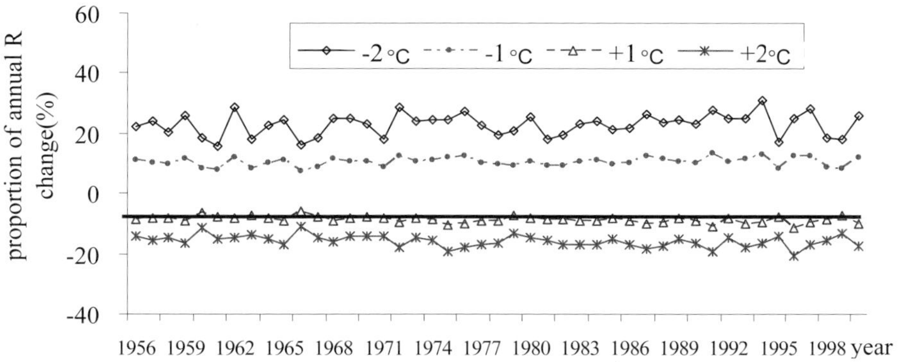

Fig. 5 Response of annual runoff to the temperature change in 45-year series.

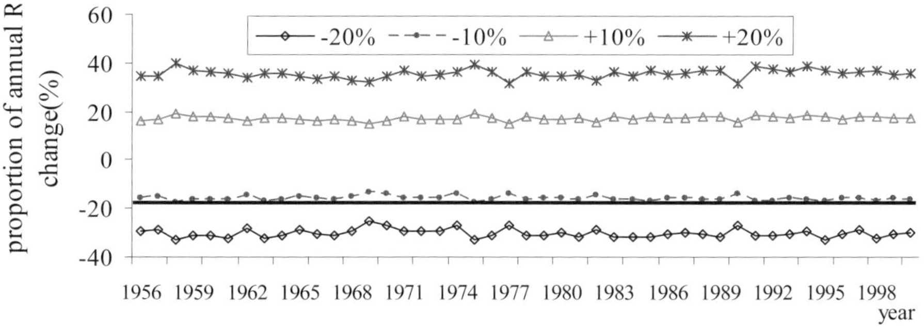

Fig. 6 Response of annual runoff to the precipitation change in 45-year series.

It can be seen that the response of runoff in the 45-year series to temperature change is accordant. Runoff decreases if the temperature increases. It is clear that in the same scenario the proportion of runoff change is at the same level, without obvious relation to precipitation.

The response of annual runoff to the precipitation is larger than that to the temperature changes. The runoff increases when the precipitation increases and in the same scenario the proportion of runoff change is at the same level.

(b) Analysis of monthly runoff change On the basis of the annual runoff analysis, the runoff change was simulated at a smaller time step. Figure 7 shows the average of monthly runoff in a 45-year series under temperature change, which is different from annual runoff.

The monthly runoff in 45-year series under temperature change is different from annual runoff. The curves in Fig. 7 intersect, which is mainly represented in the spring season every year because the headwater area is located at altiplano where there is perennial snow. When the

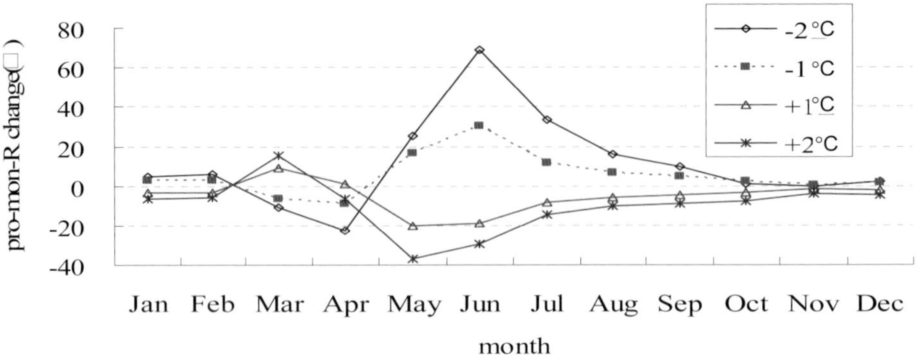

Fig. 7 Average monthly runoff under temperature change.

temperature rises to above 0°C in March or April every year, the snow melts into water, which causes spring floods. Thaw, the main part of the runoff in that season, is sensitive to temperature change, so monthly runoff increases when temperature increases in this season. The maximal proportion of change is 63.67% when temperature increased by 2°C in March, 1989. From November or December, to January or February of the next year, when the temperature is so low that the snow may not melt even under increasing temperature, the runoff response to the temperature is insensitive. In May to October every year when the actual temperature is above 0°C, temperature change deeply affects the evaporation, so monthly runoff decreases when temperature increases as the main part of runoff is surface flow.

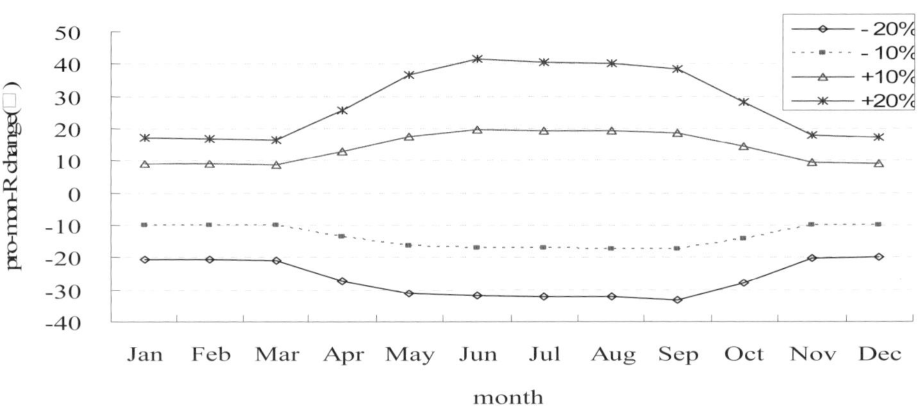

Fig. 8 Average monthly runoff under precipitation change (1956–2000).

Figure 8 shows average monthly runoff under precipitation change. We could see that the proportion of monthly runoff change in different months is different. The response to precipitation could be divided into three phases of dry season, wet season, and transitionary season. In the first phase, from November to the next March, because of the small amount of rainfall, the proportion of runoff change is maintained at the same lower level, which is close to the precipitation change proportion. During the wet season, from June to September, the change proportion is at an upper level, which is about 1.5 to 2.0 times that of the lower proportion in dry season. During three transitionary months, April, May, and October, the change proportion is between the other two.

CONCLUSIONS

In this paper, we simulated the changes of both annual runoffs and monthly ones in the headstream of the Yellow River under air temperature and precipitation changes using the WEP-L model. After calibrating the model by the runoff in 1956–1979 with a time step of day, we validated the parameters by the runoff in 1980–2000 and received good results that the relative error is not more than 4%, and the Nash coefficient is larger than 0.8. Eight different scenarios are assumed on the basis of historical observed meteorological data. The main findings are as follows:

(a) The annual runoff changes by different proportions under the temperature change. When the temperature increases, the annual runoff decreases. However, the precipitation change affects the annual runoff to different extents. When the rainfall increases, the annual runoff increases. The proportion of annual runoff change is larger than that of the precipitation change, but all of 45 years are almost at the same level under a climate change.

(b) Under temperature change, the changes of monthly runoff in different months are different during a year. In every spring season, monthly runoff increases when temperature increases. From November or December to January or February of the next year, the monthly runoff response to the temperature is insensitive. From May to October every year, monthly runoff obviously decreases with temperature increase because of the evapotranspiration increase.

(c) The monthly runoff response to precipitation change could be divided into three phases, the dry season phases, the wet season phases, and the transitionary season phases. In the first phase, from November to the next March, the proportion of runoff change is maintained at a similar lower level, which is close to the precipitation change proportion. During the wet season, from June to September, the change proportion is at an upper level, which is about 1.5 to 2.0 times of the lower proportion in the dry season. During three transitionary months, April, May, and October, the change proportion is between the other two phases.

Acknowledgements This study got financial support from the NSFC projects (50779074, 50709041), the National 973 Program of China (2006CB403404), the National 11th Five-Year Science and Technology Supporting Plan of China (2006BAB06B06) and the project of the Special Fund for Scientific Research of IWHR (ZJKF0701).

REFERENCES

Chen, D. & Gao, G. (2003) Impact of climate change on the runoffs from Hanjiang River and Ganjiang River in the Yangtze River Basin. *J. Lake Sci.* **15**(9), 105–114.

Jia, Y., Wang, H. & Wang, J. (2005) Development and verification of a distributed hydrologic model for the Yellow River Basin. *J. Nat. Resour.* **20**(2), 300–308.

Jia, Y., Wang, H., Zhou, Z., Qiu, Y., Luo, X., Wang, J., Yan, D. & Qin, D. (2006) Development of the WEP-L distributed hydrological model and dynamic assessment of water resources in the Yellow River Basin. *J. Hydrol.* **331**, 606–629.

Jiang, T., Chen, Y. & Chen, J. (2006) Study on impact of future climate change on hydrology cycle and water resources in China. *J. Zhongshan University (Natural Science)* **39**, 151–157.

Lan, Y., Wu, Y. & Kang, E. (2001) Response of runoff from the northern slope of the Qilian Mountain to global climatic changes. *Lanzhou University (Natural Sciences)* **37** (4), 121–128.

Lettenmaier, D. & Gan, T. Y. (1990) Hydrologic sensitivities of the Sacramento-San Joaquin river basin, California to global warming. *Water Resour. Res.* **26**(1), 69–86.

Li, D., Tian, Y. & Liu, C. (2004) Distributed hydrology simulation in the headwater area of the Yellow River under environment changes. *J. Geographical Sci.* **59**(4), 565–573.

Liu, C. (1997) Potential impact of climate change on hydrology and water resources in China. *Adv. Water Sci.* **9,** 220–226.

Maidment, D. R. (1992) *Handbook of Hydrology*. 1. McGraw-Hill, New York, USA.

Metoffice (2002) *PRECIS2 Update [M]*. Metoffice, London, UK.

Sellers, P. J., Dickinson, R. E., Randall, D. A., Betts, A. K., Hall, F. G., Berry, J. A., Collatz, G. J., Denning, A. S., Mooney, H. A., Nobre, C. A., Sato, N., Field, C. B. & Henderson-Sellers, A. (1997) Modeling the exchanges of energy, water, and carbon between continents and the atmosphere. *Science* **275**, 502–509.

Schaake, J. C. & Liu, C. (1989) Development and application of simple water balance models to understand the relationship between climate and water resources. In: *New Directions for Surface Water Modeling* (Proc. Symp. Third IAHS Scientific Assembly, Baltimore, May 1989), 343–352. IAHS Publ. 181. IAHS Press, Wallingford, UK.

Stockton, C. W. & Boggess, W. R. (1979) *Geohydrological Implication of Climate Change on Water Resources Development*. US Army Coastal Engineering Research Center, Fort Belvoir, Virginia, USA.

Wang, G. & Wang, Y. (2000) Sensitivity analysis about runoff of unper and middle of the Yellow River to climate change. *Water Resour. & Water Engng* **3**(11), 1–5

Yang, D., Li, C., Hu, H., Lei, Z., Yang, S., Kusuda, T., Koike, T. & Musiake, K. (2004) Analysis of water resources variability in the Yellow River of China during the last half century using historical data. *Water Resour. Res.* **40**, W06502, doi:10.1029/2003WR002763.

Yuan, F., Xie, Z. & Ren, L. (2005) Hydrological variation in Haihe River Basin due to climate change. *J. Hydraul. Engng* **36**(3), 274–279.

Ecological water resources requirement of the Ningxia natural oasis using the groundwater level variation amplitude method

XIAORONG HUANG[1], XINHAI ZHANG[2] & CAIWANG ZHAI[2]

1 *College of Water Resource & Hydropower, Sichuang University, no. 24, South Section, Yihuan Road, Chengdu 610065, China*
hxiaorong@tom.com

2 *Yellow River Engineering Consulting Co., Ltd, Zhengzhou 450004, China*

Abstract This paper presents the groundwater level variation amplitude method, which is used to calculate the Ningxia ecological water resource requirement. The information of annual groundwater level variation from observation wells is used to search for the yearly variation value of the closed groundwater flow field. The ecological water resources requirement of Ningxia natural oasis is estimated on the basis of superposition of the yearly variation value of the closed groundwater flow field, remote sensing vegetation data, and hydrogeological data. This method overcomes the insufficient quantum for ecological water demand to a certain degree, and provides a new concept for the calculation of ecological water resources requirement and for ecological protection and recovery in the northwest arid and semi-arid areas in China.

Key words hydrology; ecological water resources requirement; natural oasis; groundwater

INTRODUCTION

Ningxia is an area with a serious lack of water. The groundwater resources balance is kept mainly by the Yellow River water constantly flowing across the area, and this will go on in the future. The excess agricultural water consumption takes up much of the water for ecological environment construction. The historical "72 connected lakes" have gradually withered or even disappeared. The desertification and salinization of soil, retreat of vegetation, reduction of area and quality of farmland lead to Ningxia's small environmental capacity and poor stability of the ecosystem, and the ecological environment is in a severe worsening trend. Water is one of the most sensitive limiting factors for the combination body of nature – society – economy. In arid areas water is one of the "bottleneck" factors which limits socio-economic development and is also the key driving factor for ecological evaluation. Study of the Ningxia ecological water resources requirement based on ecological environment protection can determine the minimum value for the water supply. This is of great importance for protecting the Ningxia ecological environment, maintaining the ecological stability of oasis, even further, coordinating the reasonable distribution of economic water demand and ecological water demand, and promoting the construction of ecological water-saving society in Ningxia Province.

DISCUSSION ON CALCULATION METHODS FOR ECOLOGICAL WATER RESOURCES REQUIREMENT OF OASIS

Ecological water resources requirement has become the hot point of study. Some of the domestic scientists have carried out further and more extensive discussion on the ecological water resources requirement concept in the northwest arid area of China. For the weak ecological environment area, ecological water resources requirement shall refer to the surface and groundwater resources required for preventing the ecological environment from getting worse and for improving the ecological environment. For desert oasis, ecological water resources requirement shall refer to the water volume consumed by the system which supports the stability and development of oasis scenery, and supports sustaining and improving environment quality of oasis (Jia, 1998). For arid and semi-arid areas, ecological water resources requirement refers to the water volume consumed by the system, which satisfies the aim of ecological environment protection and supports sustaining the scenery and improving environment conditions for the purpose of maintaining its balance (Wang, 1999). In "Study of Water Resources and Ecological Environment Assessment in Northwest Arid Area in China", which is one of the important scientific brainstorm projects of the country in the 9th Five-year Plan, the operative calculation method for "ecological environment water resources requirement" is put forward for the first time definitively. With the help of GIS and remote sensing technology, Wang Fang and others calculated the vegetation water resources requirement on the base of establishing the theory of zonality and non-zonality of vegetation water requirement.

At present, there are many methods to calculate ecological water resources requirement, and they can be divided into two groups: one is called the "area-quantum method" and the other is called the "groundwater level variation amplitude method". The basic concept of the "area-quantum method" is: when the ecologic water resources requirement of a certain area is studied, the types of vegetation and the sorts of area shall be determined first, then the water demand quantum for each sort of vegetation shall be determined, and finally the ecologic water sources requirement of the area can be calculated. Though the "area-quantum method" has been widely used now, it has some problems. The first is the problem in determination of "quantum". The biological water demand of vegetation is a complicated and difficult problem in botany itself, making the accurate determination of water requirement quantum for all sorts of vegetation difficult. The quantum is usually determined with the Penman equation, which is recommended by the Food and Agriculture Organization of the United Nations (FAO) at present and the parameters can be obtained through tests. The second is the problem of scale. Even if the quantum for vegetation water resources requirement can be obtained accurately, the ecological water resources requirement laws at different scales are not the same, the water resources requirement of the same vegetation in different geographic environments is not the same, either, because of the spatial variability of underlying surface factor, hydrologic parameters, etc. This leads to the difficulty in quantum selection and it is not possible to determine various quanta for various types of vegetation or for various environments in the same area. What is more is that this kind of law is not a simple linear extension or superposition. The way to apply the study findings of small scale to large scale, the mechanism of ecological water resources requirement and the conversion of study findings of different scales have not been effectively further studied, and most of them are qualitative conclusions. The third is the problem of "practical value". Ecological water resources requirement is a variate which varies with the differences of time and location. From the view of ecological water resources requirement study at home, it can be seen that there are more qualitative descriptions while mechanism studies are fewer. It is of practical value to study the maximum duration when the ecosystem is threatened by water and the water requirement features of the ecosystem at the duration, and to calculate the water volume required for reasonably maintaining the ecosystem, particularly for arid and semi-arid areas,. Otherwise, the instructive function to the study of reasonable distribution of economic water consumption and ecological water consumption in arid and semi-arid areas and the significance levels are not great.

The basic concept of "groundwater level variation amplitude method" is to determine the groundwater level required for the growth of each sort of vegetation in the area at first, then to determine the minimum water volume for maintaining this groundwater level, and this water volume is the ecological water resources requirement in this area. "Groundwater level variation amplitude method" is more used in arid and semi-arid areas. Because of the small precipitation in this area, surface water is lacking, but if the groundwater level can be kept at the proper level, vegetation can grow normally. From this view, the "groundwater level variation amplitude method" is a scientific one. According to the actual conditions in Ningxia, it is better to adopt this method for the calculation of ecological water resources requirement.

ANALYSIS OF NINGXIA GROUNDWATER RESOURCES BALANCE

Except for in the southern mountain area and the Helan Mountain area, no runoff yield or confluence is formed, basically because of the rare precipitation in Ningxia. The groundwater

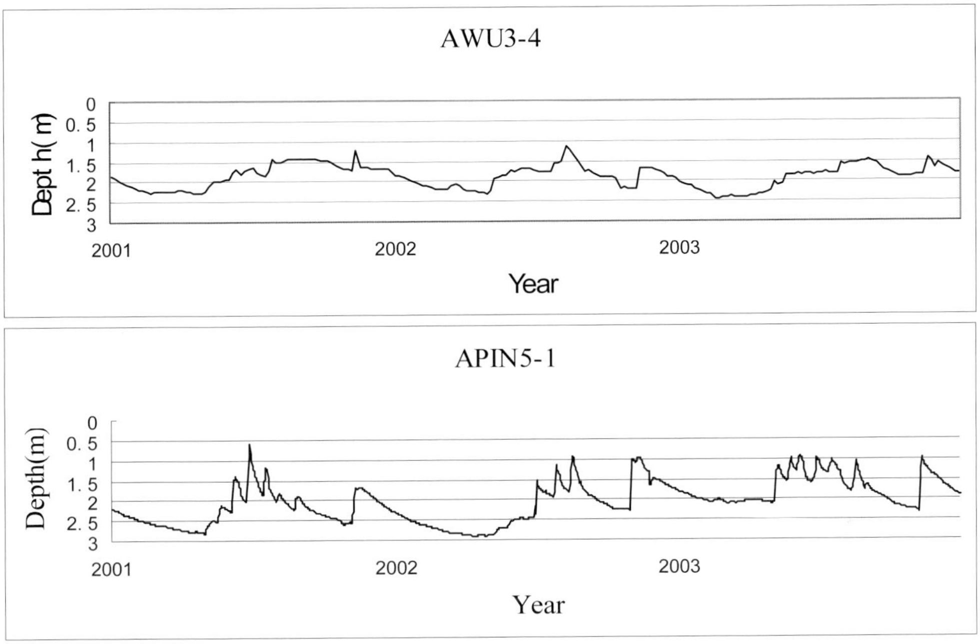

Fig. 1 Ningxia well depth hydrograph (data supplied by Ningxia Hydrological Management Bureau).

compensation mainly comes from the irrigation water from the Yellow River between April and June. The lowering and insufficiency of groundwater results from two aspects. One is phreatic water evaporation and vegetation transpiration. The other is inter-exchange and inter-movement of groundwater flow fields between the regions. Therefore, the annual mean variation of groundwater depth is analysed. The general trend is that the highest water level appears around the period between July and August and the lowest level appears between February and March, see Fig. 1.

Comparison is made between the above two figures. In Fig. 1 APIN 5-1, the large fluctuation of groundwater depth in some months in a year indicates a frequent compensation of groundwater. Besides the infiltration from periodical surface irrigation runoff, the inter-compensation between groundwater flow fields appears very frequently, which shows the groundwater flow field in the area is not closed. In Fig. 1, AWU 3-4, the fluctuation of groundwater level in a year is basically flat with only one fluctuation. This shows that the agricultural irrigation in this area is concentrated from April to June, and irrigation water is transferred to groundwater after a period of time, leading to the rising of groundwater level. The groundwater depth around July is the shallowest and the depth from the next February to March is the deepest. The regular variation of groundwater level just tallies with the regular water demand for plant growth, showing that groundwater is mainly consumed by phreatic water evaporation and vegetation transpiration, and that the groundwater flow field in this area is relatively closed. This part of groundwater consumption is considered as the ecological water resources requirement of vegetation in this area.

CALCULATION CONCEPT OF ECOLOGICAL WATER RESOURCES REQUIREMENT OF NINGXIA NATURAL OASIS

According to the general concept of ecological water resources requirement, the land utilization map on a scale of 1:100 000 was obtained from TM remote sensing information in 2000, with resolution ratio of 30 m at first, then the different types of Ningxia ecosystems (e.g. forest, brush, grass, lake, wetland, etc.) are distinguished and separated. Because the digital map may have errors from initializing itself, proper correction is conducted to each type of area in remote sensing information with the reference of area statistic results in Ningxia. According to the determined scale of ecological protection and construction, the water resources requirement of Ningxia oasis is estimated indirectly through using and analysing the observation data in years about the variation of groundwater depth on the base of yearly level variation value of the closed groundwater flow field. The processes are shown in Fig. 2.

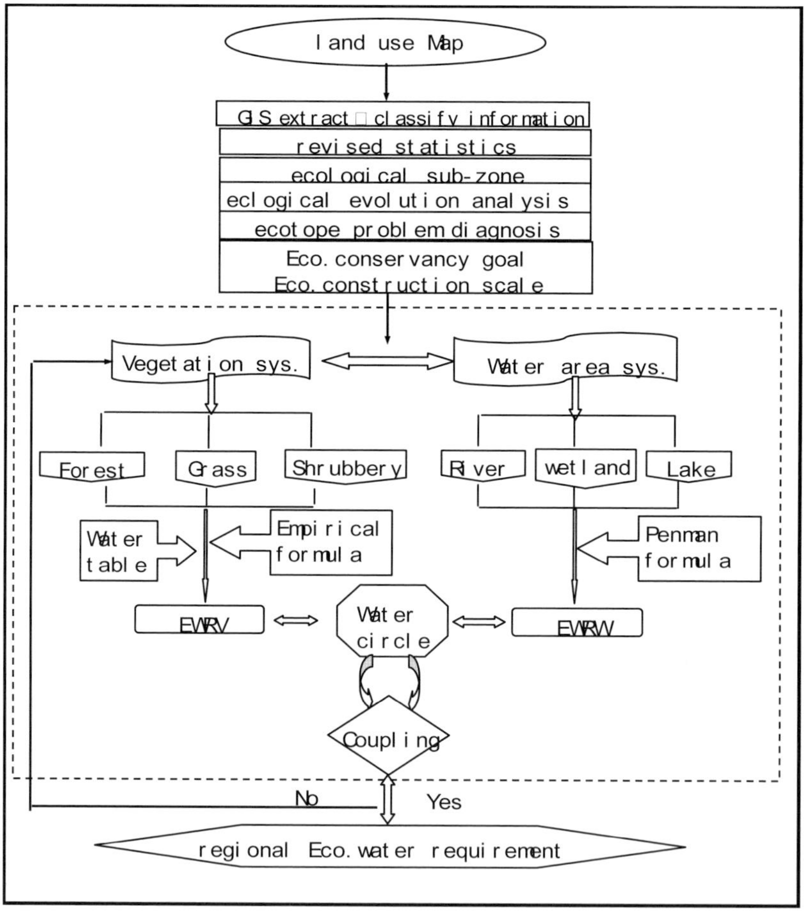

Fig. 2 Flow chart of calculation of Ningxia ecological water resources requirement. Notes: EWRV, ecological water requirement of vegetation; EWRW, ecological water requirement of water body.

CALCULATION PROCESS OF ECOLOGICAL WATER RESOURCES REQUIREMENT OF NINGXIA NATURAL OASIS

Water volume required by natural oasis

The conversion of "the four kinds of water" in Ningxia is very complicated, and it is not easy to obtain the desirable variation data for groundwater depth in the ecological sub-zone. The working procedures are: (1) locate on observation well in each ecological sub-zone; (2) collect the statistical information of annual groundwater depth from each observation well in each ecological sub-zone; (3) take the information from observation well with the flat depth variation only one time of rising and lowering trend as the effective analysis information; (4) ply the Ningxia Groundwater

Depth Curve (omitted), Distribution Map of Ningxia Groundwater Observation Wells (omitted), Distribution Map of Ningxia Aerated–zone Lithological Character (omitted) and Map of Classified Coverage of Ningxia Ecological Vegetation (omitted). The groundwater depth variation corresponding to each ecological sub-zone of Ningxia natural oasis is shown in Table 1, and the specific yield corresponding to the observation well in each ecological sub-zone is shown in Table 1. The specified yield in an observation well is determined by geological structure of the aerated zone. The artificial oasis area is deducted from the ecological sub-zone area. The ecological water resources requirement of Ningxia natural oasis vegetation can be calculated through superposition of the spatial vegetation data from remote sensing, the corresponding groundwater variation amplitude, and the specific yield, etc. The equation can be written as follows:

$$W_{EWRV} = \mu \times H_V \times F \tag{1}$$

where W_{EWRV} is ecological water resources requirement; μ is specific yield; H_v is yearly maximum variation amplitude; and F is area of ecological sub-zone.

The detailed calculation is shown in Table 2.

Table 1 Characteristics of Ningxia Annual Mean Groundwater Depth Value.

Well no.	Corresponding ecological sub-zone	Minimum value (m)	Month	Maximum value (m)	Month	Yearly maximum variation amplitude (m)
Z_{LIN-1}	Land with woods	1.89	Aug.	4.01	Apr.	2.12
S_{Y-13}	Bush	0.72	Jun.	2.7	Mar.	1.98
Z_{LING-8}	Land with scattered woods	1.51	Sept.	3.23	Mar.	1.72
S_{YIN-4}	Grassland with high coverage	1.32	Oct.	2.73	Apr.	1.41
A_{WU3-4}	Grassland with medium coverage	1.15	Sept.	2.41	Mar.	1.26
Z_{LIN-8}	Grassland with low coverage	0.82	Jun.	1.1	May	0.28

Table 2 Calculation of ecological water resources requirement of Ningxia natural oasis.

Ecological Sub-zone	Area (hm^2)	Yearly maximum variation amplitude (m)	Specific yield	Water requirement (10^8 m^3)
Land with woods	6 652	2.12	0.120	0.17
Bush	23 693	1.980	0.09	0.40
Land with scattered woods	9 080	1.72	0.106	0.17
Grassland with high coverage	57 980	1.41	0.100	0.82
Grassland with medium coverage	95 623	1.26	0.095	1.14
Grassland with low coverage	179 006	0.28	0.100	0.50
Total	372 035			3.20

Calculation of ecological water resources requirement of river and lake water bodies

The rivers in Ningxia are all seasonal except for the Yellow River and the Jinghe River in the southern mountain area. The ecological water resources requirement in channel only considers the water surface evaporation. When the water surface evaporation is more than precipitation, the difference value between channel water surface evaporation and precipitation is the water resources requirement of water surface evaporation of the water body if the channel has a water surface. When precipitation is more than evaporation, the ecological water resources requirement of evaporation is considered as zero. Therefore, the ecological water resources requirement of evaporation of each corresponding month can be calculated according to the water surface area, precipitation and water surface evaporation. The calculation equation is:

$$\begin{cases} W_E = A(E_0 - P) & \text{if} \quad E > P \\ W_E = 0 & \text{if} \quad E < P \end{cases} \tag{2}$$

in which W_E is the ecological water resources requirement of water surface evaporation; A is mean water surface area of each month (km^2); E_0 is mean evaporation capacity of each month (mm).

Total water resources requirement of Ningxia natural oasis

Based on the summary of each individual water resources requirement, the water resources requirement of natural vegetation all through the year is distributed according to its growing rules (approximately according to crop growing rules). The distribution of total ecological water resources requirement of each month is shown in Table 4.

Table 3 Study results of ecological water resources requirement for water body of Ningxia River and lake.

Month	1	2	3	4	5	6	7	8	9	10	11	12	Total
Precipitation (mm)	3.75	5.49	3.37	4.37	4.25	66.1	23	98.4	20	7.49	4.37	0	241
Evaporation (mm/d)	0.7	1.48	2.82	5.06	6.25	6.71	6.66	5.59	3.80	2.24	1.01	0.27	1300
River and lake (hm^2)	13 000	14 200	14 300	14 500	15 030	15 400	15 500	15 400	14 300	14 400	14 100	13 300	
Water requirement (10^8m)	0.02	0.04	0.12	0.16	0.29	0.13	0.28	0.12	0.09	0.09	0.03	0.01	1.37

Table 4 Ecological water resources requirement in Ningxia.

Month	1	2	3	4	5	6	7	8	9	10	11	12	Total
Water requirement 10^8m	0.07	0.15	0.36	0.56	0.86	0.52	0.79	0.46	0.33	0.29	0.12	0.05	4.57

CONCLUSION

With the consideration of high groundwater and rare precipitation in the Yinchuan Area of Ningxia, the "groundwater level variation amplitude method", which is proper to calculate Ningxia ecological water resources requirement, is put forward. The information of annual mean groundwater level variation from observation wells is used to search for the yearly variation value of the closed groundwater flow field. The ecological water resources requirement of Ningxia natural oasis is estimated on the base of superposition of the yearly variation value of closed groundwater flow field, remote sensing vegetation data, and hydrogeological data. This overcomes the insufficient quantum for ecological water demand to a certain degree and provides a new concept for calculation of ecological water resources requirement and for ecological protection and recovery in the northwest arid and semi-arid areas of China.

REFERENCES

Arnold, J. G., Srinivasan, R., Muttiah, R. S. & Williams, J. R. (1998) Large area hydrologic modeling and assessment Part I: model development. *J. Am. Water Resour. Assoc.* **34**, 73–89.

Liu, Changming, Wang, Lixian & Xia, Jun (2004) *Study of Water Resources Distribution, Ecological Environment Construction and Sustainable Development Strategy (Ecological Environment Volume)*, 140–141. Science Publishing House, China.

Wang, Fang (2002) Study of ecological water resources requirement in northwest China (2) – calculation and analysis of regional ecological water resources requirement based on technology of remote sensing and Geographical Information System. *Acad. J. Nat. Resour.* **17**(2), 129–137.

Wang, Xiqin, Liu, Changming & Yang, Zhifeng (2002) Progress and prospect of study of ecological and environmental water resources requirement. *Progr. Water Sci.* **13**(4), 507–512.

Wang, Fang, Liang, Ruiju, Yang, Xiaoliu, *et al.* (2002) A study of ecological water requirements in northwest China I: theoretical analysis. *Acad. J. Nat. Resour.* **17**(1), 1–8.

Xia, Jun & Sun, Xuetao (2002) Challenge ahead of study of ecological water resources requirement in western area, China. *Water Conservancy* **5**, 57–60.

Zhang, Li, Dong, Zengchuan & Zhao, Bin (2003) Calculation method for ecological water resources requirement of natural vegetation in arid area. *Progr. Water Sci.* **11**, 745–748.

Mapping the average annual runoff depth in Huangshui watershed using DEM

SHAOFENG JIA[1], SHIJUN LIN[1,2] & YAJUN WANG[3]

1 *Institute of Geographical Sciences and Natural Resources Research, Chinese Academy of Sciences, Beijing 100101, China*
jiasf@igsnrr.ac.cn

2 *Graduate School of the Chinese Academy of Sciences, Beijing 100039, China*

3 *Bureau of Science and Technology of Qinghai Province, Xining 81001, China*

Abstract We found that precipitation and the surface water flows generally increase with elevation, as does the vegetation cover, in Huangshui watershed. This determined that the geographical variables, such as elevation and vegetation cover, can be used to improve the precision of mapping of runoff depth. We focus on discovering an automatic approach for mapping runoff using geographical variables. Using multiple linear regression techniques we create a regression equation that relates runoff depth to altitude and relative distance from the source of water vapour. Data used for this study include runoff data measured at 13 stations, Landsat TM data and DEM data; a 50 m × 50 m grid of runoff depth was created using those models. We used the runoff depth data of 13 drainage zones for validation. Comparing with the empirical model used by local water resource survey institutions, the regression approach appears to give more precise results.

Key words runoff depth; multivariate linear regression; geographical variables; DEM

INTRODUCTION

Runoff depth is one of the most important hydrological data. Mapping runoff depth across space constitutes one of the fundamental tasks in hydrology. The only direct source of information for completing this task is hydrological observation data. However, hydrological series are sometimes too few and/or too short for a reliable determination of runoff statistic characteristics, or even not available at all in ungauged areas. According to the water balance, some studies deduce runoff from simulated spatial distribution of precipitation and evapotranspiration, using the more abundant precipitation data and a rainfall–evapotranspiration model (Morton, 1983). Some studies model the spatial distribution of runoff depth by runoff curve number based on the relation between land use and infiltration and evaporation using remote sensing data (Melesse & Shih, 2002). In recent years, distributed hydrological models using GIS and RS have undergone great development and are used in simulating the discharge of watersheds. Some of these models can obtain runoff information in spatial computation unit (grids or representative unit). TopModel (Beven, 1995) and SWAT (Neitsch *et al.*, 2000) are two typical distributed models. But all these models need vast amounts of basic data as input. In areas lacking basic data, the model can not work. Another method of spatial mapping is interpolation. A number of interpolation methods that have been used to model the spatial distribution of runoff depth include IDW, spline, kriging, etc. Usually, every subwatershed's average runoff depth is associated with the centroid of each subwatershed. Then an interpolation method is processed based on those points to model the spatial distribution of runoff depth. However, interpolation methods only consider spatial relationships among sampling points, and do not take into account other properties of the landscape (Gottschalk, 1993; Sauquet *et al.*, 2000).

This study sought to map runoff depth across a mountainous region of Huangshui catchment, in Qinghai province of China, in terms of the relationships between runoff depth and a range of geographic variables, including altitude, distance from water vapour intake, and vegetation index. Using multiple linear regression techniques, we create a model to disaggregate the mean annual streamflow measured at the gauge station to basin area. This model is compared to other mapping methods.

METHODS AND DATA

Study area

Huangshui watershed (only the part in Qinghai province, not including the Datong River watershed) is located in the east of Qinghai province. Huangshui River is a first-class tributary of the Yellow River, flowing from northwest to southeast. The Huangshui watershed covers an area of 16 120 km^2, has a shape of a leaf, located in between 36°02′–37°28′N latitude, 100°42′–103°04′E longitude. There are 13 stream-gauging stations in operation in Huangshui watershed (Fig. 1). The annual average runoff of Huangshui watershed is 2.16 × 10^9 m^3, the average flow rate is 65.7m^3/s, and the annual average runoff depth is 138 mm.

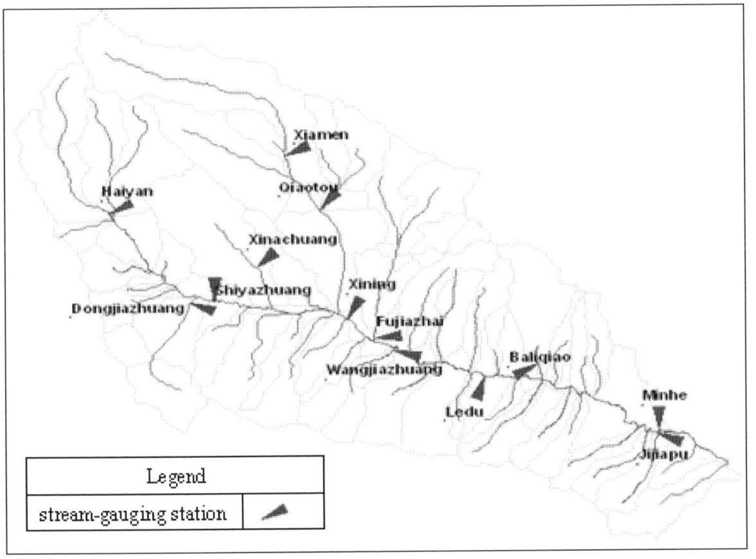

Fig. 1 Location of the study area and positions of the 13 stream-gauging stations.

Huangshui watershed is affected by a southwest water vapour, rising from the Indian Ocean Bay of Bengal. The nearest available sources of moisture are very far from Huangshui watershed. The climate there is classified as semi-humid, with an average annual precipitation of 486.6 mm. The annual precipitation distribution there is very uneven, with a Cv from 0.15 to 0.30. The upstream area receives much more precipitation than the downstream area. The annual precipitation in a valley area between Xining and Minhe varies from 250 mm to 350 mm, while more than 600 mm precipitation is received in the mountain area per year.

Huangshui watershed has a very inclining hypsography, with altitudes ranging from 4900 to 1650 m, rising gradually from southeast to northwest. The minimum elevation is located in the valley near the Gansu-Qinghai border. There are various types of topography in Huangshui watershed characterized by mountains, hills covered by loess and valley plain.

Data

Along the river, we divide the study area between every two gauges. This generates 13 nonoverlapping drainage zones. We used a period of 42 years, between 1958 and 2000, to calculate the average annual runoff depth of every drainage zone (Table 1).

In addition, a digital elevation model (DEM, 50 m × 50 m) and a Landsat TM image (Summer 2000) of Huangshui watershed were used in this study.

Table 1 Average annual runoff depth of drainage zones in Huangshui watershed.

Drainage zone	Drainage area (km^2)	Discharge ($10^9 m^3$)	Runoff depth (mm)
Xinachuan	809	1.633	201.85
Xiamen-Qiaotou	1466	2.696	183.92
Baliqiao	464	0.952	205.26
Xiamen	1308	3.617	276.55
Haiyan	715	0.473	66.21
Haiyan-Shiyazhuang	1732	1.786	103.12
Dongjiazhuang	636	0.84	132.10
Ledu-Minhe	1853	2.133	115.11
Fujiazhai	1112	2.056	184.93
Xining-Ledu	2521	2.506	99.39
Wangjiazhuang	370	0.437	118.10
Jijiabao	192	0.331	172.40
Shiyazhuang-Xining	2356	2.147	91.13

METHOD

Firstly, as in general interpolation methods, we assume that the runoff depth of the centroid point of a drainage zone equals the average runoff depth of the drainage zone and map runoff depth by using the normal interpolation methods (IDW, spline, and kriging). Then we qualitatively revised the contour based on the empirically perceived relation between runoff depth and elevation and vegetation cover. The result shows that the revised mapping is much better than the original interpolated mapping. So we want to construct a mathematical model to use the relations between runoff depth and altitude, vegetation index and other geographical variables to map runoff depth. The multivariate regression method is used in this process. SPSS and ArcInfo are used for regression analysis and spatial mapping separately.

RESULTS AND DISCUSSION

Interpolation of runoff depth

There are three kinds of interpolation methods available in GIS software: IDW, spline, and kriging. Those interpolation methods have been applied to Huangshui watershed by using ArcInfo. The standard deviation of residuals from interpolation methods are shown in Table 2, which shows the spline interpolation, obtains the best result.

Table 2 SD and SDORR of IDW, spline and kriging.

Interpolation method	SD (mm)	SDORR (%)
IDW	27.60	23.05
Spline	17.58	13.42
Kriging	27.36	23.29

SD, the standard deviation of residuals from estimated model. SDORR, the standard deviation of relative residuals from estimated model.

Revision of contour lines based on DEM and Landsat TM

Based on the spline contour, we revised the runoff depth contour by using DEM data. We use a criterion to revise the runoff depth contour that a runoff depth contour line should be nearly paralleled with the elevation contour. The revised runoff depth contour is shown as Fig. 2.

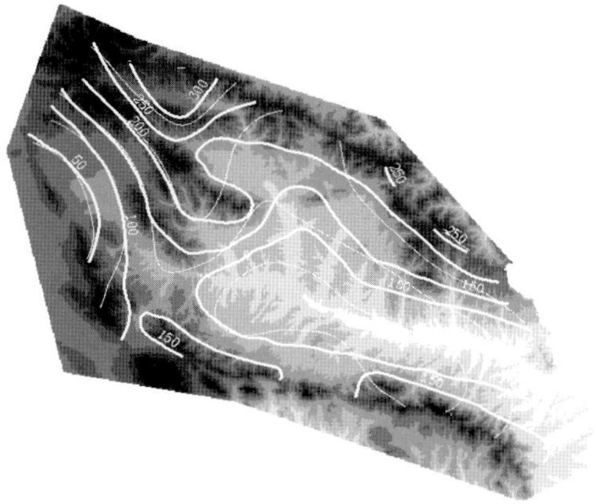

Fig. 2 Revised based on DEM. White line: after revised; white thin line: before revised.

Table 3 SD and SDORR after revision based on DEM.

Runoff mapping method	SD (mm)	SDORR (%)
Spline	17.58	13.42
Revised based on DEM	17.97	12.14

After the contour lines were revised based on the DEM, the residuals became much smaller than those in the spline interpolation model (Table 3). The standard deviation of relative residuals dropped from 13.42% to 12.14%, though the standard deviation of residuals was increased a little (from 17.58 to 17.97). The precision of the estimated values of the relative small runoff depth is improved, apparently at the cost of a little less precision of the relative big runoff depths.

Vegetation is a very important part of the environment. The vegetation can usually reflect the local landform, soil, geology, climate and elevation. Generally, the vegetation cover is thick in the places where precipitation is abundant, and is sparse in the areas lacking precipitation. So, we can assume that more precipitation is expected in the area where the vegetation cover is thick. And we can qualitatively determine that in an area where the vegetation cover is thick, more precipitation is received and runoff is also abundant; and *vice versa*. Based on the vegetation information on Landsat TM images, we revised the runoff depth contour further. Let the distribution of the contour be consistent with the vegetation coverage and get the revised runoff depth contour as shown in Fig. 3. The validation results shows that the revised contour improved further (Table 4).

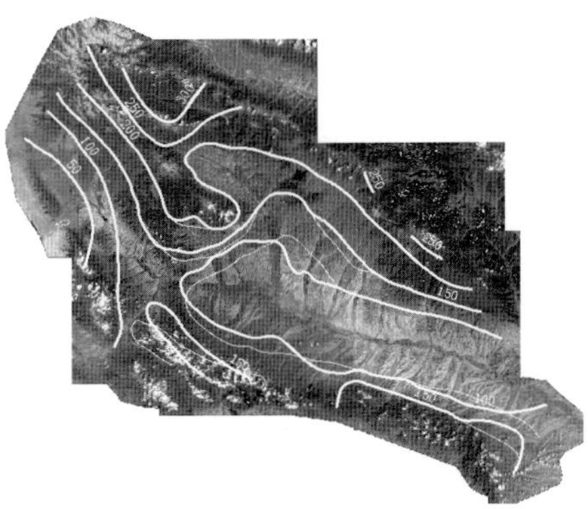

Fig. 3 Revised based on Landsat TM. White line: after revised; white thin line: before revised.

Table 4 SD and SDORR after revision based on Landsat.

Runoff mapping method	SD (mm)	SDORR (%)
Spline	17.58	13.42
Revised based on DEM	17.97	12.14
Revised based on Landsat TM	16.52	11.74

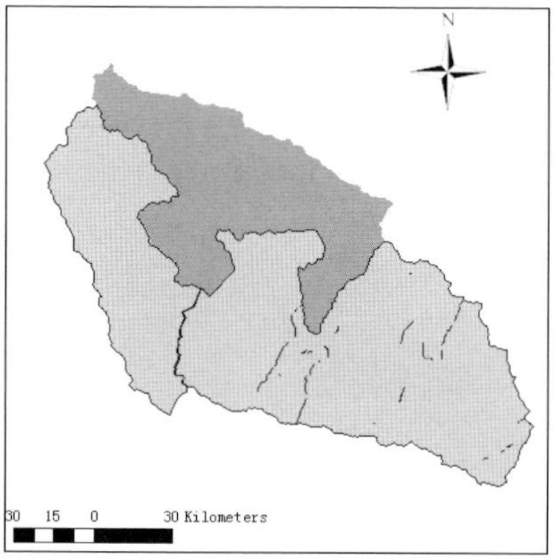

Fig. 4 divided the study area into two parts.

Regression model

According to the statistical characteristics of the runoff depth, we divided Huangshui watershed into two parts: (1) the southern part; (2) the northern part (Fig. 4). The northern part includes Xinachuan, Xiamen-Qiaotou, Xiamen and Fujiazhai drainage zones. The others are included in the southern part. In the regression analysis we model the spatial distribution of runoff depth for the northern part and the southern part, respectively.

The relationships between runoff depth and the independent topographic variables generally fit squared or cubic or higher-power models better than linear ones. So we introduced the squared, cubic and fourth order values of the independent variables to achieve a better fit of the theoretical curve to the data.

We used DEM to provide estimates of elevation, and the Landsat TM data to derive vegetation index. We computed the distance from a drainage zone's centroid to the south side of the watershed (dsouth (m)), and to the east side (deast (m)). The average runoff of the centroids of the 13 subwatersheds described previously was used for multivariate regression analysis using SPSS software.

We took the average runoff depth of a drainage zone as the independent variable, and took the average elevation of a drainage zone (H), the higher power of H (H^2, H^3, and H^4), the average vegetation index of a drainage zone (V), deast, $deast^2$, $deast^3$, $deast^4$, dsouth, $dsouth^2$, $dsouth^3$ and $dsouth^4$ as dependent variables.

For the regression, we followed the criterion of the stepwise approach, selecting the variables backwards. This method has the advantage that a variable selected in one step can be eliminated in another later step. In this way, initially all the variables are introduced in a single step and then they are discarded one by one, based on the outset criteria.

Pearson correlation coefficients for the relationships between runoff depth and the independent variables in the southern area are shown in Table 5. The best runoff depth was estimated with model 2 in Table 5, which is expressed as equation (1).

$$R = 1.71E - 12*H^4 - 0.001*deast.73.891 \quad (1)$$

For the north area, the final regression model is expressed as equation (2).

$$R = -5.2E - 19*dsouth^4.132.42 \quad (2)$$

Statistic test values for this model are shown in Table 6.

Table 5 Statistic values of regression models of southern area.

Model	R	R square	Adjusted R square	Std error of the estimate
1	0.830(a)	0.690	0.651	34.50673
2	0.910(b)	0.829	0.780	27.39156

Table 6 Statistic values of regression model of northern area.

Model	R	R square	Adjusted R square	Std error of the estimate
1	0.999(a)	0.998	0.996	2.12078

Table 7 Estimated average annual runoff depth by regression model.

Drainage zone	Observed (mm)	Estimated (mm)	Difference (mm)	Difference (%)
Xining–Ledu	99.39	118	–18.15	–18.26
Dongjiazhuang	132.1	132	0.52	0.39
Wangjiazhuang	118.1	105	13.39	11.34
Baliqiao	205.26	234	–29.07	–14.16
Xiamen	276.55	250	26.56	9.60
Haiyan	66.21	62	4.21	6.36
Xinachuan	201.85	151	50.95	25.24
Haiyan–Shiyazhuang	103.12	97	6.42	6.23
Shiyazhuang–Xining	91.13	78	13.37	14.67
Jijiabao	172.43	132	40.32	23.39
Xiamen–Qiaotou	183.92	148	35.90	19.52
Ledu–Minhe	115.11	135	–19.83	–17.22
Fujiazhai	184.93	144	41.31	22.34
SD			25.48	15.26

We have noticed that there is no apparent relationship between R and H in the north part. The reason may be that the altitude of the north changes more slowly than the south, the distance from the source of water vapour becomes the major factor of runoff generation in the north part.

Initially, some vegetation variables were included in the regression analysis. But all the vegetation variants were finally eliminated. There are two reasons for the absence of vegetation variables. First, vegetation is strongly correlative with elevation. So when elevation is used in the regression model, the vegetation variables are no longer necessary. Secondly, the main vegetation in the valley plain area is crop, and the crop is hardly correlative with runoff depth.

Residual analysis

The resolution of the DEM used in this study is 50 m. So we can create a 50 m × 50 m grid of runoff depth using the regression model. The value of each grid cell represents the estimated runoff depth of a special 50 m × 50 m area. The observed runoff depth of each 50 m × 50 m area is not available. We can not use the grid to assess the validity of the models directly. But the observed runoff depth data of the 13 drainage zones are available. And the estimated runoff depth of each drainage zone can be easily calculated by using the grid values. So we can use the runoff depth of the 13 drainage zones for validation.

The comparison of the estimated runoff depth of the 13 drainage zones using the regression model and the observed ones are shown in Table 7. The standard deviation of residuals is 25.48 mm, and the standard deviation of relative residuals is 15.26%.

Table 8 SD and SDORR of the six runoff mapping methods.

Runoff mapping method	SD (mm)	SDORR (%)
IDW	27.60	23.05
Spline	17.58	13.42
Kriging	27.36	23.29
Regression	25.48	15.26
Revised manually	16.52	11.74
Empirical model used by local institutions	29.01	18.52

When the runoff depth contour is available, we can create a grid of runoff depth by using GIS software. In order to compare the interpolation models with the regression model, we created a 50 m × 50 m grid of runoff from the runoff depth contour. And each grid created was used to validate the performance of each model. The standard deviation of residuals and relative residuals in the six methods have been calculated and are presented in Table 8.

CONCLUSION

In this study, we succeed in improving the precision of mapping of runoff depth using DEM and related geographic data. By comparing various models we can reach the following conclusions:

(a) The mapping results of runoff depth, computed by using the normal interpolation methods included in GIS software, appear to be better than the empirical mapping which is still used in the local hydrological bureau, according to SD (mm). This indicated that using the GIS software to map runoff depth can not only improve the efficiency, but can also get a more reliable result.

(b) The geographical variables, such as elevation and vegetation cover, can be used to improve the precision of mapping of runoff depth.

(c) Using multiple linear regression techniques we create a regression model that relates runoff depth to altitude and distance from the source of water vapour. And the regression model can achieve a similar precision as that from the normal interpolation methods. But it fails to greatly improve the precision. The reason is that the relation between runoff depth and geographical characteristics is very complicated and it is not always possible to use one regression function to show these complex relations.

(d) One potential merit of the regression model is that it may be extended and used to predict runoff depth in ungauged basins with similar geographical conditions, while interpolation methods do not have this function. Because of the geographical difference between the north and the south in the study area, we used two different equations even in the same basin. This shows that it is very difficult to assess the geographical similarity between various areas, and it needs more effort and study to find a feasible method to decide whether an area's runoff depth can be estimated by the runoff depth equation of another area.

Acknowledgements The research is supported by the National Natural Science Foundation of China (Grant 90302009), and the Institute of Geographical Sciences and Natural Resources Research, Chinese Academy of Sciences (CXIOG-A05-13).

REFERENCES

Beven, K. J. (1995) TOPMODEL. In: *Computer Models of Watershed Hydrology*, Chapter 18 (ed. by V. P. Singh). Water Resources Publications, Littleton, Colorado, USA.

Gottschalk, L. (1993a) Correlation and 5 covariance of runoff. *Stochastic Hydrol. Hydraul.* **7**, 85–101.

Gottschalk, L. (1993b) Interpolation of runoff applying objective methods. *Stochastic Hydrol. Hydraul.* **7**, 269–281.

Melesse, Assefa, M. S. & Shih, F. (2002) Spatially distributed storm runoff depth estimation using Landsat images and GIS. *Computers and Electronics in Agriculture* **37**, 173–183.

Morton, F. I. (1983) Operational estimate of areal evapotranspiration and their significance to the science and practise of hydrology. *J. Hydrol.* **66**(1–4), 1–76.

Neitsch, S. L., Arnold, J. G., Kiniryj, R. *et al.* (2000) Soil and Water Assessment Tool Theoretical Documentation / Version 2000 [EB.OL] http://www.brc.tamus.edu/swat/.

Sauquet, E., Gottschalk, L. & Leblois, E. (2000) Mapping average annual runoff: a hierarchical approach applying a stochastic interpolation scheme. *Hydrol. Sci. J.* **45**(6), 799–815.

2-D hybrid approach to storm flood modelling in the Shalan catchment

CHONG LI[1], JIKANG LU[2] & BAISHENG YE[3]

1 *Department of Water Environment, China Institute of Water Resources and Hydropower Research (IWHR), Beijing 100038, China*
lichong@iwhr.com

2 *Department of Flood Disaster, China Institute of Water Resources and Hydropower Research, Beijing 100038, China*

3 *Cold and Arid Regions Environmental and Engineering Research Institute, Chinese Academy of Sciences, Lanzhou Gansu 730000, China*

Abstract A hybrid approach to storm flood modelling has been introduced and applied to a flood simulation in the Shalan catchment in Heilongjiang Province, northeast China, by combining a simplified hydrological rainfall model with a 2-D hydraulic runoff model. When combining, the scale matching and numerical instability should be taken into account and solved. In the application a rainfall process has been constructed based on field surveys and quantified flooding facts. One of the parameters in its expression should be inversely determined. Digital Elevation Models (DEM) and their extracted river network provide a strong geological base to enhance rainfall and runoff simulation. The practical simulation results agree well with collected flood data in Shalan. It turns out to be one of the most useful tools to simulate storm floods in flat plains or uneven mountain areas.

Key words DEM; hybrid approach; rainfall; runoff; Shalan catchment

INTRODUCTION

A heavy storm occurred in Shalan valley, a very small mountainous area in Heilongjiang Province, northeast China, on 10 June 2005, causing the most severe disaster in 100 years of this area's history, often referred to as a "6.10" storm. A significant flood suddenly formed and surged into a local village and a primary school. A total of 109 people lost their lives in the tragedy, including 105 children. They had no chance to escape from the violent current because the school was located in a lower part of the valley terrain and maximum water depth at the school reached up to 2 m. The administration was shocked and made a series of field investigations to find out the main reason for this tragedy and how to avoid it in the future (IWHR Survey, 2005).

In the provincial primary investigation report, "6.10" storm is identified as a local severe storm, occurring once every 100 years. The area-averaged rainfall was 123 mm in 3 hours and the maximum rainfall was over 200 mm in some places. Since it is an ungauged mountain area, the hydrological rainfall approach was based on surface runoff analyses during the flood investigation.

In general, the latter can be done if sufficient runoff data are available, and a proper hydrological model is adopted to simulate local storm floods (Singh, 2000b).

In this paper a hybrid approach to that problem has been studied, i.e. for rainfall, hydrological analysis based on meteorological observed data, and quantified narrative facts. For runoff, a hydraulic analysis was applied to simulate surface flow along the valley. In the study, a completed 2-D unsteady flow model was applied, instead of using a simplified kinematic wave model or a Muskingum-Cunge model. A hydraulic model needs a numerical description of the valley, so a Digital Elevation Model (DEM) was applied.

Firstly a 2-D hydraulic model is introduced in this paper, which has previously been successfully used to simulate flood evolution in flood plains in China. The advantage is that it is easy to display the velocity field and water depth distribution based on its output results. Subsequently, a simpler rainfall mode was introduced. To couple the rainfall mode with the hydraulic model, some difficulties needed to be overcome such as their scale matching and numerical stability caused by geological unevenness and complex flow pattern. Based on DEM, a river network in the Shalan area was extracted, and rainfall subregions were classified. An explicit expression of the rainfall process based on field surveys and quantified flooding facts have also been introduced, as well as its parameters. One of the parameters should be determined inversely. Finally, based on the simulation results, some facts were revealed, which will be discussed in some detail.

HYBRID APPROACH USING SIMPLIFIED HYDROLOGIC PROCESSES AND THE 2-D HYDRAULIC MODEL

2-D hydraulic model

The applied model is often referred to as a 2-D unsteady model, derived from depth averaged Saint-Venant equations. Under Cartesian coordinates, its continuity and moment equations could be given as follows:

$$\frac{\partial z}{\partial t} + \frac{\partial (uh)}{\partial x} + \frac{\partial (vh)}{\partial y} = q \tag{1}$$

$$\frac{\partial u}{\partial t} + u\frac{\partial u}{\partial x} + v\frac{\partial u}{\partial y} + g\frac{\partial z}{\partial x} + g\frac{n^2 u\sqrt{u^2+v^2}}{h^{4/3}} - k\sqrt{u^2+v^2}\frac{\partial u}{\partial x} = 0 \tag{2}$$

$$\frac{\partial v}{\partial t} + u\frac{\partial v}{\partial x} + v\frac{\partial v}{\partial y} + g\frac{\partial z}{\partial y} + g\frac{n^2 v\sqrt{u^2+v^2}}{h^{4/3}} - k\sqrt{u^2+v^2}\frac{\partial v}{\partial y} = 0 \tag{3}$$

where Z is water elevation; h is water depth; u and v are depth averaged Cartesian velocity components; q is the external water discharge, such as rainfall or laterally coming flow, etc.; n is the manning coefficient; t is the time(s); $g\frac{n^2 v\sqrt{u^2+v^2}}{h^{4/3}}$, $g\frac{n^2 v\sqrt{u^2+v^2}}{h^{4/3}}$ is the flow resistance in x or y direction, respectively; $k\sqrt{u^2+v^2}\frac{\partial u}{\partial x}$, $k\sqrt{u^2+v^2}\frac{\partial v}{\partial x}$ is the local flow resistance in x or y direction, respectively; k has a positive value when the oncoming flow velocity is larger than local flow velocity in the calculated cell, otherwise k is zero. In this study, k is set to be 0.4.

Once the above equations were discretized, convection terms neglected and characteristic theory applied, the momentum algorithm forms, were developed by Li *et al.* (2005); they are shown in equation (4). The discretized scheme is shown in Fig. 1, where the flow simulation field is divided into cells and cell C stands for the currently calculated cell with all variables subscripted by C and neighboring cells are, according to their directions, labelled as E, S, W, N, ES, WS, WN, and EN. All variables with superscript n and $n + 1$ signify their value is taken at nth time and $n +$ 1th time, and stand for the known and unknown, respectively. So V_E^n, V_S^n, V_W^n, V_N^n, V_{ES}^n, V_{WS}^n, V_{WN}^n, V_{EN}^n are velocities at the current time step in the corresponding cells, and $V_{E,b}^{n+1}$, $V_{S,b}^{n+1}$, $V_{W,b}^{n+1}$, $V_{N,b}^{n+1}$ are velocities at the next time step in four borders of cell C.

$$\begin{aligned}
V_{i,b}^{n+1} = {} & \lambda^+ \cdot \{(\frac{V_c^n + V_i^n}{2}) + [sign(c,i)\cdot g\frac{Z_i^n - Z_c^n}{\Delta D} \\
& + \mathrm{k}\frac{\sqrt{(V_c^n + V_{i,A}^n)_x^2 + (V_c^n + V_{i,A}^n)_y^2}}{4}\cdot\frac{V_{i,A}^n - V_c^n}{\Delta D} \\
& + V_{i,b}^{n+1}\sqrt{(V_{i,b,x}^{n+1})^2 + (V_{i,b,y}^{n+1})^2}(g\cdot\frac{n^2}{(\frac{h_i+h_i}{2})^{4/3}})]\Delta t\} \\
& + \lambda^- \cdot \{(\frac{V_c^n + V_j^n}{2}) + [sign(c,j)\cdot g\frac{Z_j^n - Z_c^n}{\Delta D} \\
& + k\frac{\sqrt{(V_c^n + V_{j,A}^n)_x^2 + (V_c^n + V_{j,A}^n)_y^2}}{4}\cdot\frac{V_{j,A}^n - V_c^n}{\Delta D} \\
& + V_{j,b}^{n+1}\sqrt{(V_{j,b,x}^{n+1})^2 + (V_{j,b,y}^{n+1})^2}(g\cdot\frac{n^2}{(\frac{h_j+h_c}{2})^{4/3}})]\Delta t\}
\end{aligned} \tag{4}$$

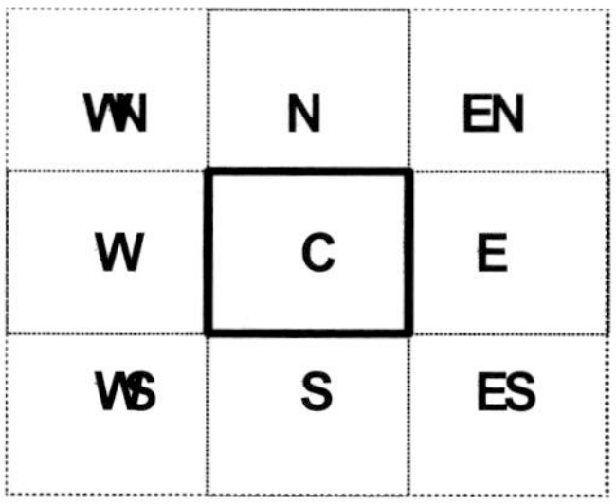

Fig. 1 Discretized scheme in grids.

where $i,j = E,S,W,N$, $\Delta D \begin{cases} = \Delta X, if\ i = E,W \\ = \Delta Y, if\ i = S,N \end{cases}$ and *sign*(*c*,*i*) are sign function, its positive or negative value depends on cell *i*th location compared with cell *C*; a positive value is taken when cell *i* is located on the upper or left side of cell *C*.

For the continuity equation, a central difference discretized scheme is adopted, Z_c^{n+1} is expressed as:

$$\begin{aligned} Z_c^{n+1} = Z_c^n + [&\frac{(V_{W,b}^n + V_{W,b}^{n+1})(h_W^n + h_W^{n+1}) - (V_{E,b}^n + V_{E,b}^{n+1})(h_E^n + h_E^{n+1})}{2\Delta x} \\ &+ \frac{(V_{N,b}^n + V_{N,b}^{n+1})(h_N^n + h_N^{n+1}) - (V_{S,b}^n + V_{S,b}^{n+1})(h_S^n + h_S^{n+1})}{2\Delta y}]\Delta t \end{aligned} \quad (5)$$

Simplified hydrological process

It is often assumed that the hydrological cycle has neither a beginning nor end, and that it processes continuously (Chow *et al.*, 1988). Precipitation might come down into soil, and precipitated water might be intercepted by vegetation, forming surface flow. The infiltrated flow through soil might come out as surface runoff and runoff might infiltrate into the soil again. Evaporated water becomes part of the atmosphere, then falls down as rain. Though the runoff–rainfall expression could be described in Horton's mode or as saturation overland flow mode, its complexity still remains due to a certain amount of spatio-temporal difference in climate and soil. If flood-runoff is caused by a heavy storm, i.e. the rainfall intensity is much stronger than the soil infiltration capacity, some factors in the water cycle, such as evaporation, interception, infiltration, or subsurface flow, might be insignificant. Therefore, hydrological processes influencing flood–runoff can be simplified as net rainfall, which could be embedded into an hydraulic model. If detailed rainfall data is obtained, the land can be subdivided into different rainfall zones. If additional detailed geographic data and the vegetation distribution are acquired, suitable hydrological parameters can be set.

In this paper, for each cell, a saturation overland flow mode is adopted to get q in equation (1), where q is defined as $q = (P - E) - (W_m^{'} - W^{'})$, and P and E stand for precipitation and evaporation, respectively. Here $W_m^{'}$ is the water storage capacity (mm), $W^{'}$ is the actual water storage at the beginning (mm).

Though precipitation can be viewed as an input to a hydraulic model, there are still some challenges met in the hybrid approach. These are discussed as follows:

(a) Scale matching problem. Once a hydrological mode is set to couple with a hydraulic model, scale matching becomes a challenge. It can be classified as either temporal scale matching or spatial scale matching. For the rain process, its time step takes at least several minutes. But for the runoff process, its calculation time step is bound by the model and its mesh size, as well as what flow situation is encountered. In some cases this could be within 60 or 30 s. Another problem is that for the hydraulic model, water depth in a cell should be more than 0.001 m, otherwise the calculation cannot be completed. However, in a single time step, the cell only gets a net rain amount of less then 0.001 m, and this cannot be neglected. To maintain a water quantity balance is therefore very important in the calculation. To solve this issue, the concept of the memory cell is introduced in our study. The calculation cell possesses a memory function; it automatically receives net rainfall, no matter how small it is. And it will not let the program run until its accumulated water amount reaches a limit, e.g. 0.001 m. In the memory function design, it not only remembers how much net rainfall there is into or out of cells, but also takes water exchange between cells into account to maintain the water balance.

Spatial scale in rainfall models is also different from that in runoff models. Their entire simulated areas are different in size; usually the simulated area in rainfall models is much larger than one in runoff models, and the former covers the latter. To simply solve the problem, cells can also be marked with a subregion number of a rainfall region.

(b) Numerical instability caused by topographic unevenness. Topographic unevenness, such as sharp slopes and banks, concave or protruding places, as well as dry-wet joints often makes hydraulic model running numerically unstable. Such a kind of numerical instability occurs especially with flood calculations in mountainous areas.

To overcome such numerical instability, some algorithms techniques have been used in our programming, such as adding a local energy dispersion term, taking characteristic algorithms forms and partly using try-correct techniques.

SIMULATION OF THE "6.10" STORM FLOOD

Spatial information derived from DEM

The objective zone was located at the upper reach of the Shalan River, above Shalan town in An'ning City, Heilongjiang Province. Based on the 2004 (3) Arc Second SRTM elevation, digital elevation data with 90-m resolution, the river network of the Shalan River basin was extracted using an eight-level matrix (D8) approach. The matrix approach is developed as software for extracting drainage networks from raster DEM using the Pfafstetter scheme. The river network could be marked as the main river and its tributaries (Li *et al.*, 2004). The river network and related information of the upper basin of the Shalan River is shown in Fig. 2.

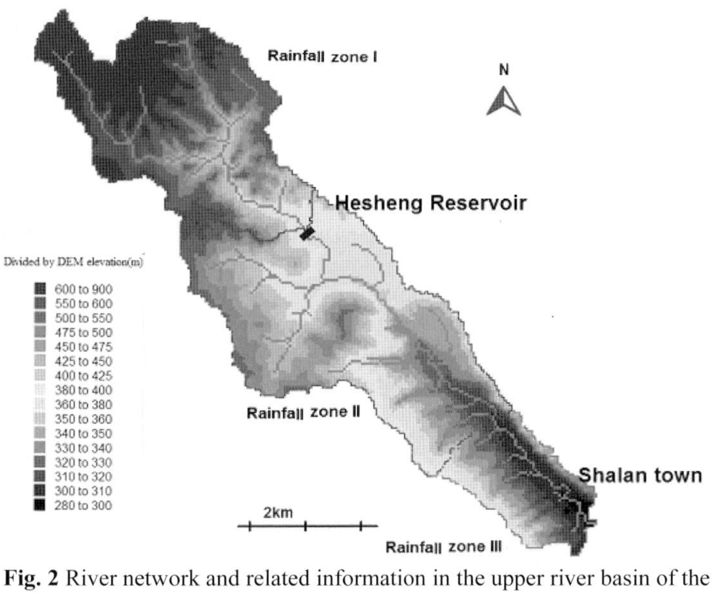

Fig. 2 River network and related information in the upper river basin of the Shalan River.

The extracted drainage area is 114.3 km^2, a narrow strip running from northwest to southeast, with a length of 30 km and width of 3~6 km. The elevation in the northwest mountain area is 500~900 m, whereas in the southeast valley area it is 300~400 m. In a previous study (Cheng, 2005), the real drainage area above Shalan village is 115 km^2, including the Hesheng reservoir catchment of about 45 km^2, which is covered with rich forest vegetation. The riverbed average slope is 6‰, with an elevation difference of 84 m. Based on the above information and DEM, the entire rainfall region is divided into three subregions: (i) Hesheng reservoir catchment with a drainage area of 42.6 km^2, (ii) part of the valley between the reservoir and valley outlet with drainage area of 49.9 km^2, and (iii) the river basin between the valley outlet and Shalan town with a drainage area of 21.8 km^2. The total elevation difference is 82 m. All the data measured in DEM is calibrated, and is very close to that in the previous study (Cheng, 2005).

Rainfall process

After checking the provincial primary investigation report and all available field survey data, it came to mind that the rainfall peak appeared to shift along the valley and become higher and higher. Consequently, the matter of how to express this kind of phenomenon seems to be a key subject in this study. One of the distribution patterns of rainfall processes can be expressed as follows:

$$F(x,t) = \left(1 + \frac{\alpha x}{X_0}\right) F_0\left(x, \beta\left(t - x/_{V}\right)\right) \tag{6}$$

where X_0 is distance from the reservoir to the valley outlet; x is also distance measured from the reservoir to a specific rainfall subregion; α and β are distribution parameters, α has a constant value, β is set to be 1 at the reservoir, otherwise it takes the value from statistically analysed rainfall duration in a specific rainfall subregion; v is shift velocity of the rainfall peak.

Before running the model, some parameters should be determined based on field surveys and statistical analysis. In our case study, peak shift velocity is V = 1.8 m s^{-1}, and for all subregions rainfall duration retains the same value of β = 1. One of the parameters, α, is obtained through a series of calculations, its value continuously reset until the simulation results agree well with the observed flooding facts. The final α value is about 4.8 mm km^{-1}, indicating that the maximum rainfall intensity had increased to 4.8 mm while passing a distance of 1 km downstream.

Initial and boundary conditions

In the study, rectangular grids of 90 m are adopted for the hybrid model. There are a total of 14 110 grids in the simulated area, and the maximum elevation difference in adjacent grids was 45 m. Roughness coefficient is expressed as the Manning coefficient.

The simulated start time is the same as the actual storm, which started at 12:30 h, and ended at 17:00 h. Initially the valley is saturated with earlier rainfall, and its streamflow is set at 20 $m^3 s^{-1}$. The upstream reservoir, the Hesheng Reservoir, is small and during the rainfall no flood had been released from the reservoir (only regular flow out downstream to keep the reservoir water at a fairly constant level). Therefore, for runoff simulation, it is assumed that flow out of the reservoir is at a regular flow, without considering the reservoir regulation. Of course it can also serve as an upstream boundary condition. The downstream condition has been set at the valley outlet, because there is a bridge across the stream and its design flood discharge is known. The flow over the bridge is assumed to be flow over a board weir. The rainfall process at the reservoir is assumed to follow a specific curve of normal distribution; its duration is about 2.5 h as cited from other reports. The accumulated rainfall, which is a definite integral of rain intensity over its duration, is verified to be nearly equal to the measured volume in the reservoir pluviometer.

Simulation results

As a kind of inverse problem, the rainfall shift parameter α, is determined through computation, with a value of 4.8 mm km^{-1}. Therefore, the downstream rainfall process at subregions can be

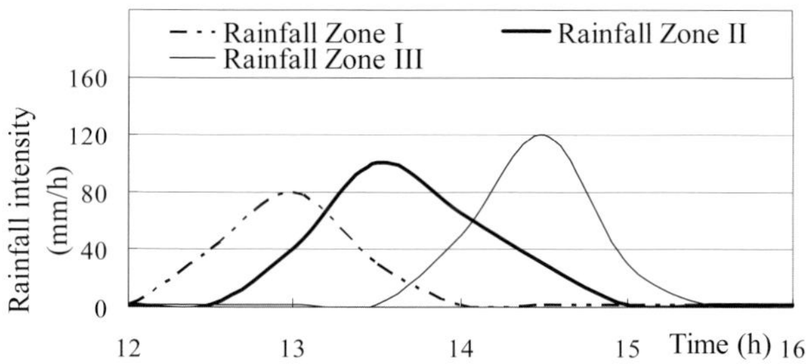

Fig. 3 Rainfall in three rainfall subregions.

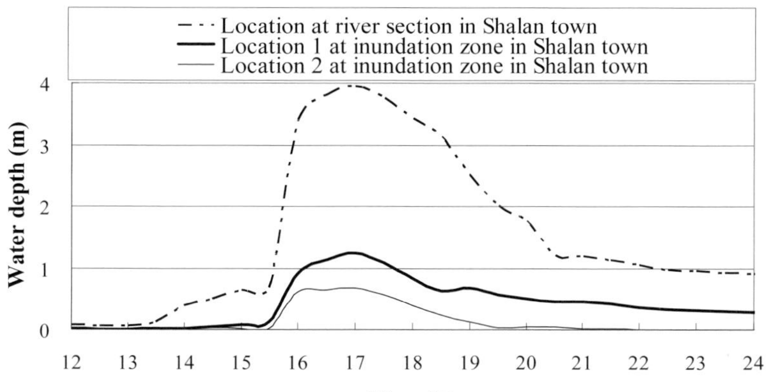

Fig. 4 Flood process in three locations along the Shalan River.

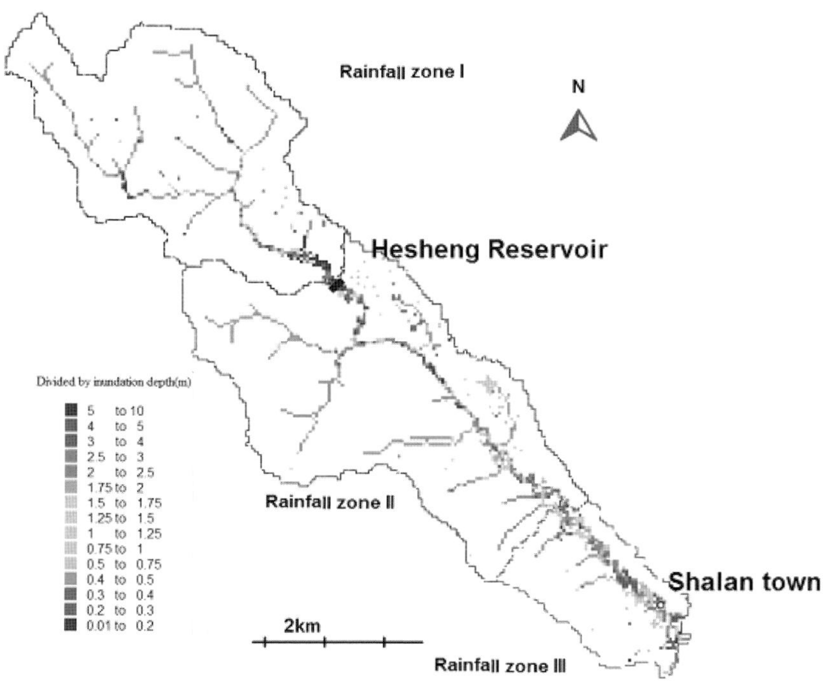

Fig. 5 Inundation fields distribution in Shalan River Basin.

revealed from upstream rainfall processes. The results are shown in Fig. 3. The rainfall intensity in subregion II and subregion III is about 100 mm h^{-1} and 120 mm h^{-1}, respectively, and the rainfall amount in the three subregions is about 75 mm, 94 mm and 112.5 mm, respectively.

The results also show that a total of 9 million m^3 runoff water is yielded over a 5 h period of rainfall. The maximum flood discharge through the valley outlet was 415 $m^3 s^{-1}$, which is beyond the bridge design capacity, and caused some water to overtop the bridge with a water depth of 0.5 m. The maximum water depth nearby the Shalan town was 1.3 m, and at the primary school 2 m and subsided to half these values in 3 h (Fig. 4). Most inundated fields ranged within an extension of 500 m along the river (Fig. 5).

DISCUSSION AND CONCLUSION

A hydraulic runoff model can be coupled to a hydrological rainfall model if their scale matching and numerical stability problems are carefully overcome. Such a hybrid approach can serve as one of the helpful tools to solve rainfall floods in mountainous areas.

DEM and its extracted flow networks offer a strong geological base to build hydraulic runoff models. For some ungauged areas the rainfall process can be expressed based on field surveys and quantified flooding facts.

The rain flood in the Shalan valley resulted in serious loss of life and property. Through our simulation, the following facts are revealed:

(a) The "6.10" storm rain was less strong than that presented in the provincial primary investigation report.
(b) The bridge and some buildings at its two abutments at the valley outlet reduced the discharge capacity and caused more serious inundation in lower lying upstream areas, such as the village and the primary school. To derive rainfall data from runoff data should be carefully carried out; if there is something constructed in the flow passage or outlet, the derived rainfall might be exaggerated.
(c) Another reason for getting more serious floods, is that the rainfall peak was travelling along the valley in the same direction of its forming flow.

Acknowledgements This research was sponsored and supported by the Chinese National Nature Science Foundation (no. 30490231), and received a lot of help from the Information Division, Department of Flood Disaster, IWHR. Thanks to all of them.

REFERENCES

Cheng, Xiaotao (2005) *Wish the disaster in Shalan goes away*. www.cws.net.cn (in Chinese).

China Institute of Water Resources and Hydropower Research (IWHR Survey) (2005) *Survey on flood disaster in Shalan River Basin*. http://www.hwcc.com.cn (in Chinese).

Chow, Vente, Maidment, D. R. & Mays, L. W. (1988) *Applied Hydrology*. McGraw-Hill Book Company, New York, USA.

Department of Flood Disaster, IWHR (2005) *Study on the flood risk map and emergency regulation map in 7 main rivers in China*. IWHR report (in Chinese).

Li, Chong & Yang, Dawen (2004), Deriving drainage networks and catchment boundaries from grid DEM. *J. China Institute of Water Resources and Hydropower Research* **2**(3), 208–214 (in Chinese).

Li, Chong, Lu, Jikang & Ye, Baisheng (2005) Simulation of storm flood basing on deriving drainage networks from grid DEM and 2-D hybrid approach in Shalan River Basin. *J. China Institute of Water Resources and Hydropower Research* **3**(3), 161–167 (in Chinese).

Singh, Vijay P. (2000a) *Hydrologic Systems, Watershed Modeling*. The Yellow River Conservancy Press, China (Zhao weiming translation, in Chinese).

Singh, Vijay P. (2000b) *Hydrologic Systems, Runoff–Rainfall Simulation*. The Yellow River Conservancy Press, China (Zhao weiming translation, in Chinese).

The uncertainty analysis of the wetted perimeter method via axis scaling for setting minimum ecological in-stream flow requirements

SUXIA LIU[1], XINGGUO MO[2], JUN XIA[1], CHANGMING LIU[1] & LINA JI[1,3]

1 *Key Lab of Water Cycle & Related Land Surface Processes, Institute of Geographical Sciences and Natural Resources Research (IGSNRR), Chinese Academy of Sciences (CAS), Beijing 100101, China*
liusx@igsnrr.ac.cn

2 *Key Lab. of Ecological Net Observation and Modeling, IGSNRR, CAS, Beijing 100101, China*

3 *Beijing Management Division of North Grand Canal, Beijing 101100, China*

Abstract The minimum ecological in-stream flow requirement (MEIFR) is the in-stream flow requirement necessary to guarantee the basic ecological function of a river. This paper discusses the uncertainty involved in one of the MEIFR evaluation techniques, the Wetted Perimeter (WP) method, due to axis scaling based on an analytical solution of MEIFR, under the assumption of simple triangular cross-section. It is clearly shown that the solution of MEIFR based on original variables is stable and independent of the data series. That based on scaled variables, however, is highly related to the data series, time period, and scaling factor. The maximum difference of MEIFR based on the scaled variables via Maximum, Z-score and Summation scaling schemes relative to that based on original variables reaches 36%. These results suggest that in order to decrease the uncertainty of the WP method, axis scaling should be used with caution.

Key words minimum ecological in-stream flow requirements; wetted perimeter method; analytical solution; axis scaling; water resources management

INTRODUCTION

The minimum ecological in-stream flow requirement (MEIFR) is the in-stream flow requirement (IFR, King & Louw, 1998; Rowntree & Wadeson, 1998; Hughes, 2001; Houghes & Hannart, 2003; Levite *et al.*, 2003; Symphorian *et al.*, 2003) necessary to guarantee the basic ecological functions of a river. Estimates of MEIFR are becoming an important index of water resources management in China and all over the world. Local governments, state and federal water resources management agencies and consultants are all interested in determining the MEIFR. Similarly, stakeholders have a need to acknowledge the impacts of estimates of MEIFR on decision-making. To date, an impediment to an accurate estimation of MEIFR is the uncertainty involved in MEIFR evaluation techniques. The lack of observed MEIFR for validation makes determination of the reliability of estimation methods impossible.

The wetted perimeter (WP) method (Annear & Conder, 1984; Gordon *et al.*, 2004), as a hydraulic method, is one of several popular methods to estimate MEIFR (Thoms & Sheldon, 2002). It uses the critical point on the relationship curve between wetted perimeter and discharge to determine MEIFR. If the discharge is less than the critical minimum discharge corresponding to the critical point, the wetted perimeter declines rapidly. With decreasing flow, assuming the aquatic habitat is given by the wetted perimeter, a small decrease in flow will result in a significant decrease in the available aquatic habitat, increasing the living pressure of the biome. Above the critical minimum discharge, the wetted perimeter declines slowly, and a large decrease in discharge will produce only a small decrease in wetted perimeter. By defining the wetted perimeter as an index sensitive to habitat quality, the critical minimum discharge corresponding to the critical point is defined as MEIFR in the meaning that if a river keeps the discharge at the level of MEIFR, the optimal ecological function of the river, including aspects of physical, chemical and biological components and their interactions (Acreman, 2005), can be guaranteed.

Unlike the Tennant method (Tennant, 1976), which uses a subjective percentage of historic annual discharge to determine MEIFR, the WP approach has a relatively clear mathematical definition of MEIFR. Unlike habitat, holistic and hybrid methods, which need large amounts of detailed ecological data (see the review in Liu *et al.*, 2006), the WP method just needs the data of streamflow and channel geometry. These features perhaps make WP the least uncertain MEIFR evaluation technique among others. However, the WP method has considerable uncertainty in the determination of the critical point. In the WP method, the critical point was originally determined subjectively by eye from a graph (Collings, 1974; Cochnauer, 1976; Nelson, 1980). Gippel & Stewardson (1998) pointed out that it is not possible to select the breakpoint on wetted perimeter–discharge curves reliably by eye, since the appearance of the slope of the curve is strongly dependent on the relative scaling of the axes. They therefore proposed that axes must be scaled before estimating MEIFR by defining the critical point where the slope equals 1 (slope technique), or where the curvature is maximized (curvature technique). Typically, a logarithmic or power law function is fitted to the observed relationship between flow and wetted perimeter.

Scaling the axes, or standardizing the variables is a common technique (e.g. Liu *et al.*, 2001; Ouarda *et al.*, 2001; Mo & Beven, 2004; Riad *et al.*, 2004). There are several approaches to scale axes. One is to linearly rescale the data to lie in the interval between 0 and 1 by subtracting the minimum, and then dividing by the range; or dividing the data by the maximum value; or by dividing by the summation (Mo & Beven, 2004) of all coordinates. Another approach is to scale individual variables to standard normal, by subtracting the mean and dividing by the standard deviation (Liu. *et al.*, 2001). There are many other options. Application of such transformations prior to the analysis attempts to remove the influence of scaling effects from the analysis, with the aim of obtaining a simpler expression (Ouarda *et al.*, 2001). Furthermore, when plotting the results for a large number of catchments, variables with the largest means tend to dominate the display. As discussed by Friendly & Kwan (2003), scaling axes can effectively result in an incoherent display in which no systematic trends or relations can be seen.

This paper discusses the uncertainty in the WP method for estimating the MEIFR due to axis scaling. Based on the MEFIR deduced in Liu *et al.* (2006) for original variables, analytical solutions of MEIFR based on scaled variables by three scaling schemes are first deduced. Then the dependency of MEIFR on the data series, scaling scheme and climate factor is discussed. Finally conclusions are given.

An analytical solution of MEIFR based on original and SCALED VARIABLES

Liu *et al.* (2006) deduced an analytical solution of MEIFR, as shown in the second row of Table 1, based on original variables by the slope technique and curvature technique for a hypothetical triangular cross-section with water depth, D, channel width over water surface W and the angle between banks 2θ. Similarly, MEIFR based on scaled variables via three scaling schemes is deduced in this paper. The first scheme is to define a relative wetted perimeter, p, and relative discharge, q, by dividing discharge Q and wetted perimeter P by a quantity to make Q and P dimensionless. The quantity can be the maximum value of the data series (hence called Maximum Scheme), or median or mean. For the Maximum Scheme, this gives (Table 1):

Table 1 MEIFR based on original and scaled variables via curvature and slope technique.

Variables	Curvature (C)	Slope (S)
Original (o)	$MEIFR_{Co}=a^{\frac{1}{1-b}}b^{\frac{b}{1-b}}(\frac{2-b}{1-2b})^{\frac{b}{2b-2}}$	$MEIFR_{So}=a^{\frac{1}{1-b}}b^{\frac{b}{1-b}}$
Maximum (m)	$MEIFR_{Cm}=(\frac{P_{max}}{Q_{max}})^{\frac{b}{1-b}}a^{\frac{1}{1-b}}b^{\frac{b}{1-b}}(\frac{2-b}{1-2b})^{\frac{b}{2b-2}}$	$MEIFR_{Sm}=(\frac{P_{max}}{Q_{max}})^{\frac{b}{1-b}}a^{\frac{1}{1-b}}b^{\frac{b}{1-b}}$
Z-score (z)	$MEIFR_{Cz}=(\frac{\sigma_P}{\sigma_Q})^{\frac{b}{1-b}}a^{\frac{1}{1-b}}b^{\frac{b}{1-b}}(\frac{2-b}{1-2b})^{\frac{b}{2b-2}}$	$MEIFR_{Sz}=(\frac{\sigma_P}{\sigma_Q})^{\frac{b}{1-b}}a^{\frac{1}{1-b}}b^{\frac{b}{1-b}}$
Summation (s)	$MEIFR_{Cs}=(\frac{S_P}{S_Q})^{\frac{b}{1-b}}a^{\frac{1}{1-b}}b^{\frac{b}{1-b}}(\frac{2-b}{1-2b})^{\frac{b}{2b-2}}$	$MEIFR_{Ss}=(\frac{S_P}{S_Q})^{\frac{b}{1-b}}a^{\frac{1}{1-b}}b^{\frac{b}{1-b}}$

Note: $a = (\sqrt{S}/32/n)(\sin 2\theta)^{5/3}$, b = 8/3, n is Manning roughness coefficient with the unit of $m^{-1/3}$ s, S is the energy grade slope (non-dimensional), 2θ is the angle between the riverbanks (see Liu *et al.*, 2006).

$$q = Q/Q_{max}, \qquad p = P/P_{max} \tag{1}$$

where Q_{max} and P_{max} are the maximum values of discharge and wetted perimeter, respectively.

Similarly, by defining:

$$q = (Q - \overline{Q})/\sigma_Q, \qquad p = (P - \overline{P})/\sigma_P \tag{2}$$

where σ_Q and σ_P are the standard deviations of Q and P, $\overline{Q}$ and $\overline{P}$ are the averages of Q and P, we get the Z-score scheme. By defining:

$$q = Q/S_Q, \qquad p = P/S_P \tag{3}$$

where S_Q and S_P are the summations of data series of Q and P, we get Summation scheme. Results of MEIFR deduced based on the three scaling schemes are shown in Table 1.

DISCUSSIONS

Via different values of data series

From Table 1, it is obvious that the solution of MEIFR based on original variables is independent of the data series (maximum wetted perimeter and discharge). However, the solution of MEIFR based on scaled variables is highly related to the data series. Taking the Maximum scheme as an example, assuming θ = 60°, n = 0.044 and S = 0.0032, the results of MEIFR based on the hypothetic data set with different values of the maximum wetted perimeter and discharge are compared in Fig. 1. It is shown that both the slope and curvature methods of estimating the MEIFR based on original variable are stable; that is, not sensitive to variations in the value of the maximum wetted perimeter and discharge. However, MEIFR estimates based on scaled variables are sensitive to this scaling factor for both techniques. Because the maximum values are highly dependent on the time period, the results will depend on the selected period, introducing more uncertainty in the MEIFR estimates than found without scaling.

With different schemes of axis scaling

The role of axis scaling is further assessed by testing different scaling schemes. If the MEIFR results based on the different scaling schemes vary, the additional axis scaling methods will also

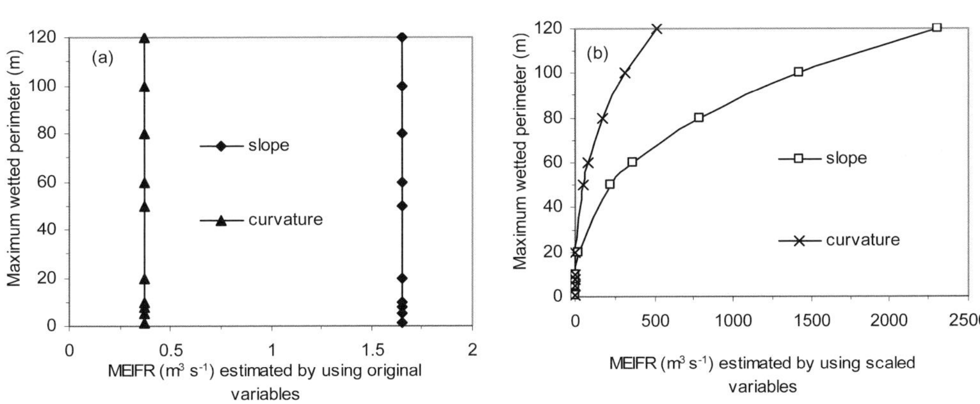

Fig. 1 The change of MEIFR with the change of the maximum value of the wetted perimeter data by using, the (a) original and (b) scaled variables via Maximum scheme, respectively.

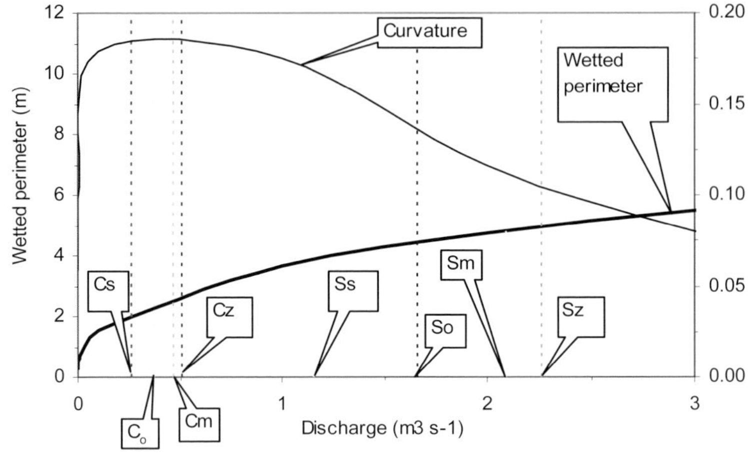

Fig 2 The MEIFR estimated by different schemes (pointed by arrows, the explanation of Cs, Co, Cm, Cz, Ss, So, Sz, see Table 1).

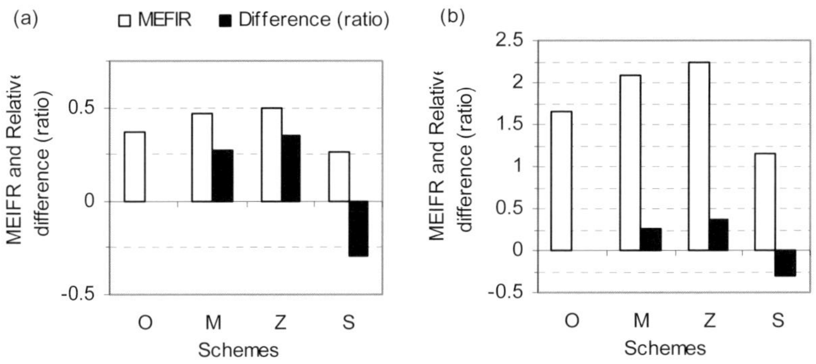

Fig. 3 The specific values of MEIFR estimated via: (a) curvature and (b) slope techniques based on original (noted as o in x-axis) and scaled variables by using three scaling schemes (noted as M, Z and S, respectively) for a hypothetical river.

result in increased uncertainty. As in the above section, assuming $\theta = 60°$, $n = 0.044$ and $S = 0.0032$, based on the hypothetical data set with the value of the maximum wetted perimeter and discharge being 8.66 m and 10 $m^3 s^{-1}$, respectively, the original and scale variables with the three scaling schemes can be calculated based on Table 1. The results are shown and explained in Fig. 2.

Generally speaking, the results via Maximum scheme and Z-score scheme are not obviously different; all are larger than the results with original variables. The summation scheme brings a smaller result than the results with original variables. The maximum difference of MEIFR based on the scaled variables relative to that based on original variables reaches 36%, as shown in Fig. 3. Therefore, in order to get a reasonable estimation of MEIFR, axis scaling should be used with caution.

Should MEIFR be independent of the climate?

While it is obvious that scaled variables may bring more uncertainties than the original variables as shown above, we also have to admit that even using original variables, WP still displays uncertainty. The fact that the left-hand-side of the MEIFR equations in the second row of Table 1 are not the same tells us that determination of which technique (slope or curvature) is more accurate needs further study. Furthermore, it shows that MEIFR based on original variables is only related to channel geometries, independent of the climate. Surely from both a hydrological and ecological view, rivers in different climate zones will have different MEIFR values, even based on the same channel structure. Further, how the scaling of the axes affects other channel profiles needs to be determined. More research will be necessary on the principles of the WP method itself, with more general study on various cross-section channels.

CONCLUSIONS

This paper analyses the uncertainty of the WP method due to axis scaling. Based on wetted perimeter method under the assumption of a triangular cross-section, an analytical solution to the estimation of MEIFR suggests that in order to decrease the uncertainty of WP, axis scaling should be used with caution. Whether the unscaled formulation is generally independent of the climate signal needs further study. While this study is focused on the estimation of MEIFR for channels with a triangular cross section, it does have applicability for actual catchments. This kind of channel is not unusual in headwater regions, where most of the rivers are ungauged (Liu *et al.*, 2006). This work is expected to be valuable for predicting MEIFR in headwater areas as one of the components of the Prediction in Ungauged Basins initiative of the International Association of Hydrological Sciences (Sivapalan *et al.*, 2003).

Acknowledgements This work is financially aided by the National Natural Science Foundation (no. 40671032, 40671033), the Chinese Ministry of Science and Technology for "863" agriculture high technology project (2006AA10Z228)) and the Key Innovation Project of the Institute of Geographical Sciences and Natural Resources Research, Chinese Academy of Sciences (no. CXIOG-A04-12). Many thanks to two anonymous reviewers who give pertinent comments and Dr Barry Croke of The Australian National University who assisted with editing the paper.

REFERENCES

Acreman, M. (2005) Linking science and decision-making: features and experience from environmental river flow setting. *Environ. Modelling Software* **20**, 99–109.

Annear, T. C. & Conder, A. L. (1984) Relative bias of several fisheries instream flow methods. *N. Am. J. Fish. Manage.* **4**, 531–539.

Cochnauer, T. (1976) Instream flow techniques for large rivers. In: *Proceedings, Symposium and Special Conference on Instream Flow Needs* (ed. by J. F. Osborn & C. H. Allman) Vol. II, 387–399 (3–6 May, Boise, Idaho. American Fisheries Society, Bethesda, Maryland, USA.

Collings, M. R. (1974) Generalisation of spawning and rearing discharges for several Pacific salmon species in western Washington. *USGS Open File Report.*

Gippel, C. J. & Stewardson, M. J. (1998) Use of wetted perimeter in defining minimum environmental flows. *Regulated Rivers: Res. & Manage.* **14**(1), 53–67.

Gordon, N. D., McMahon, T. A., Finlayson, B. L., Gordon, Nancy D., McMahon, Thomas A., Gippel, C. J. & Nathan R. J. (2004) *Stream Hydrology–An Introduction for Ecologists*, second edn. John Wiley & Sons, Chichester, UK.

Friendly, M & Kwan, E. (2003) Effect ordering for data displays. *Computational Statistics & Data Analysis* **43**, 509–539.

Hughes, D. A. (2001) Providing hydrological information and data analysis tools for the determination of ecological instream flow. *J. Hydrol.* **241**(1–2), 140–151.

Hughes, D. A. & Hannart, P. (2003) A desktop model used to provide an initial estimate of the ecological instream flow requirements of rivers in South Africa. *J. Hydrol.* **270**(3–4), 167–181.

King, J. & Louw, D. (1998) Instream flow assessments for regulated rivers in South Africa using the Building Block Methodology. *Aquatic Ecosystem Health & Manage.* **1**(2), 109–124.

Levite, H., Sally, H. & Cour, J. (2003) Testing water demand management scenarios in a water-stressed basin in South Africa: application of the WEAP model. *Physics and Chemistry of the Earth, Parts A/B/C* **28**(20–27), 779–786.

Liu, S., Mo, X., Li, H., Peng, G. & Robock, A. (2001) The spatial variation of soil moisture in China: geostatistical characteristics. *J. Meteorol. Soc. Japan* **79**(2B), 555–574.

Liu, S., Mo, X., Xia, J., Liu, C., Lin, Z., Men, B. & Ji, L. (2006) Estimating the minimum in-stream flow requirements via wetted perimeter method based on curvature and slope techniques. *J. Geographical Sci.* **16**(2), 242–250.

Mo, X. & Beven, K. (2004) Multi-objective parameter conditioning of a three-source wheat canopy model. *Agric. & Forest Met.* **122,** 39–63.

Nelson, F. A. (1980) Evaluation of four instream flow methods applied to four trout rivers in southwest Montana. *Draft Report to US Fisheries and Wildlife Service*. Montana Department of Fish Wildlife and Parks. p. 105.

Ouarda, T. B. M. J., Girard, C., Cavadias, G. S. & Bobee, B. (2001) Regional flood frequency estimation with canonical correlation analysis. *J. Hydrol.* **254**(1–4), 157–173.

Riad, S., Mania, J., Bouchaou, L. & Najjar, Y. (2004) Rainfall–runoff model using an artificial neural network approach. *Mathematical & Computer Modelling* **40**(7–8), 839–846.

Rowntree, K. & Wadeson, R. (1998) A geomorphological framework for the assessment of instream flow requirements. *Aquatic Ecosystem Health & Manage.* **1**(2), 125–141.

Sivapalan, M., Takeuchi, K., Franks, S. W., Gupta, V. K., Karambiri, H., Lakshmi, V., Liang, X., McDonnell, J. J., Mendiondo, E. M., O'Connell, P. E., Oki, T., Pomeroy, J. W., Schertzer, D., Uhlenbrook, S. & Zehe. E. (2003) IAHS decade on Predictions in Ungauged Basins (PUB), 2003–2012: shaping an exciting future for the hydrological sciences. *Hydrol. Sci. J.* **48**(6), 857–880.

Symphorian, G. R., Madamombe, E. & van der Zaag, P. (2003) Dam operation for environmental water releases; the case of Osborne dam, Save catchment, Zimbabwe. *Physics and Chemistry of the Earth, Parts A/B/C* **28**(20–27), 985–993.

Tennant, D. L. (1976) Instream flow regimens for fish, wildlife, recreation and related environmental resources. In: *Proceedings, Symposium and Special Conference on Instream Flow Needs* (ed. by J. F. Osborn & C. H. Allman) Vol. II, 359–373 (3–6 May, Boise, Idaho. American Fisheries Society, Bethesda, Maryland, USA.

Thoms, M. C. & Sheldon F. (2002) An ecosystem approach for determining environmental water allocations in Australian dryland river systems: the role of geomorphology. *Geomorphology* **47**, 153–168.

Temporal and spatial variations of δ^{18}O along the main stem of the Yangtze River, China

BAOHONG LU[1], TINGTING SUN[1], CHUNYAN WANG[1], SU DAI[1,2], JIAN KUANG[3] & JIYANG WANG[1]

1 *State Key Laboratory of Hydrology-Water Resources and Hydraulic Engineering, Hohai University, Nanjing 210098, China*
lubaohong@126.com; lubaohong@hotmail.com

2 *Flood-control Division, Taihu Basin Authority of MWR, Shanghai 200434, China*

3 *Hydrology Bureau of MWR, Beijing 10053, China*

Abstract The isotope compositions of δD and δ^{18}O in river water are very useful tools for interpreting hydrological processes related to climate changes and anthropogenic activities. Since 2003, 170 water samples recovered from the first water campaign and 1-year regular sampling at four stations along the main stem of the Yangtze were analysed for δD and δ^{18}O composition. The results revealed that temporal and spatial variations in the oxygen- and hydrogen-isotopes of water samples along the main stem of the Yangtze strongly relies on the isotope pattern of the regional precipitation. Secondary signals deriving from influx of evaporatively enriched waters from several major lakes or reservoirs along the system are also apparent. The peaks of the river water isotopic temporal variations are clear, corresponding to the boundary of the beginning or ending for annual flooding period at-site. It can used as an indicator to split flooding or low water standing period at a given location for a water year.

Key words flooding period; indicator; river water isotopic monitoring; temporal and spatial variations; water campaign; Yangtze River

INTRODUCTION

The Yangtze River is the largest river in China, the third largest in the world. The river snakes its way 6397 km from western China's Qinghai-Tibet Plateau to the East China Sea near the city of Shanghai, crossing nine provinces and spanning 90°33′ to 122°25′E and 24°30′ to 35°45′N. Its watershed has an area of about 179.93 × 104 km^2, including approx. 20% of China's total land area and 25% of its total farming land area. About 350 million people live near the Yangtze River and its 700 tributaries.

The river from origin to estuary crosses three main physiographic regions in China corresponding to the three main reaches, which are upstream, mid-stream and downstream. From the headwaters to Chongqing, the upstream portion of the river flow is located on the uplands of the Tibetan Plateau. The downstream portion of the river extends from the mouth of Poyang Lake to the estuary in the lowland region. The portion of the river in between these regions is located on an intermediate step and is referred to as the mid-stream. The headwaters of the Yangtze are situated at an elevation of about 4876.8 m in the Kunlun Mountains in the southwestern section of Qinghai, in which, north of the Himalayas at the Yangtze's origin, the Tibetan Plateau has mighty glaciers and continuous snow cover that melt into the River. The famous three gorges are located in the medium-stream; all four of the largest freshwater lakes of China are distributed in the medium and low reaches of the river.

Flood inundation, water shortages caused by environmental pollution, soil erosion and geological disaster are a few of the major issues facing the Yangtze River basin because of intensive human activities and unsuitable land use. To ensure sustainable economic and social development, a better understanding of hydrological processes and the water cycle is very important at the watershed scale. Watershed hydrology requires integrating knowledge of the hydrosphere, biosphere and atmosphere. Previous studies have shown that the commonly-used hydrological approaches cannot completely recognize intrinsic characteristics and mechanisms of hydrological processes and the water cycle.

River discharge is composed of snow melt, surface runoff and groundwater seepage, and is a very important linkage of the global hydrological cycle. It integrates all of the information on all three disciplines mentioned above which impact on the watershed. Much valuable information is contained in the isotopic composition of river discharge. Isotopic signatures in the river discharge can potentially provide information about how the hydrological cycle is impacted by both climate change and land use. Stable isotope measurements of river discharge are now used in the studies of watershed hydrology, e.g. to describe major flow pathways of groundwater, to trace water mixing history; to determine mean residence time, surface water and groundwater exchange and renewal rates, as well as evaporation–transpiration partitioning; to split hydrograph. Dansgaard (1964) first proposed the use of the value d, to characterize the deuterium excess in global precipitation. The value d is defined for a slope of 8, and is calculated for any precipitation sample as $d = \delta^2H - 8\delta^{18}O$. On a global basis, d averages about 10. It changes due to variations in humidity, wind speed and sea surface temperature during primary evaporation (Clark & Fritz, 1997). The parameter d has been widely used to estimate the impacts of global climate changes and anthropogenic activities on the large scale water cycle.

IAEA has accumulated lots of rainfall isotopic data. However, application of isotope techniques to the study of large-scale rivers has been limited by the lack of discharge isotopic data. To support isotope techniques applied in the large-scale watershed, a global research project entitled "Design criteria for a network to monitor isotope composition of runoff in large rivers" was started in 2002 by IAEA. "The isotopic tracing of hydrology process of the Yangtze River basin" is one cooperative research program of the project. The main purpose of the project is to better understand the interaction of surface and groundwater, the contributions of the groundwater to stream flow and the water balance, and identification of the impact of human activities and climate changes on river runoff.

Apparently, coupling this information into atmospheric GCMs and hydrological models equipped with isotope tracers should improve the accuracy of forecasts of basin floods and water resources. The measurement of isotopic fluxes and volumetric discharge is very useful for the development, utilization and protection of water resources in a large-scale watershed.

WATER SAMPLING

The first water campaign combined with water pollution investigation was carried out by the Environmental Monitoring Center of the Water Resources Committee of the Yangtze River. It started on 13 January 2003, and was completed at the end of January 2003. A total of 74 water samples were collected for the first water campaign, 4 samples were collected for each cross-

Fig. 1 Sampling sites for the 1st water campaign and the 4 regular monitoring sites (black dot and triangle represent the sampling site for the first water campaign and for the regular monitoring site, respectively).

section within the Three Gorges Reservoir, in total 11 transects and 1 sample per transect without the reservoir, in total 30 transects. At 4 regular monitoring stations, water was sampled at 8:00–8:30 h on the first day and the 15th day of the month from 1 October 2003 until 30 September 2005. Water samples were taken by using the boat measurement method at the 0.1 relative depths below the river water surface and at the 0.3 relative width apart from the right (or left) banks for each transect of the whole river. The sampling sites are demonstrated in Fig. 1.

All samples were analysed for $\delta^{18}O$ and δD in the Environmental Isotope Laboratory of the University of Waterloo (Canada) and Hydrology Isotope Laboratory of IAEA (Vienna), as well as in the Key Laboratory of IGCS (China).

RESULTS AND DISCUSSION

Relationships between runoff and precipitation isotopic compositions

The relationships between $\delta^{18}O$ and δD for precipitation and river water are demonstrated in Fig. 2. As illustrated, the real lines represented the relationships for the first water campaign in the January and a few weeks in 2003, and for 4 regular sampling sites about one year from October 2003. The corresponding trend line equations are $\delta D = 7.64\delta^{18}O + 6.93$ and $\delta D = 7.68\delta^{18}O + 5.61$, respectively. It can be seen that the local meteoric water line (LMWL: $\delta D = 7.62\delta^{18}O + 8.20$) around the whole Yangtze River is very close to the global meteoric water line (GMWL: $\delta D = 8\delta^{18}O + 10$) and appreciably higher than it (Rozanski *et al.*, 1993). The trend lines for river water along the main stem in the Yangtze River basin are also close to LMWL, and nearly parallel to the LMWL. It reflects that the river water isotopic variations along the main stem strongly depend on precipitation isotope fields. As observed, the differences of the trend lines for river water behaved as d-excess. It proves the evaporation effect because river water experiences multiple evaporation, including surface water evaporation and groundwater evaporation after the precipitation water is converted into river water, which results in a higher d-excess value of the river water than the precipitation water (Araguas-Araguas *et al.*, 2000; Martinelli *et al.*, 2004).

Spatial variations of stable isotope ^{18}O

The variation of $\delta^{18}O$ composition for the first water campaign with the distance from the headwater along the main stem of Yangtze River is shown in Fig. 3.

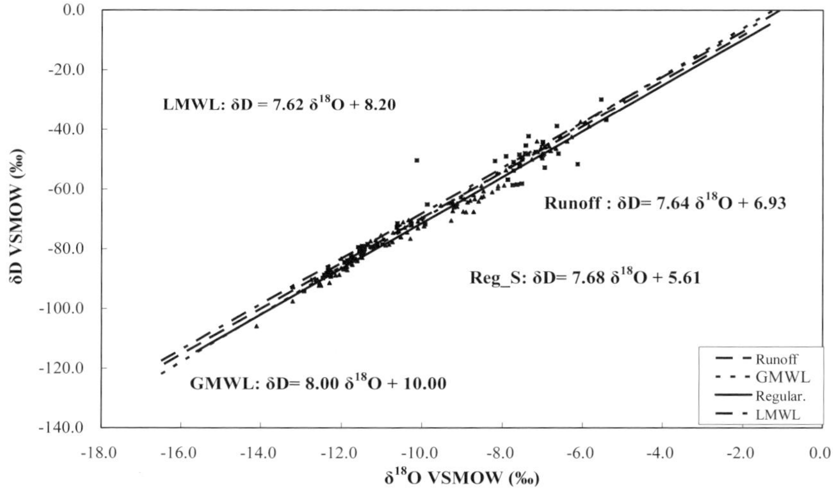

Fig. 2 D and ^{18}O relationship of the runoff and precipitation isotope in Yangtze River.

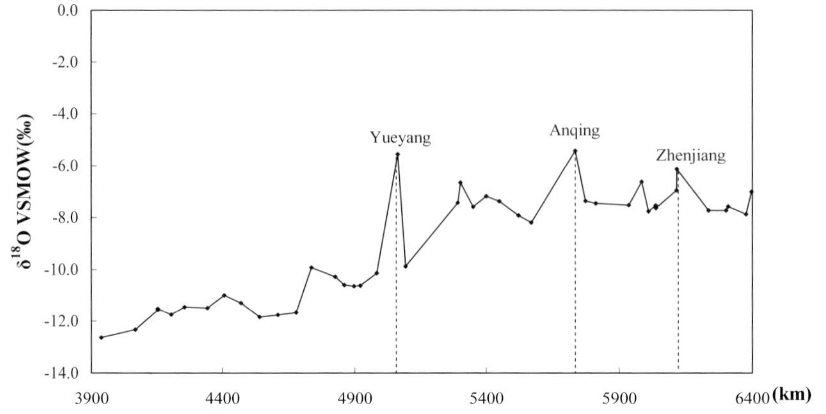

Fig. 3 Stable Isotope ^{18}O compositions changed with distance apart from the headwater of the Yangtze River.

As observed, the isotopic composition of the river water is progressively increasing along the main stem and there are three major peaks in Fig. 3. The increase of the δ^{18}O along the river course is because of the evaporation effect enhancing, due to the long distance. Along the flow path, the isotopic compositions of the water regulated by lakes or reservoirs are more positive than those unregulated waters of neighbour sites due to the evaporation, and the isotopic compositions between former and later cross-sections are obviously different, which also reveals that the isotope pattern depends on the precipitation isotope field. The largest δ^{18}O peak showing the greatest evaporative enrichment (first δ^{18}O peak) is caused by the water sample directly collected from Dongting Lake, which may consist of recharge from rice plant fields (Su *et al.*, 2003; Gibson *et al.*, 2005). The following peak is because the water sample was collected in the mixing region of Poyang Lake and the main stem of Yangtze River. The last δ^{18}O peak is at Zhengjiang where the collected water sample was a composite consisting of water from the main channel of the Yangtze River and runoff from Hongze Lake.

Temporal variations of stable isotope ^{18}O

The temporal variation of the δ^{18}O via discharge change with time is reflected in Figs 4–5.

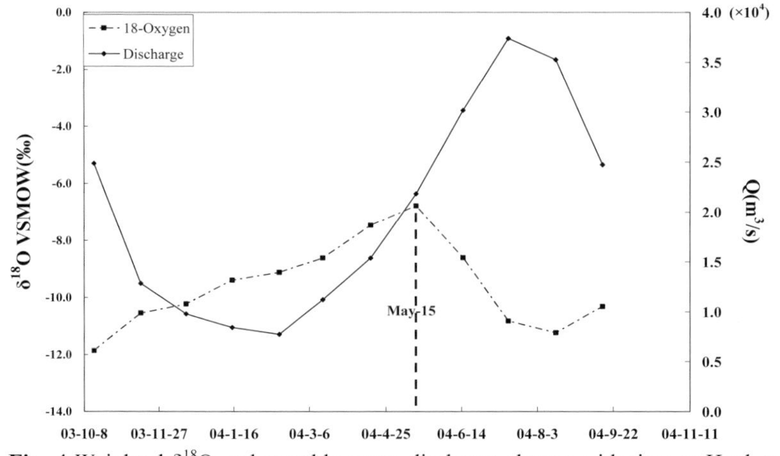

Fig. 4 Weighted δ^{18}O and monthly mean discharge change with time at Hankou Hydrological Gauge Station.

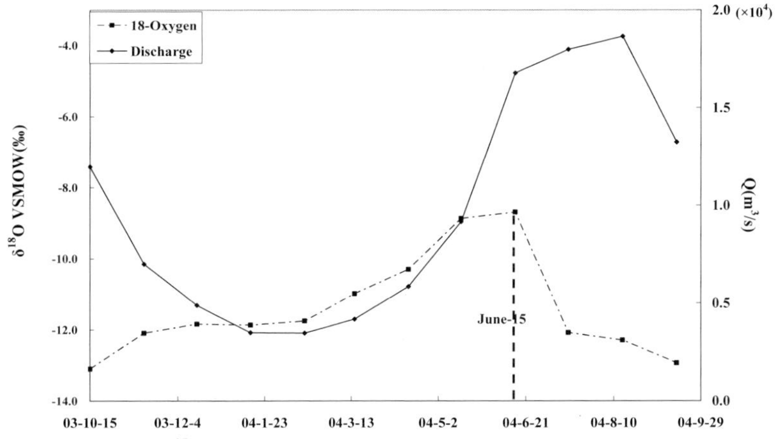

Fig. 5 Weighted $\delta^{18}O$ and monthly mean discharge change with time at Chongqing Hydrological Station.

In Figs 4 and 5, monthly average $\delta^{18}O$ is weighted by river discharge using the following equation:

$$\overline{\delta^{18}O_i} = \frac{\left(\delta^{18}O_f \times Q_f + \delta^{18}O_m \times Q_m + \delta^{18}O_e \times Q_e\right)}{\left(Q_f + Q_m + Q_e\right)}$$

where $\overline{\delta^{18}O_i}$ represents average $\delta^{18}O$ composition of the *i*th month, $\delta^{18}O_f$, $\delta^{18}O_m$, $\delta^{18}O_e$ and Q_f, Q_m, Q_e represent $\delta^{18}O$ values and river discharge at the first, the medium, and the end of the *i*th month, respectively.

River water is composed of melt water, surface water, and groundwater. At the low water standing period, the flux of river water is dominated by groundwater because of the decreasing precipitation contribution to river water.

As shown in Figs 4 and 5, the $\delta^{18}O$ peak appears before the flooding period in summer at the two sites, and the time interval before the peak could be divided into two phases: (1) winter, mainly including the last month of 2003 and the first month of 2004, river water is dominated by groundwater with higher $\delta^{18}O$ value because of the little rainwater in winter. The $\delta^{18}O$ value of the river water is more positive in the period and the $\delta^{18}O$ value is about –10‰ in Hankou station, –12‰ in Chongqing station. (2) Old water gradually drained out from the basin becomes the main composition in the river water, starting from February and ending at the $\delta^{18}O$ peak. Besides, the stronger evaporation caused by the increasing temperature also makes the $\delta^{18}O$ value become higher and higher. Eventually the value reaches the highest point at the end of the period. The flood period comes after the isotopic peak. The $\delta^{18}O$ value declines because the new water has become the main composition of the river water.

More attention should be paid to the occurrence time of the $\delta^{18}O$ peak that corresponds well to the beginning of the flooding period at the corresponding region, just as 15 June for Chongqing station, and 15 May for Hankou station. The feature could be used to split a water year into the flooding and the low water standing periods.

SUMMARY AND CONCLUSIONS

Based on the study of water stable isotopes along the main stem of the Yangtze River, pertinent conclusions can be drawn as follows:

– Temporal and spatial variations in the oxygen- and hydrogen-isotopes of water along the main stem of the Yangtze River strongly reflect the isotope pattern of the regional precipitation.

- Lakes or reservoirs influence on the isotopic compositions of the river water could be very large. They directly result in d-excess values increasing.
- Stable isotope composition peak or valley of river water is a good indicator of split flooding period and low water standing period at a given location for a water year.

Acknowledgements The study was sponsored by the National Natural Science Foundation Committee of China (NSFC 50379008) and International Atomic Energy Agency (IAEA-CRP12285), as well as the Hydrology Bureau of the Ministry of Water Resources of China and Hohai University Academician Foundation. The authors acknowledge all those who have been concerned with the study of water isotopic monitoring along the main stem of the Yangtze River.

REFERENCES

Araguas-Araguas, L., Froehlich, K. & Rozanski, K. (2000) Deuterium and oxygen-18 isotope composition of precipitation and atmospheric moisture. *Hydrol. Processes* **14**, 1341–1355.

Clark, I. D. & Fritz, P. (1997) *Environmental Isotopes in Hydrogeology*, 43–61. Lewis Publishers of CRC Press, USA.

Gibson, J. J., Edwards, T. W. D., Birks, S. J., St Amour, N. A., Buhay, W. M., McEachern, P., Wolfe, B. B. & Peters, D. L. (2005) Progress in isotope tracer hydrology in Canada. *Hydrol. Progresses* **19**, 303–327.

Martinelli, L. A., Gat, J. R., De Camargo, P. B., Lara, L. L. & Ometto, J. P. H. B. (2004) The Piracicaba River basin: isotope hydrology of a tropical river basin under anthropogenic stress. *Isotope in Environmental and Health Studies* **40**(1), 45–56.

Rozanski, K., Araguas-Araguas, L. & Gonfiantini, R. (1993) *Isotopic Pattern in Modern Global Precipitation*. American Geophysical Union, USA.

Su, Xiaosi, Lin, Xueyu, Liao, Zisheng & Wang, Jinsheng (2003) Spatial variations of Isotope ^{18}O, ^{2}H, and ^{3}H in Yellow River and the causing factors study. *Geochemica* **32**(4).

Genetic algorithm based combined evaluation model for regional water security evaluation: a case study

FANG TONG & ZENGCHUAN DONG

The State Key Laboratory of Water Resources-Hydrology and Hydraulic Engineering, Hohai University, Nanjing 210098, China

eyre521@126.com; tongfang824@163.com

Abstract The water security crisis is one of the most serious challenges facing sustainable water development in the world. The North River basin in Guangdong Province of China is one such area facing a water security crisis. The case study in the North River basin applied a novel Combined Evaluation Model based on Genetic Algorithm (CEM-GA) developed for regional water security evaluation. The study method considered the pros and cons of both the subjective and objective weighting evaluation methods, and integrated four single evaluation methods with the Minimizing Difference Degree Model based on Nash Equilibrium in Game Theory and Genetic Algorithm as the coordinated objective function. The case study of the North River basin illustrated the results of this methodology. The results suggest that CEM-GA, as a practical method, can be widely used for the quantitative evaluation and comparison of water security status in different regions.

Key words water security; combined evaluation model; genetic algorithm; Nash equilibrium; Minimization Difference Degree Integration Model

INTRODUCTION

Water, one of the most important natural resources, is widely considered as the key to human health and wealth (Johnson *et al.*, 2001; Brown, 2002). Water security, affected by natural factors as well as human actions, has increasingly become a major threat to food security, human health, economic development and natural ecosystems on a global level (Rockström, 2000; Xia *et al.*, 2006). Water security evaluation bridges water security scenario analysis and water strategic decision-making. Hence the study of precise and practical regional water security evaluation methods is the foundation of regional water resources management and planning (Lundqvist *et al.*, 2000).

The crux of water security evaluation is to determine the weights of the evaluation indicators. The existing weighting methods mainly consist of two categories: (1) the subjective weighting methods and (2) the objective weighting methods (Gu, 1990). The limitation of each weighting method, as well as each single comprehensive evaluation method, can not be neglected. Under this backdrop, a combined evaluation concept was conceived. This novel method mainly focuses on the reinforcement of the merits of each single evaluation method to fully explore the feature information of the evaluation objective from different angles (Gregory, 1996; Tang, 2001).

The purpose of this study is to set up a combined model for regional water security evaluation based on Nash Equilibrium in Game Theory and Genetic Algorithm to solve problems of subjective and objective weighting methods, as well as defects of single evaluation methods. Further, stability and sensibility analysis on this combined model, significant for water security evaluation in ungauged basins, is also discussed.

METHODOLOGY OF THE CEM-GA

The standard processing of the evaluation indicator sample data

Define $\{x^*(i,j)\}$ (i ranges from 1 to m, j ranges from 1 to n) as the evaluating object vector sample set, where $x^*(i,j)$ is the jth evaluation indicator value of the ith evaluating object vector, m and n are the numbers of the evaluating object vector and indicators, respectively.

The standard processing of the original evaluation sample series should be implemented to eliminate the influences of the units and dimension of the evaluation sample series. Use equation (1) for the effective-type evaluation indicators (the bigger the better), and equation (2) for the cost-type evaluation indicators (the smaller the better).

$$x(i,j) = x^*(i,j) / x^*_{\max}(j) \tag{1}$$

$$x(i,j) = x^*_{\min}(j) / x^*(i,j) \tag{2}$$

where, $x^*_{\min,j}$ and $x^*_{\max,j}$ are the minimum and maximum of the jth indicator in sample series, and $x^*(i,j)$ is the unified effective-type indicator value.

Selecting the comparable evaluation method set *M*

Several comparable comprehensive evaluation methods should be selected. Weights of indicators, on the one hand, reflect the bias of the decision-maker (the subjective aspect), and on the other hand, present the various importance of each indicator value during decision-making (the objective facet). Therefore, the Genetic Algorithm improved Projection Pursuit (GAPP), Ideal Solution Point Method (TOPSIS), Fuzzy Comprehensive Evaluation (FCE) and Analytic Hierarchy Process (AHP) evaluation methods are selected as the comparable evaluation method set. The former two methods, belonging to objective weighting method, can explore the whole information diversity of the evaluation indicators of different evaluating object vectors; and the latter two methods, belonging to subjective weighting method, can reflect the local diversity information of the evaluation indicators of the same evaluating object vector:

$$M = \{m_1, m_2, \cdots, m_l\} \tag{3}$$

where l is the number of compatible evaluation methods, and l equals 4 in this research.

The determination of the compatible evaluation method set M_1

(a) Implement the four selected comparable evaluation methods within the comparable evaluation method set M to evaluate the relative evaluation indicator values of each evaluating object vector $x(i,j)$, and the comprehensive indicator value vector of each evaluating object vector $Z(i,p)$ (where i ranges from 1 to m, and p ranges from 1 to l), can be calculated by equation (4). The detailed calculation process can be referred in the literature concerned (Friedman & Turkey, 1974; Saaty, 1980; Chen & Hwang, 1992; Ozernoy, 1992).

$$Z(i,p) = \sum_{j=1}^{n} w(j,p)x(i,j) \quad i = 1 \sim m, j = 1 \sim n, p = 1 \sim l \tag{4}$$

where $Z(i,p)$ is the comprehensive indicator value of the ith evaluating object vector with the pth method, $w(j,p)$ is the weight of the jth indicator with the pth method and l is the number of the selected comparable methods.

(b) Apply Fuzzy Clustering Analysis (FCA) to the evaluation comprehensive indicator value vector, $Z(i,p)$; then divide the comparable evaluation method set M into two groups with the fixed threshold value ξ: the first group is only one vector $Z(i,k)$ (where k belongs to the interval of $[1,l]$), and the other group is the remaining other methods except $Z(i,k)$.

(c) Calculate the mean vector of the second evaluation group $MZ(i)$, by equation (5):

$$MZ(i) = \left[\sum_{p=1, p \neq k}^{l} Z(i,p) \right] / (l-1) \quad i = 1 \sim m, p = 1 \sim l \tag{5}$$

Then calculate the correlation coefficient of $Z(i,k)$ and $MZ(i)$ by equation (6):

$$r_{z,mz} = \frac{\sum_{i=1}^{n}(Z(i,k)-\overline{Z(i,k)})(MZ(i)-\overline{MZ(i)})}{[\sum_{i=1}^{n}(Z(i,k)-\overline{Z(i,k)})^2\sum_{i=1}^{n}(MZ(i)-\overline{MZ(i)})^2]^{0.5}} \quad (6)$$

And the compatibility between the kth evaluation method and the other remaining methods can be obtained through testing the correlation of $Z(i,k)$ and $MZ(i)$ using equation (6) with the required significance level.

(d) If the testing result presents significant correlation between $Z(i,k)$ and $MZ(i)$, the selected l comparable methods are compatible with each other, that is to say, M is the compatible method set M_l; otherwise, the kth evaluation method is the incompatible method from M, and should be excluded from the group. Then implement these steps narrated above to the remaining methods, and the compatible method set M_l can be obtained as equation (7).

$$M_l = \{m_1, m_2, \ldots, m_q\} \quad (7)$$

where q is the number of selected compatible evaluation methods.

The establishment of the combined comprehensive evaluation indicator function

The combined comprehensive evaluation indicator function can be set up as equation (8).

$$ZZ(i) = \sum_{p=1}^{q} a(p)Z(i,p) \quad i = 1 \sim m, p = 1 \sim q \quad (8)$$

where $ZZ(i)$ is the combined comprehensive evaluation indicator of the ith evaluating object vector, $a(p)$ is the weight of the pth evaluation method, and q is number of selected compatible evaluation methods.

The establishment of the minimizing difference degree integration model based on Nash equilibrium of game theory

The competitive and yet coordinative relationship between different evaluation methods can be analysed through Game Theory, and Nash Equilibrium can be applied into the integration model of single evaluation methods as the coordinative objective. Hence the optimal value of $a(p)$ can be solved from equation (9). The optimization objective for this problem is to minimize the Difference Degree of the results of selected single evaluation methods within the M_l set and the combined comprehensive evaluation indicator value.

$$\begin{cases} Minf(a) = \sum_{i=1}^{n}\sum_{p=1}^{q}\|Z(i,p)-ZZ(i)\|_2 \\ s.t. \sum_{p=1}^{q} a(p) = 1, a(p) \geq 0, \end{cases} \quad (9)$$

This is a complicated nonlinear optimal problem with $a(p)$ as the optimal variables, and it is difficult to be solved by general approaches. Real coding Accelerating Genetic Algorithm (RAGA), as an overall optimal method, can easily solve this optimal problem (Jin & Ding, 2000).

The ranking evaluating results of the combined evaluation model

Apply the optimal weight vector $a^*(q)$ into equation (8), and the combined evaluation indicator values of evaluating object vector $ZZ^*(m)$ can be obtained. Then rank the $ZZ^*(m)$ values, and the ranking results of the evaluating object vector sample can be obtained.

A CASE STUDY

The lower reaches and the delta region of the North River, with an area of 32 000 km^2 and a population of 835 940 000, is one of the socio-economically robust regions in Guangdong Province in China. Water security, reflecting water availability in terms of quality and quantity, has become one of the key factors challenging the sustainable socio-economic development in this region.

The indicator system for water security evaluation in the North River basin is established based on the Pressure-State-Response (P-S-R) framework and its application in selecting indicators for sustainability assessment (OECD, 1993). The P-S-R mechanism for regional water security assessment can be obtained and is shown in Fig. 1. The indicator system and the indicator values of the status quo and the planning level years are presented in Table 1.

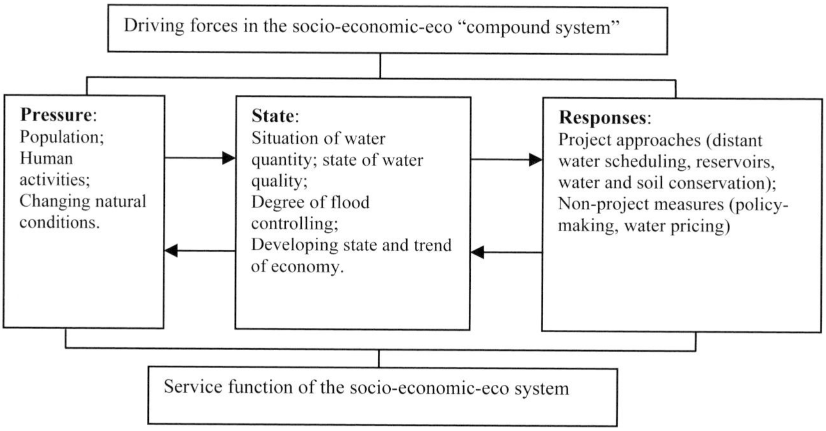

Fig. 1 The Pressure-State-Response (P-S-R) framework for regional water security assessment.

Table 1 the evaluation indicator values of water security situation in North River basin of the year of the status quo and the planning level years.

Time Indicator	2000	2005	2010	2015	2020	2025	2030	2035	2040	2045	2050
I_1	1	0.9678	0.9485	1	1	1	1	1	1	0.9808	0.9763
I_2	1	0.9347	0.893	1	1	1	1	1	1	0.9565	0.9454
I_3	1	1	1	1	1	1	1	1	1	1	1
I_4	0.0188	0.0181	0.0168	0.0156	0.0147	0.0137	0.0112	0.0085	0.0075	0.0066	0.0063
I_5	0.4906	0.5553	0.6832	0.9369	1	1	1	1	1	1	1
I_6	0.3092	0.3194	0.3449	0.3513	0.3522	0.355	0.3672	0.3879	0.4141	0.4484	0.492
I_7	0.1079	0.0894	0.0835	0.0854	0.0655	0.0602	0.0551	0.0513	0.0516	0.0472	0.0466
I_8	0.0059	0.0077	0.0082	0.0046	0.0023	0.0031	0.0040	0.0047	0.0046	0.0055	0.0056

I_1: ratio of industrial water supply (%); I_2: ratio of agricultural water supply (%);
I_3: ratio of ecological water supply (%); I_4: ratio of flood disaster loss (%);
I_5: over-standard ratio of released water (%); I_6: utilization ratio of water resources (%);
I_7: GDP ratio of increase (%); I_8: proportion of investment on water resources construction (%).

Implement the selected four comparable comprehensive evaluation methods to the water security evaluation indicator sample series in Table 1, and the comprehensive indicator value vector $Z(11,4)$ and the ranking results are shown in Table 2.

Then carry out Fuzzy Clustering Analysis to the four evaluation vectors in Table 2, then with the required threshold value of ξ (ξ equals to 0.95), divide it into two groups. The first group is

Table 2 the comprehensive indicator value vector of water security situation in North River basin of the year of the status quo and the planning level years with four comparable evaluation methods.

year	GAPP Indicator values	Ranking results	TOPSIS Indicator values	Ranking results	FCE Indicator values	Ranking results	AHP Indicator values	Ranking results
2000	0.5350	11	0.3635	10	0.1621	11	0.6725	10
2005	0.5600	10	0.3620	11	0.4352	8	0.6695	11
2010	0.6130	9	0.3665	9	0.3368	10	0.6870	9
2015	0.7425	8	0.4231	6	0.4251	9	0.7405	1
2020	0.7590	6	0.3895	8	0.4668	7	0.7155	8
2025	0.7570	7	0.3925	7	0.5986	5	0.7165	7
2030	0.7695	5	0.4365	5	0.6125	4	0.7201	5
2035	0.7710	3	0.532	4	0.5698	6	0.7265	2
2040	0.7735	1	0.9951	2	0.7521	3	0.7175	6
2045	0.7700	4	0.754	3	0.9421	1	0.7245	3
2050	0.7720	2	0.9963	1	0.8655	2	0.7225	4

the evaluation vector of GAPP, $Z(11,1)$ equals to (0.5350, 0.5600, 0.6130, 0.7425, 0.7590, 0.7695, 0.7710, 0.7735, 0.7700, 0.7720, 0.7710), and the second group is the remaining three evaluation vectors. Then through equation (5), calculate the mean vector of the second evaluation group MZ (11), MZ (11) equals to (0.4302, 0.5067, 0.5008, 0.4751, 0.5827, 0.6157, 0.6347, 0.6498, 0.8096, 0.8391, 0.8381); and then apply equation (6) to calculate $r_{z,mz}$, the correlation coefficient of $Z(11,1)$ and MZ (11). In this case study, $r_{z,mz}$ equals to 0.869 and is larger than $r_{0.01}$ ($r_{0.01}$ equals to 0.735 with the sample volume of 11); hence the GAPP evaluation method is compatible with the three other methods, and the compatible evaluation method set M_1 contains PP, TOPSIS, FCA and AHP.

Then the Minimizing Difference Degree Integration Model of evaluation methods can be defined as follows:

$$\begin{cases} Minf(a) = \sum_{i=1}^{11}\sum_{p=1}^{4}\|Z(i,p) - ZZ(i)\|_2 \\ s.t. \sum_{p=1}^{4} a(p) = 1, a(p) \geq 0 \end{cases}$$

Apply RAGA to solve this optimization problem, and the optimal weight vector $a^*(4)$ can be obtained: $a^*(4)$ equals to (0.3120, 0.2963, 0.1952, 0.1965); then put $a^*(4)$ into equation (8), and the combined comprehensive evaluation indicator values of water security situation in North River basin from 2000 to 2050, $ZZ^*(11)$, can be obtained. The comprehensive evaluation indicators results, as well as the final ranking results of the water security situation in the North River basin from 2000 to 2050, are presented in Table 3.

The result in Table 3 shows that the water security level after implementation of water resources planning gradually increases in the course of the planning level years. Therefore, it proves that CEM-GA can be effectively used for qualitative analysis and quantitative evaluation of water security level of a single region, as well as for comparison of different regions.

Table 3 The Combined Comprehensive Evaluation Indicator (CCEI) values and the final rank of water security situation in North River basin of the status quo and the planning level years.

Time (year)	2000	2005	2010	2015	2020	2025	2030	2035	2040	2045	2050
CCEIV	0.4627	0.5600	0.5356	0.5580	0.6374	0.6595	0.6765	0.6770	0.6881	0.7973	0.8183
Final rank	11	8	10	9	7	6	5	4	3	2	1

CCEIV: Combined Comprehensive Evaluation Indicator Value.

DISCUSSIONS

About the selection of regional water security comprehensive evaluation indicators

Selection of regional water security evaluation indicators is the first and also one of the key steps of the whole modelling process. The P-S-R framework was applied in setting up an indicator system in the case study of the North River basin. However, the selection process of the eight indicators was still quite qualitative, due in part to the limitations of the available data.

A large quantity of indicators from various driving forces can be available if full-scale and detailed data material is accessible. Hence it could be a demanding job to establish an optimal evaluation indicator system, since this involves both qualitative and quantitative "evaluation" of each indicator. Theoretical research and case studies of this problem require further study.

About the sensitivity of standardization method of the evaluation sample series

In theory, standardization processing for the evaluation indicator sample series is very important; however, it is more or less neglected in practice. Three common linear standardization methods are Maximum-and-Minimum method, Difference of Maximum and Minimum method and Mean value method, and the calculation expressions are presented in equation (10), (11) and (12), respectively.

$$\begin{cases} x(i,j) = x^*(i,j)/x^*_{\max}(j) \\ x(i,j) = 1.0 - x^*(i,j)/x^*_{\max}(j) \end{cases} \tag{10}$$

$$\begin{cases} x(i,j) = (x^*(i,j) - x^*_{\min}(j))/(x^*_{\max}(j) - x^*_{\min}(j)) \\ x(i,j) = 1.0 - (x^*(i,j) - x^*_{\min}(j))/(x^*_{\max}(j) - x^*_{\min}(j)) \end{cases} \tag{11}$$

$$\begin{cases} x(i,j) = x^*(i,j)/\bar{x}(j) \\ x(i,j) = 1.0 - x^*(i,j)/\bar{x}(j) \end{cases} \tag{12}$$

In the case study of the North River basin, the sensitivity of the indicator standardization procedures is influential to the evaluation precision for Genetic Algorithm improved Projection Pursuit method (GAPP). The three standardization methods mentioned above were applied into the Projection Pursuit evaluation method to further study of the regulation of this problem, and the experimental error results of the three methods are shown in Table 4.

Table 4 Comparison error results of the three general standardization methods on GAPP.

	PEAV (%)				PERV (%)				AAE	ACE
	[0~0.1]	[0~0.2]	[0~0.25]	[0~0.3]	[0~0.04]	[0~0.07]	[0~0.09]	[0~0.15]		(%)
Equation(10)	42.01	61.52	71.12	100	40.42	61.65	86.04	99.63	0.17	4.13
Equation(11)	5.63	15.96	36.85	40.21	3.65	10.53	22.63	30.36	0.45	17.85
Equation(12)	12.65	43.35	58.21	71.32	13.65	32.63	42.30	61.85	0.33	9.63

PEAV: Percentage of Absolute Error Values within the intervals;
PERV: Percentage of Relative Error Values within the intervals;
AAE: Average Absolute Error; ACE: Average Relative Error.

In analysis, equation (11) enlarges the comparative differences among the original indicator sample series. Therefore, it greatly changes the data structure feature of the indicator sample series, hence resulting in lower precision of the evaluation model than the other two equations.

About the stability of the integration method for the Combined Evaluation Model

The stability experimental study on the result of this integration model was designed as follows: choose four major indicators, namely, GDP ratio of increase, over-standard ratio of released water, utilization ratio of water resources and flood disaster loss ratio, to implement the result of sensitivity with the designed change rate of ±10% of each indicator sample series. Table 5 presents the results.

Table 5 The stability experimental results on the result of CEM-GA.

Indicator Selected	Change (%)	Change rate of the comprehensive indicator values *ZZ*(*i*) of 11 evaluation years of North River basin (%)										
		2000	2005	2010	2015	2020	2025	2030	2035	2040	2045	2050
IS_1	10	−1.32	−3.20	−1.63	0.96	−2.85	−2.74	−1.96	−1.32	−1.85	−2.63	−1.08
	−10	2.01	1.36	2.12	1.63	2.63	1.78	1.96	0.85	1.36	1.64	1.29
IS_2	10	1.22	1.36	1.88	1.02	1.74	1.66	1.59	1.89	2.01	2.03	1.65
	−10	−1.01	−1.36	−1.42	−1.50	−2.18	0.99	−2.09	−1.85	−1.47	−1.63	−2.07
IS_3	10	2.63	2.85	3.01	2.45	2.87	1.99	2.46	3.05	2.03	3.12	3.78
	−10	−2.78	−2.60	−2.11	−3.01	−2.96	−2.17	−3.00	−1.99	−2.65	−2.18	−2.09
IS_4	10	−5.56	−3.96	−4.56	−4.01	−3.01	−4.12	−2.96	−5.56	−2.64	−4.87	−3.65
	−10	4.63	3.78	3.41	3.10	3.85	4.14	4.69	2.99	3.18	3.89	4.67

IS_1: GDP ratio of increase; IS_2: over-standard ratio of released water;
IS_3: utilization ratio of water resources; IS_4: flood disaster loss ratio.

As shown in Table 5, the biggest change rate of the comprehensive indicator value ZZ(*i*) is 4.69%, and the smallest is 0.96%, with the designed change rate of ±10% of each indicator sample series. The favourable stability of the combined evaluation model is largely attributed to the combined effect of decreasing evaluation risks and the integration of the Genetic Algorithm characterized by its excellent self-adaptation, fault-tolerance and robustness. This feature can be significant for water security evaluation in ungauged basins: indicator values through analogy from neighbouring regions can still work on this model with the inaccuracy to some tolerable degree, and the evaluation could still be applicable and constructive for further analysis and study.

CONCLUSIONS

At a global level, water security is emerging as one of the highest priorities on the development agenda. In this study, a Combined Evaluation Model for regional water security evaluation was set up to use both subjective requirements of decision-maker and objective evaluation information. The results of the case study of the North River basin in South China suggest that CEM-GA combines information on both subjective and objective weights, hence objective evaluation information and the requirements of the decision-maker can be well balanced. Besides, the favourable stability of the combined evaluation model presents great significance for water security evaluation in ungauged basins. Therefore this model can be extensively applied into water security evaluation of various regions.

REFERENCES

Brown, K. (2002) Water scarcity: forecasting the future with spotty data. *Science* **297**, 926–927.

Chen, S. J. & Hwang, C. L. (1992) *Fuzzy Multiple Attribute Decision Making*, 163–287. Springer Press, Berlin, Germany.

Friedman, J. H. & Turkey, J. W. (1974) A projection pursuit algorithm for exploratory data analysis. *IEEE Tran On Computer* **23**(9), 881–890.

Gregory, A. J. (1996) The road to integration: reflections on the development of organizational evaluation theory and practice. *Int. J. Manage. Sci.* **24**(3), 295–370.

Gu, J. F. (1990) *Methods of Comprehensive Evaluation*. 22–26. Science Technology Press of China Beijing, China (in Chinese).

Jin, J. L. & Ding, J. (2000) *Genetic Algorithm and Its Applications to Water Science*. Sichuan University Press, Chengdu, China (in Chinese).

Johnson, N., Revenga, C. & Eeheverria, J. (2001) Managing water for people and nature. *Science* **292**, 1071–1072.

Lundqvist, J., Falkenmark, M. & Folke, C. *et al.* (2000) *New Dimensions in Water Security*. FAO, UN Food and Agriculture Organization, Rome, Italy.

OECD (1993) *OECD Core Set of Indicators for Environmental Performance Reviews*. Paris, France.

Ozernoy, V. M. (1992) Choosing the "best" multiple criteria decision-making method. *INFOR* **30**, 159–171.

Rockström, J. (2000) Green water security for the food makers of tomorrow: windows of opportunity in drought prone savannahs. In: *Proc. Stockholm Water Symposium 2000: Water Security for the 21st Century—Innovative Approaches. Water Science & Technology* **43**(4).

Saaty, T. L. (1980) *The Analytic Hierarchy Process*. McGraw-Hill Company, New York, USA.

Tang, X. J. (2001) Model integration. *J. Systems Engng* **16**(5), 322–329 (in Chinese).

Xia, J., Zhang, L., Liu, C. M. & Yu, J. J. (2006) Towards better water security in North China. *Water Resour. Manage*. **6**, 122–126.

Comparing univariate ARMA and ARFIMA model for forecasting daily streamflows

WEN WANG[1], XI CHEN[1], JUN MA[2] & SHIQING HUO[2]

1 *State Key Laboratory of Hydrology-Water Resources and Hydraulic Engineering, Hohai University, Nanjing 210098, China*
wangwen@hhu.edu.cn

2 *Yellow River Conservancy Commission, Hydrology Bureau, Zhengzhou 450004, China*

Abstract The performances of two types of univariate time series model, i.e. the autoregressive moving average (ARMA) model and the long-memory fractionally integrated autoregressive moving average model (ARFIMA) model, for forecasting daily streamflows, are investigated in the present study. Both models are applied to four daily average discharge series of medium-sized watersheds in cold mountainous regions, and built on the basis of deseasonalized series. The result shows that both the ARMA model and the ARFIMA model work well for forecasting short-term daily average discharges, and the performance of the ARFIMA model is generally slightly better than that of the ARMA model. The use of the ARFIMA model is therefore recommended as an alternative to the conventional ARMA model for modelling univariate daily streamflow processes for the cold mountainous regions from the perspective of forecasting.

Key words streamflow process; ARMA model; long memory; ARFIMA model; time series analysis; hydrological forecast

INTRODUCTION

When modelling streamflow processes in ungauged basins, we may sometimes have to deal with univariate streamflow processes. The most commonly used model for modelling univariate hydrological time series is the autoregressive moving average (ARMA) model (see Hipel & McLeod, 1994), which is a short-memory model, whose autocorrelation function (ACF) decays exponentially. However, since the early work of Hurst (1951), it has been well recognized that many time series, in diverse fields of application, may exhibit the phenomenon of long-memory or long-range dependence. In the hydrology community, many studies have been carried out on the test for long-memory in streamflow processes (e.g. Burlando *et al.*, 1996; Montanari *et al.*, 1997; Rao & Bhattacharya, 1999; Ooms & Franses, 2001; Wang *et al.*, 2007). Burlando *et al.* (1996) and Montanari *et al.* (1997) applied the ARFIMA mode to the modelling of the daily water inflow to Lake Maggiore, and showed that the fit with an ARFIMA model is much better than the one provided by more traditional ARIMA models. But their result is based on the one single streamflow process, and the model performance comparison is based on the model-fitting residuals, without considering the performance for out-of-sample forecasting.

In the present study, the performances of ARMA and ARFIMA model for making 1- to 7-days ahead out-of-sample forecasts are compared for four daily streamflow processes in the cold mountainous regions of British Columbia in Canada.

METHODS OF MODELLING UNIVARIATE TIME SERIES

Autoregressive moving average (ARMA) model

The ARMA model is one of the most important and highly popularized time series models, and has a long history of being applied to streamflow forecasting problems (see Hipel & McLeod, 1994). An ARIMA(p, q) model is given by:

$$(1-\phi_1 B-\cdots-\phi_p B^p)x_t=(1-\theta_1 B-\cdots-\theta_q B^q)\varepsilon_t \tag{1}$$

where B is the backshift operator, that is, $Bx_t = x_{t-1}$; ε_t is a white noise process with zero mean and variance σ^2. When $q = 0$, then the ARMA(p,q) model reduces to an AR(p) model.

Fractionally integrated autoregressive moving average (ARFIMA) model

The ARFIMA model has not been widely accepted in the hydrology community so far. The general form of an ARFIMA(p,d,q) model is given by:

$$(1-\phi_1 B-\cdots-\phi_p B^p)\nabla^d x_t = (1-\theta_1 B-\cdots-\theta_q B^q)\varepsilon_t,\ |d|<0.5 \qquad (2)$$

where $\nabla^d = (1-B)^d$ can be given by means of the binomial expansion, and d denotes the fractional differencing parameter. A larger value of d indicated a stronger intensity of long-memory. When $d = 0$, equation (2) reduces to the classical ARMA(p,q) model. The conventional ARMA process is often referred to as a short memory process because its autocorrelation function (ACF) $\rho(h)$ decays exponentially as $h\to\infty$, whereas the long-memory ARFIMA process is characterized by a hyperbolically decaying autocorrelation function. In fact, it decays so slowly that the autocorrelations can not be summed.

Many methods for estimating the long-memory parameter d are available, which can be categorized into two classes: the heuristic and the parametric methods. The heuristic methods are useful to find a first estimate of d and to test if a long-range dependence in the data exists. The parametric methods obtain consistent estimators of d via maximum likelihood estimation of parametric long-memory models. They give a more accurate estimate of d, but generally require knowledge of the true model. The S-Plus function *arima.fracdiff* fits ARFIMA models to univariate time series data through the approximate Gaussian maximum likelihood algorithm of Haslett & Raftery (1989).

When forecasting from the univariate ARFIMA($p,d,0$) model, we can express the process $(1-\phi_1 B-\cdots-\phi_p B^p)\nabla^d x_t = \varepsilon_t$ in the form of an infinite autoregression, where $\nabla^d = \sum_{j=0}^{\infty}\left[\frac{\Gamma(j-d)}{\Gamma(j+1)\Gamma(-d)}B^j\right]$. The infinite autoregression may be truncated at a suitably high lag (say, 200), and the forecasts are then made recursively for multi-steps ahead.

DAILY STREAMFLOW DATA USED

Daily average discharge series of four watersheds in British Columbia, Canada, are used in the present study. All the four watersheds are located in cold mountainous regions where snow-melt water makes a considerable contribution to runoff. The description of the eight gauging stations is given in Table 1. The data are selected data so that the series should be approximately stationary, at least by visual inspection, and the time-span of the data ranges from 23 to 30 years. All the four daily series exhibit strong and regular seasonal variations.

MODEL BUILDING FOR DAILY STREAMFLOW SERIES

ARMA model

The procedure of fitting deseasonalized ARMA models to daily flow series includes two steps: deseasonalization and ARMA model construction. We first log-transform the data series, and

Table 1 Daily streamflow data used.

Gauging stations	Area (km^2)	Latitude	Longitude	Period	Avg. discharge (m^3/s)	ACF(1)
Canoe River below Kimmel Creek	298	52.728	−119.408	1972–1994	14.5	0.9294
Columbia River Near Fairmont Hot Springs	891	50.324	−115.863	1946–1975	11.1	0.9676
Columbia River at Nicholson	6660	51.244	−116.912	1933–1962	107.5	0.9778
Fraser River at Mcbride	6890	53.286	−120.113	1959–1988	197.3	0.9582

Note: The ACF(1) is calculated based on log-transformed and deseasonalized series.

deseasonalize the series by subtracting the seasonal (e.g. daily or monthly) mean values and dividing by the seasonal standard deviations of the log-transformed series. To alleviate the stochastic fluctuations of the daily means and standard deviations, we smooth them using the first eight Fourier harmonics.

In the present study, for simplicity, we considered using the AR(p) model, that is, $q = 0$ in the ARMA(p,q) model, when using the ARMA model. The AR(p) model structure (listed in Table 2) for all the daily flow series are determined according to the Alkaike Information Criterion (AIC). The goodness of fit of the models are examined by inspecting the ACF of the residuals from the models, which indicates that there is generally no significant autocorrelation left in the residuals from all the AR models.

Table 2 AR(p) and ARFIMA(p,d,0) model structure for the streamflow series.

Gauging stations	AR(p)	ARFIMA(p,d,0)	
	p	p	d
Canoe River below Kimmel Creek	10	3	0.31
Columbia River at Nicholson	35	3	0.4392
Columbia River Near Fairmont Hot Springs	6	8	0.4213
Fraser River at Mcbride	9	7	0.1886

ARFIMA model

Before fitting the ARFIMA model to a daily flow series, the existence of long-memory should be confirmed. The first heuristic indication of the presence of long-memory comes from the slow decaying ACF structure of the four deseasonalized daily series, as shown in Fig. 1. In addition, it has been shown (Wang *et al.*, 2007) that all the daily flow series under consideration have significant long-memory.

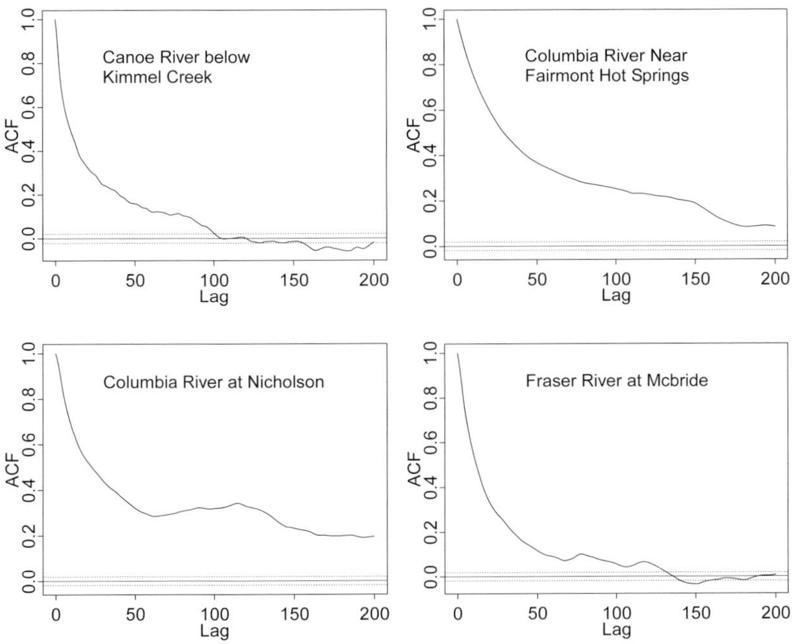

Fig. 1 ACF of the log-transformed and deseasonalized daily mean discharge series.

Similar to constructing the ARMA model to the daily streamflow series, the procedure of fitting ARFIMA models also includes two steps, i.e. firstly deseasonalize the data, and then construct the ARFIMA model. We only consider using the ARFIMA model of the structure $(p,d,0)$. The deseasonalization is done in the same way as that for building the ARMA model. The order of the AR component and estimates of d for the fitted ARFIMA model of all the deseasonalized daily series are also listed in Table 2. The order p of the AR component in the ARFIMA model is determined according to AIC, and the parameters of the AR component and value of d are estimated using the function *arfima.fracdiff* in S-Plus. The examination of the ACF of the residuals from the ARFIMA models shows that no significant autocorrelations are observed in the residual series.

COMPARISON OF MODEL PERFORMANCES

Model performance measures

There is extensive literature on model forecasting evaluation indices. Despite its crudeness and identified weaknesses, the Coefficient of Efficiency (*CE*) introduced by Nash & Sutcliffe (1970) is still one of the most widely used criteria for the assessment of model performance. However, *CE* is a global measure for comparing the predicted value with the overall mean value; it is not efficient enough to evaluate the predictions for those series whose mean values change with seasons, which is almost always the case for hydrological processes. Therefore, a Season-Adjusted Coefficient of efficiency (*SACE*) (Wang *et al.*, 2004) is used here for evaluating the model performance. *SACE* is calculated by:

$$SACE = 1 - \frac{\sum_{i=1}^{n}(Q_i - \hat{Q}_i)^2}{\sum_{i=1}^{n}(Q_i - \bar{Q}_m)^2} \tag{3}$$

where Q_i is the observed value; $\hat{Q}_i$ is the predicted value; $\bar{Q}_m$ is the mean value of season m; $m = i \bmod S$ (mod is the operator calculating the remainder), ranging from 0 to $S-1$; and S is the total number of "*season*". Note that, a "*season*" here is not a real season. It may be a month or a day, depending on the timescale of the time series. For daily streamflow series, one season is one day. Furthermore, for daily series, the mean values of 365 days are smoothed using the first eight Fourier harmonics.

Besides the above two performance measures, the root mean squared error (*RMSE*) is also used in the present study.

Comparison of model performances

For each daily flow series, we make 1- to 7-day ahead forecasts for the last 5 years of data to evaluate the performance of the fitted model. Both the AR and ARFIMA models are built on a rolling-forward basis. For instance, we use the daily flow data of the Columbia River near Fairmont Springs in 1946–1970 to fit an AR(6) model, and make forecasts for 1971; then use the data in 1946–1971 to fit another AR(6) model, and make forecasts for 1972; etc. Data are log-transformed and deseasonalized before being used. The evaluation results for 1- to 7-day ahead forecasts are reported in Table 3. According to the values of performance measures, we see that:

(a) The measure CE shows that both models give good forecasts (i.e. the value of CE > 0.8) for 7-day ahead forecasts. But SACE shows that, although for all the cases both models can give forecasts significantly better than seasonal mean values for all lead times (i.e. SACE > 0), satisfying forecasts (i.e. SACE > 0.8) are available only for 1- to 3-days ahead for the Columbia River at Nicholson and Columbia River near Fairmont Hot Springs, for 1-day ahead for the Fraser River at Mcbride, and no satisfying forecasts are available, even for 1-day

Table 3 Multi-step forecasting performances of ARMA and ARFIMA models.

Lead time (days)	Measure	Model	1	2	3	4	5	6	7
Canoe River below Kimmel Creek	RMSE	ARMA	4.119	6.044	6.696	7.151	7.419	7.543	7.607
		ARFIMA	4.122	6.046	***6.692***	***7.143***	***7.406***	***7.525***	***7.583***
	CE	ARMA	0.944	0.879	0.851	0.830	0.817	0.811	0.808
		ARFIMA	0.944	0.879	0.851	0.831	0.818	0.812	0.809
	SACE	ARMA	0.719	0.394	0.256	0.152	0.087	0.056	0.040
		ARFIMA	0.718	0.394	0.257	0.154	0.090	0.061	0.046
Columbia River Near Fairmont Hot Springs	RMSE	ARMA	0.851	1.475	1.984	2.423	2.795	3.113	3.399
		ARFIMA	***0.848***	***1.468***	***1.971***	***2.405***	***2.774***	***3.089***	***3.372***
	CE	ARMA	0.995	0.986	0.975	0.963	0.951	0.939	0.927
		ARFIMA	0.995	0.986	0.975	0.963	0.951	0.940	0.928
	SACE	ARMA	0.974	0.921	0.858	0.788	0.717	0.649	0.582
		ARFIMA	0.974	0.922	0.859	0.791	0.722	0.655	0.588
Columbia River at Nicholson	RMSE	ARMA	6.157	11.962	17.263	21.944	25.997	29.498	32.469
		ARFIMA	***6.102***	***11.844***	***17.084***	***21.729***	***25.723***	***29.220***	***32.220***
	CE	ARMA	0.998	0.991	0.982	0.970	0.958	0.946	0.935
		ARFIMA	0.998	0.991	0.982	0.971	0.959	0.947	0.936
	SACE	ARMA	0.983	0.936	0.868	0.786	0.700	0.613	0.531
		ARFIMA	0.983	0.938	0.870	0.790	0.706	0.620	0.538
Fraser River at Mcbride	RMSE	ARMA	30.893	52.331	66.262	75.012	81.011	85.403	88.747
		ARFIMA	***30.885***	***52.286***	***66.153***	***74.813***	***80.684***	***84.939***	***88.178***
	CE	ARMA	0.979	0.939	0.902	0.875	0.854	0.838	0.825
		ARFIMA	0.979	0.939	0.903	0.875	0.855	0.839	0.827
	SACE	ARMA	0.888	0.677	0.483	0.337	0.227	0.141	0.072
		ARFIMA	0.888	0.678	0.485	0.341	0.233	0.150	0.084

Note: the bold italic values denote better performances of ARFIMA model against ARMA model in terms of *RMSE*.

ahead for the Canoe River. The results illustrate that the measure CE exaggerates the performance of the forecasting models in cases where the streamflow processes have strong seasonality, whereas the SACE is a better alternative.

(b) Among the four cases, the ARFIMA model outperforms the ARMA model in three cases (the Columbia River at Nicholson, the Columbia River near Fairmont Hot Springs and the Fraser River at Mcbride) for all lead times, and outperforms the ARMA model for 3-days and longer lead time forecasts for the Canoe River, indicating its good modelling performance for the four cases under consideration in the present study.

(c) According to the performance measures CE and SACE, we see that the ARFIMA model and ARMA perform much better for the streamflows measured at Nicholson and near Fairmont Hot Springs for the Columbia River than those measured below Kimmel Creek for the Canoe River and at Mcbride for the Fraser River. This is probably because the streamflow processes for the Columbia River not only have significant snow-melt water contributions, but are also regulated by two lakes, i.e. Lake Koocanusa and Columbia Lake, which lead to stronger temporal dependence and stronger intensity of long-memory, as indicated by larger values of ACF(1) (Table 1) and larger values of the long-memory parameter (Table 2) than those for the Fraser River and Canoe River.

The scatter plot of the one-day ahead predicted values *versus* the observed values for the four streamflow series with the ARFIMA models are shown in Fig. 2. From Fig. 2 we see that there is no systematic bias in the forecasts, and the results are generally satisfying, especially for the Columbia River.

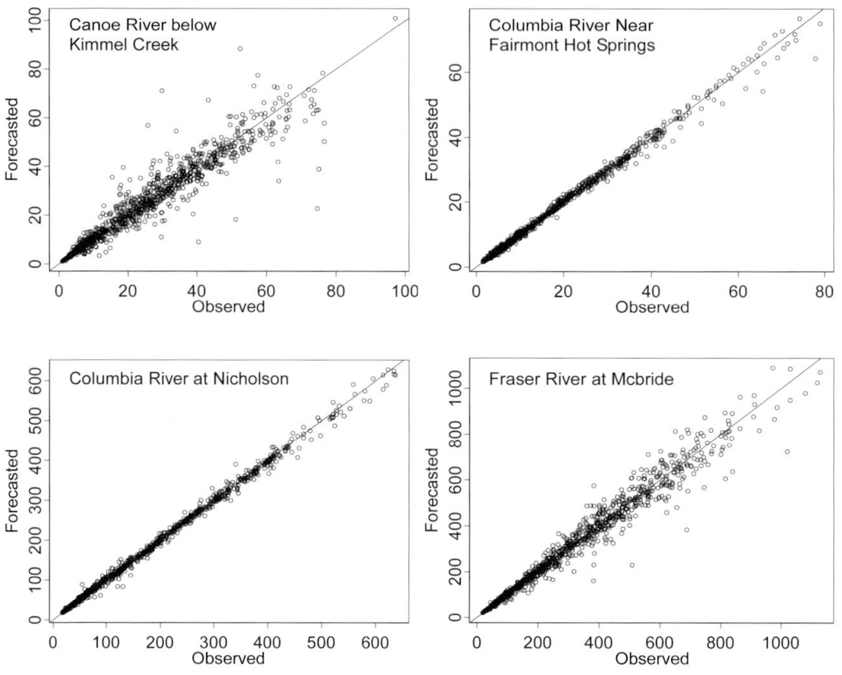

Fig. 2 Observed discharges *versus* 1-day ahead forecasts with ARFIMA models.

CONCLUSIONS

When modelling streamflow processes in ungauged basins, we may sometimes have to deal with univariate streamflow processes. In the present study, two univariate time series models, i.e. the autoregressive moving average (ARMA) model and the fractionally integrated ARMA (ARFIMA) model are applied to four medium-sized watersheds. The result shows that univariate time series models work very well for forecasting daily streamflow processes of the watersheds in cold mountain areas where snow-melt water makes a considerable contribution to runoff. According to the values of coefficient of efficiency (CE), satisfying 7-day ahead forecasts (CE > 0.8) are available for the four cases. According to the season-adjusted coefficient of efficiency (SACE), satisfying forecasts (i.e. SACE > 0.8) can be made for a lead time of three days for two out of four cases (i.e. for the Columbia River at Nicholson and the Columbia River near Fairmont Hot Springs), and one-day ahead satisfying forecasts are available for one case (the Fraser River at Mcbride). And the ARFIMA model built on the basis of deseasonalized daily streamflow series outperforms the ARMA model in all cases and all lead times except for 1- and 2-day ahead forecasts for the Canoe River. The use of the ARFIMA model can therefore be recommended as an alternative to the conventional ARMA model for modelling univariate daily streamflow processes from the perspective of forecasting for the watersheds where snow-melt water contributes considerably to runoff.

Acknowledgements The financial support from the National Science Foundation of China (Project no. 40771039) and the Research Fund for the Doctoral Program of Higher Education (Project no. 20060294003) are gratefully acknowledged.

REFERENCES

Burlando, P., Montanari, A. & Rosso, R. (1996) Modelling hydrological data with and without long memory. *Meccanica* **31**(1), 87–101.

Haslett, J. & Raftery, A. E. (1989) Space-time modelling with long-memory dependence: assessing Ireland's wind power resource (with discussion). *Applied Statistics* **38**, 1–50.

Hipel, K. W. & McLeod, A. I. (1994) *Time Series Modelling of Water Resources and Environmental Systems*. Elsevier, Amsterdam, The Netherlands.

Montanari, A., Rosso, R. & Taqqu, M. S. (1997) Fractionally differenced ARIMA models applied to hydrological time series: Identification, estimation, and simulation. *Water Resour. Res.* **33**(5), 1035–1044.

Nash, J. E. & Foley, J. J. (1982) Linear models of rainfall–runoff systems. In: *Proc. Int. Symp. on Rainfall-Runoff Modelling*. (ed. by V. P. Singh), 51–66. Water Resources Publications, Mississippi State University, USA.

Ooms, M. & Franses, P. H. (2001) A seasonal periodic long memory model for monthly river flows. *Environ. Modelling & Software* **16**, 559–569.

Rao, A. R. & Bhattacharya, D. (1999) Hypothesis testing for long-term memory in hydrological series. *J. Hydrol.* **216**(3–4), 183–196.

Wang, W., van Gelder P. H. A. J. M., Vrijling, J. K. & Ma, J. (2004) Predictability of streamflow processes of the Yellow River. In: *Proc. 6th Int. Conf. on Hydroinformatics*, 1261–1268. World Scientific, Singapore.

Wang, W., van Gelder P. H. A. J. M., Vrijling, J. K. & Chen, X. (2007) Detecting long-memory: Monte Carlo simulations and application to daily streamflow processes. *Hydrol. Earth Syst. Sci.* **11**, 851–862.

Effects of land use changes on hydrological responses of the Mekong River

SHAN XIE[1,2], TIANQI AO[1], LINGLING WU[1] & WEI FENG[1]

1 *Department of Hydrology and Water Resources, College of Water Resources & Hydropower, Sichuan University, Chengdu, Sichuan 610065 China*
shanxie_bj@163.com

2 *Wuhan Water Conservancy Planning & Designing Research Institute, Wuhan, Hubei 430015 China*

Abstract In hydrological modelling, one of the challenges is how to reflect the effects of physical basin features on hydrological responses. This paper focuses on investigating the effects of land use/land cover changes on hydrological responses for the Mekong River. In simulation, the strategy of gridded parameterization is explored for the physically based distributed hydrological model: block-wise use of the TOPMODEL with the Muskingum-Cunge routing method (BTOPMC); the basin is subdivided into 737 natural sub-basins; land use of the upper Mekong basin is changed into impervious area; rainfall and discharge data sets from 1980–1982 were used for optimization by the SCE-UA algorithm; data sets of 1985 were used for validation. The results indicate that land use changes will effect hydrograph and flow components.

Key words BTOPMC model; hydrological responses; land use changes; Mekong River

INTRODUCTION

Since the 20th century, the urbanization process in many countries has become quicker, especially in developing countries; concrete land cover is continually increasing and vegetation area is comparatively decreasing. Such human-made land-use changes will result in influences to the hydrological responses of a watershed, especially to the flood-runoff process. Quantitative studies on the effects of physical basin features on hydrological responses have significant value in protecting life, property and the natural environment in cities and downstream areas. Existing research indicates that parameters of physically-based distributed hydrological models can represent different physical basin features separately, such as vegetation, soil type, land use, etc. This kind of rainfall–runoff model has great potential for simulating hydrological responses resulting from natural and man-made land-use changes.

In this study, the Mekong River 3-year (1980–1982) hydrometeorological data sets are used for parameter estimation through calibration/optimization, in which the physically-based distributed hydrological model, block-wise use of the TOPMODEL with the Muskingum-Cunge routing method (BTOPMC), is used. Obtained parameters are used to simulate 1985 daily runoff of the five study sites in the lower Mekong mainstream. In simulation, land use/land cover of upper Mekong is assumed as concrete to investigate the effects on hydrological responses.

BTOPMC MODEL STRUCTURE AND PARAMETERS

BTOPMC is a physically-based distributed hydrological model which consists of several sub-models. It has the advantages of both lumped and distributed models (Ao *et al.*, 1999; Takeuchi *et al.*, 1999). There are only five parameters that need to be specified and they are briefly introduced as follows:

Topographic sub-model and its parameters

Automatic topographic analysis tools are used in BTOPMC to get all topographic variables. The watershed is described by drainage networks extracted from digital elevation maps (DEMs), in which all pits are filled with calculated small elevation increments (Ao *et al.*, 2001, 2003a). The topographic index, λ_i, of any grid cell, *i* (Beven & Kirkby, 1979; Quinn *et al.* 1995) is calculated by the following equation:

$$\lambda_i = \ln(a_i / \tan \beta_i) \tag{1}$$

where a_i is the drainage area per contour length, $\tan\beta_i$ denotes the slope of grid cell, i.

The basin is then divided into sub-basins, whose average topographic indices $\lambda(k)$ are calculated as:

$$\lambda(k) = \frac{1}{N_p(k)} \sum_{i=1}^{N_p(k)} \lambda_i \tag{2}$$

where k is sub-basin code and $N_p(k)$ is the total number of grid cells belonging to sub-basin, k. Equations (1) and (2) demonstrate the topographic parameters in BTOPMC. These are specified by DEM and subdivision level (Ao *et al.*, 2002, 2003b).

Runoff generation sub-model and its parameters

The assumptions and concepts of TOPMODEL are used for runoff generation for each grid cell by applying to each sub-basin. The effective precipitation $R_e(i,t)$ stored in the root zone at time step t is represented as:

$$R_e(i,t) = R_o(i,t) - S_{rmax}(i) - E_p(i,t) \tag{3}$$

where $E_p(i,t)$ denotes evapotranspiration; $R_o(i,t)$ is total precipitation approximated by the Thiessen method; S_{rmax} is maximum storage capacity of the root zone and is a parameter reflecting the effects of vegetation interception and soil moisture. The storage in the unsaturated zone is calculated as:

$$S_{uz}(i,t) = S_{uz}(i,t-1) + R_e(i,t) - q_v(i,t) \tag{4}$$

where $q_v(i,t)$ is the recharge to groundwater calculated by:

$$q_v(i,t) = T_0(k)\exp[-S(i,t)/m(k)] \tag{5}$$

where T_0 is the saturated soil transmissivity in m^2/h and m is the decay factor (in metres) of T_0. $S(i,t)$ is the local saturation deficit in metres that can be thought of as the distance from groundwater level to ground surface. It is determined by:

$$S(i,t) = S_{bar}(k,t) + m(k)[\lambda(k) - \lambda_i] \tag{6}$$

where $S_{bar}(k,t)$ is the average saturation deficit in meters of sub-basin k and calculated as:

$$S_{bar}(k,t) = S_{bar}(k,t-1) + \frac{1}{N_p(k)} \sum_{i=1}^{N_p(k)} (q_b(i,t) - \frac{1}{N_p(k)} \sum_{i=1}^{N_p(k)} q_v(i,t) \tag{7}$$

where $q_b(i,t)$ is the baseflow flowing into streams as calculated by:

$$q_b(i,t) = T_0(k)\exp[-S(i,t)/m(k)]\tan\beta_i \tag{8}$$

For $t = 1$, the initial value of the average saturation deficit, $S_{bar0}(k)$, which is a parameter to be calibrated, indicates $S_{bar}(k,0)$. As shown in equations (3), (5) and (7), in the runoff generation sub-model there are four parameters in total to be calibrated: $S_{rmax}(i)$, $T_0(K)$, $m(k)$ and $S_{bar0}(k)$.

Flow routing sub-model and its parameters

The Muskingum-Cunge (M-C) method (Cunge, 1969) is adopted for the flow routing sub-model. The equivalent Manning roughness coefficient for each grid cell is estimated/calibrated according to its land cover and soil type.

Evaluation criteria and objective function

Model performance is evaluated by the Nash-Sutcliffe Efficiency (E) measure and the Volume Ratio (V_r) of simulated to observed discharges. The objective function for optimization is designed as below:

$$F = 1 - E \; ; \; E = \sum_{i=1}^{N} w_i E_i \tag{9}$$

where N denotes the total number of calibration sites in study basin, w is the weighting factor and the sum of w_i equals 1.0. Here, each w_i equals 0.2 (Ao, 2004, 2005).

BASIN DESCRIPTION AND CALCULATION CONDITIONS

Basin description

The Mekong Basin, which is one of the largest international river basins in the world, is located in southeast Asia. Its headwater originates from the Tibetan highlands (elevation 4968 m) of China, and then flows through six countries: China, Myanmar, Laos, Thailand, Kampuchea and Vietnam, continuing in a southeasterly direction to the South China Sea. The basin has a drainage area of about 795 500 km^2. The length of the river is about 4620 km (excluding tributaries), and the basin is shaped like a violin; the upper basin being narrow and the lower part being wide. The basin's mean annual rainfall is about 1672 mm, and the average annual discharge is about 14 000 $m^3 s^{-1}$, measured at a site located approx. 545 km from the outlet (Hori, 1996). The five sites for simulating simultaneously along the lower portion of the mainstream are Chiang Saen, Luang Prabang, Vientiane, Mukdahan and Pakse, of which the total drainage area is 189 000–545 000 km^2, as shown in Fig. 1.

Topographic analysis

The DEM of the Mekong Basin was downloaded from the GTOPO30 database. For the purpose of shortening computational time, the grid cell size was enlarged from 1 × 1 km to 5.3 × 5.54 km by

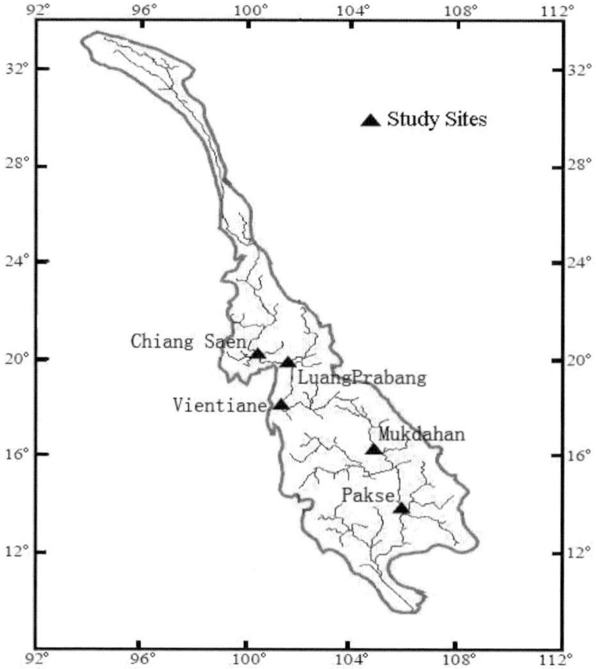

Fig. 1 Digital stream network of the Mekong River and the five study sites.

using the minimum elevation method. The digital drainage network of this basin is extracted from the pre-processed DEM. The whole basin was divided into 737 sub-basins with natural boundaries.

Data used

The 4-year (1980–1982 and 1985) hydro-climatic data sets are used, which include daily distributed basin precipitation of 43 raingauges obtained from the NOAA CD-ROM and the Yearbook of the Lower Mekong Basin, and daily observed stream discharges for the study sites attained from the Yearbook of the Lower Mekong Basin published in Thailand. The land cover map of IGBP Version 2 (USGS) and FAO soil map are used to get distributed data of land cover and soil types for each grid cell. Land cover is classified into four kinds. Three soil types which are clay, sand and silt have 108 combinations. The annual potential evapotranspiration was assumed to be 1700 mm.

Optimization description

The optimization is conducted by the SCE-UA algorithm (Duan *et al.*, 1994). Based on physical meaning of BTOPMC model parameters and modelling experience, all the five parameters of BTOPMC are optimized for each type of soil, land cover and topography. The three-year low flow rate during 1980–1982 was used to get parameter values, as shown in Table 1.

Table 1 Optimized model parameter values.

Parameter and unit	Physical interpretation	Physical properties	Final value
T_0 (m^2/s)	Saturated soil transmissivity	Clay	99.625
		Sand	298.098
		Silt	198.244
m (m)	The decay factor of T_0	Clay	0.004
		Sand	0.229
		Silt	0.027
$S_{r\max}$ (m)	Maximum storage capacity of the root zone	Deep rooted	0.017
		Shallow rooted	0.025
		Shallow rooted and irrigated	0.020
		Impervious area	0.003
S_{bar0} (m)	Initial value of averaged soil saturation deficit	$0 \leq \lambda \leq 10$	0.545
		$11 \leq \lambda \leq 20$	0.261
		$21 \leq \lambda \leq 30$	0.698
		$31 \leq \lambda \leq 40$	0.484
n_0	Equivalent Manning roughness coefficient	River mesh	0.015
		Built up area	0.002
		clay	0.043
		sand	0.043
		silt	0.044

Calculation strategy

To investigate the effects of land-use changes on hydrological responses, two cases are detailed here, and the same parameter values for each kind of physical basin features are shown in Table 1. Case 1 is to use original land cover and vegetation conditions for simulation of daily discharge of 1985. In case 2, land use/land cover of all the upper Mekong Basin, in which latitude ranges from 23 to 34, is assumed as impervious area. The hydrographs of the two cases are compared to investigate the hydrological effects of land-use changes.

Table 2 Comparison of hydrological responses of the two cases for the five study sites (validation).

Indices		Study sites	Chiang Saen	Luang Prabang	Vientiane	Mukdahan	Pakse
Average daily discharge ($m^3 s^{-1}$)	Surface	Case1	294	1071	1141	3123	4751
		Case2	896	1749	1836	3929	5784
	Subsurface	Case1	1606	2695	2834	3941	4525
		Case2	402	1441	1570	2632	3160
	Total	Case1	1900	3766	3975	7067	9276
		Case2	1298	3190	3406	6561	8944
Peak discharge ($m^3 s^{-1}$)	Surface	Case1	1764	5988	6109	19234	23543
		Case2	3520	7680	7896	19103	23657
	Subsurface	Case1	2725	5138	5482	8778	9811
		Case2	772	3683	3988	7185	8124
	Total	Case1	3615	10850	11430	25141	30457
		Case2	4248	11178	11830	23824	29315

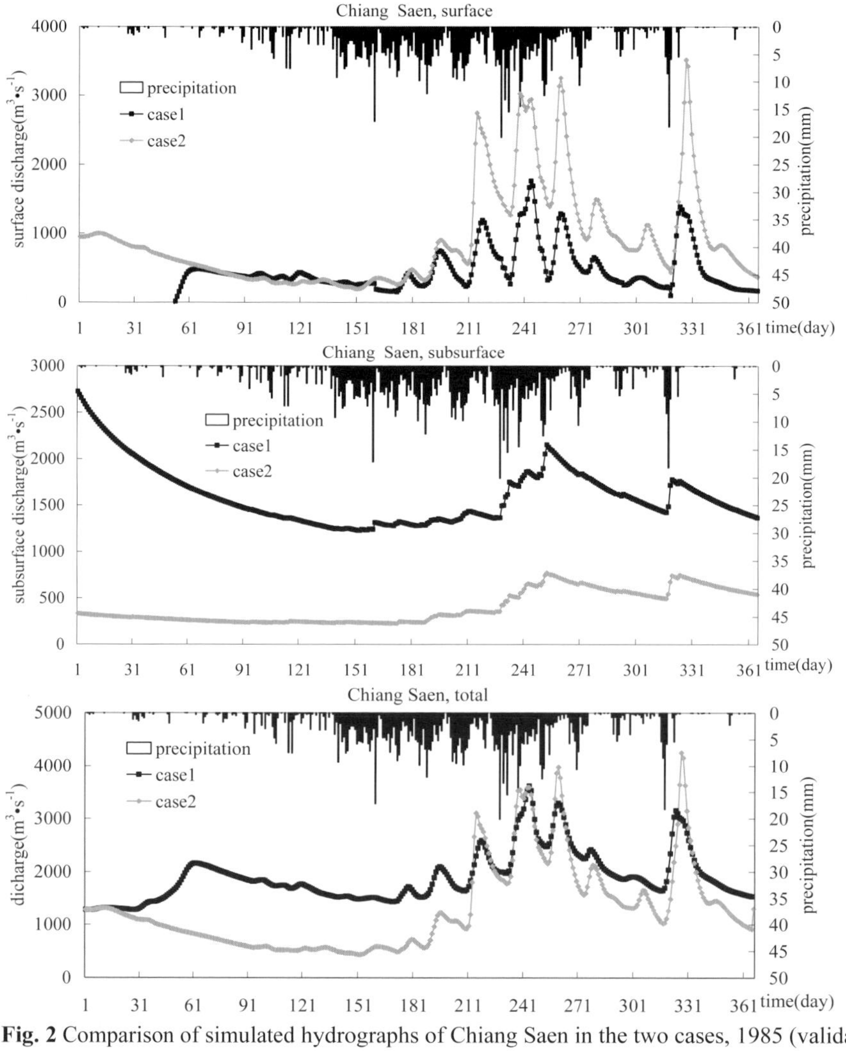

Fig. 2 Comparison of simulated hydrographs of Chiang Saen in the two cases, 1985 (validation).

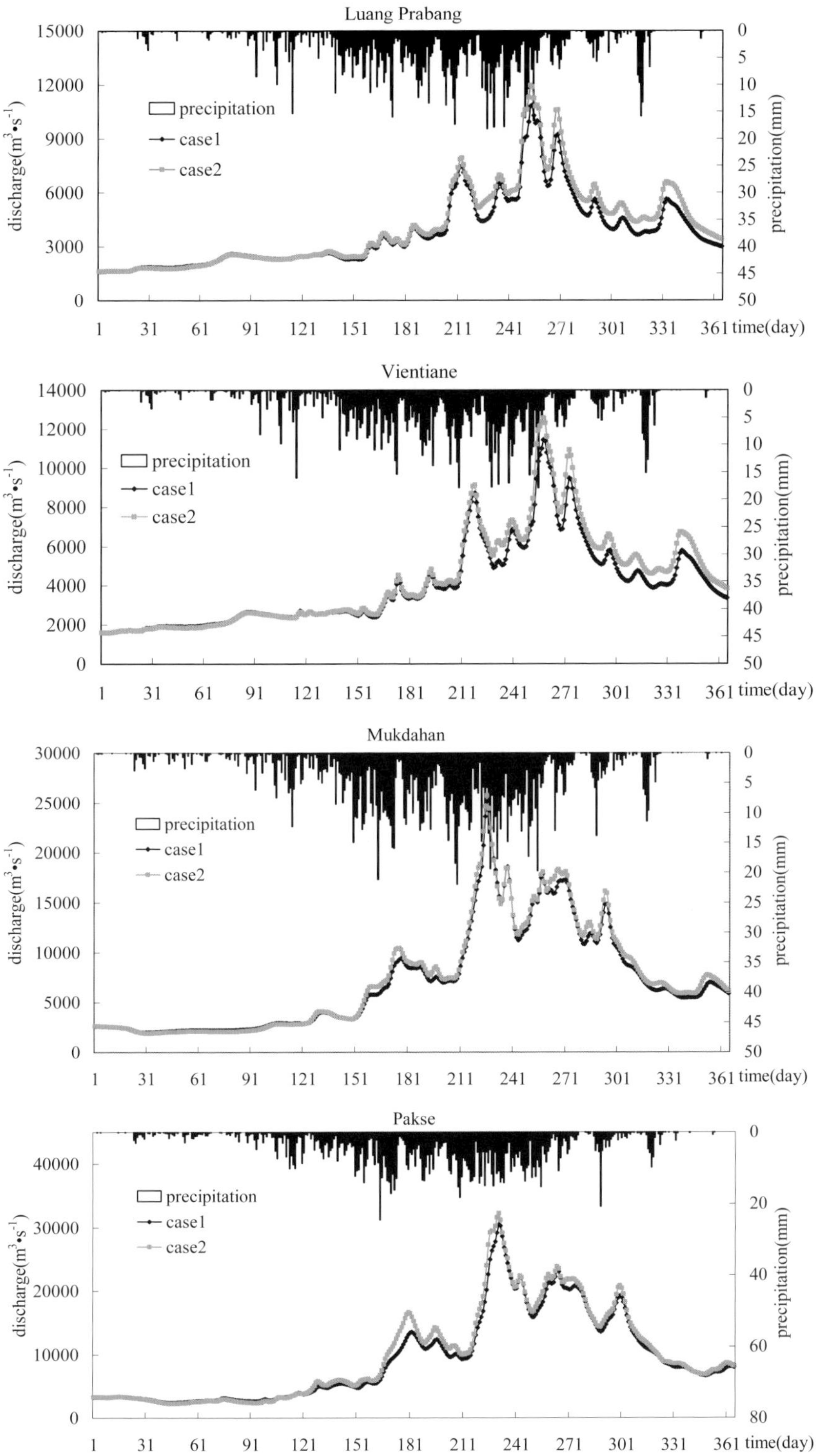

Fig. 3 Comparison of simulated hydrographs of the other four study sites in the two cases, 1985 (validation).

RESULTS AND ANALYSIS

Using the calculation strategy described above, two sets of BTOPMC validation performances corresponding to the two cases are summarized in Table 2. As validation results, the hydrograph of the first study site, Chiang Saen, of which all drainage area is changed into impervious area, is shown in Fig. 2, and the hydrograph of the other four study sites are shown in Fig. 3. From Table 2 and Figs 2 and 3, it can be seen that:

- In Table 1, optimized model parameters can be thought of as physically reasonable, these parameter values approximately reflected their corresponding physical basin features. Furthermore, from Table 2 and Figs 2 and 3, it can been seen that different land use leads to different hydrological responses, implying that BTOPMC and its parameters have the potential to reflect the effects of basin changes on its hydrological responses.
- Table 2 indicates that, for all the five study sites, land use changes result in changes of flow component. In detail, as the concrete area increases the ration of surface flow is increased while subsurface flow is decreased. This is reasonable because in the former case less water is available for infiltration.
- Table 2, Figs 2 and 3 also indicate that land use changes will influence peak flow. Peak flow increases with the increase of concrete area in the upper drainage area. This can be explained as concrete area generates more runoff and makes water flow fast.
- From Figs 2 and 3, it can be seen that for different study sites the degree of the effect is different. In this study, the upper study sites get more influence of hydrological responses than lower study sites. This can be considered as the effects of multiple factors, such as the distribution of rainfall, branches, reservoirs, etc.
- Table 2 indicates that land-use changes will influence the total amount of annual runoff. In this study, the bigger the ratio of concrete drainage area, the smaller the total runoff. This could be partly considered as the effects of evaporation, namely, more water is evaporated in the case of having a larger concrete drainage area.

CONCLUSIONS

In this study, physically-based distributed hydrological model BTOPMC was applied to the Mekong River to explore the effects of land use/land cover changes on hydrological responses. The preliminary main findings are summarized as follows:

- Model parameters obtained from gridded optimization can approximately reflect corresponding physical basin features such as land use/land cover, vegetation and soil types.
- Assuming land use of the upper Mekong as impervious concrete area, hydrographs and flow components at the lower Mekong River are different. This implies that BTOPMC and its parameters have the potential to simulate the effects of physical basin feature changes on the basin's hydrological responses.

However, due to the limitation of data set, scale problem, etc., further study is needed in the future to get more reasonable and accurate simulation results.

REFERENCES

Ao, T. Q., Ishidaira, H. & Takeuchi, K. (1999) Study of distributed runoff simulation model based on block type TOPMODEL and Muskingum-Cunge method. *Ann. J. Hydraul. Engng, JSCE* **43**, 7–12 (in Japanese).

Ao, T. Q., Takeuchi. K. & Ishidaira, H. (2001) On the method of generating artificial stream networks of large river basins and its effect on runoff simulation. *Ann. J. Hydraul. Engng, JSCE* **45**, 139–144 (in Japanese).

Ao, T. Q., Yoshitani, J., Fukami, K., Mutsuura, T., Takeuchi, K. & Ishidaira, H. (2002) Topographic analysis for distributed runoff model BTOPMC when applied to large river basins (in Japanese). *Ann. J. Hydraul. Engng, JSCE* **47**, 193–198.

Ao, T. Q., Takeuchi. K., Ishidaira, H., Yoshitani, J. & Fukami, K. (2003a) Development and application of a new algorithm for automated pit removal for grid DEMs. *Hydrol. Sci. J.* **48**(6), 985–997.

Ao, T. Q., Yoshitani, J., Takeuchi, K., Fukami, K., Mutsuura, T., Ishidaira, H. (2003b) Effects of sub-basin scale on runoff simulation in distributed hydrological model: BTOPMC. In: *Weather Radar Information & Distributed Hydrological* (ed.

by Y. Tachikawa, B. E. Vieux, K. P. Georgakakos & E. Nakakita), 227–233. IAHS Publ. 282. IAHS Press, Wallingford, UK.

Ao, Tianqi, Takeuchi, Kuniyoshi, Ishidaira, Hiroshi, Zhou, Maichun, Hapuarachchi, H. A. P., Hirabayashi, Yukiko & Magome, Jun (2004) On the effects of calibration locations on parameter transfer functions for the Mekong River. In: *Advances in Integrated Mekong River Management* (Proc. Int. Conf. Vientiane, the Lao PDR, 25–27 October), 34–39.

Beven, K. J. & Kirkby, M. J. (1979) A physically-based, variable contributing area model of hydrology. *Hydrol. Sci. Bull.* **24**(1), 43–69.

Cunge, J. A. (1969) On the subject of a flood propagation computation method (Muskingum method). *J. Hydraul. Res.* **7**(2), 205–230.

Duan, Q., Sorooshian, S. & Gupta, V. K. (1994) Optimal use of the SCE-UA global optimization method for calibrating watershed models. *J. Hydrol.* **158**, 265–284.

Dunn, S. M. & Lilly, A. (2001) Investigation of the relationship between a siols classification and the spatial parameters of a conceptual catchment-scale hydrological model. *J. Hydrol.* **252**, 157–173.

Hori, H. (1996) *The mekong: The Development and its Environmental Effects*. Published by KOKON-SHOIN Publishing Co., Ltd, Japan (in Japanese).

Quinn, P. F., Beven, K. J. & Lamb, R. (1995) The ln(a/tanβ) Index: how to calculate it and how to use it within the TOPMODEL framework. *Hydrol. Processes* **9**, 161–182.

Takeuchi, K., Ao, T. Q., Ishidaira, H. (1999) For hydro-environmental simulation of a large ungauged basin – introduction of block-wise use of TOPMODEL and Muskingum-Cunge method. *Hydrol. Sci. J.* **44**(4), 633–646.

Analysis of water resources variability in the Yellow River of China using a distributed hydrological model

DAWEN YANG & WEIWEI SHAO

State Key Laboratory of Hydro-Science and Engineering, Department of Hydraulic Engineering, Tsinghua University, Beijing 100084, China

yangdw@tsinghua.edu.cn

Abstract Drying up of the main river along the lower reach of the Yellow River has occurred since 1972, and the situation has become more and more serious over the last 30 years. By incorporating historical meteorological data and the available geographic information related to the land surface conditions, a distributed hydrological model (GBHM) has been employed to simulate the natural runoff of this basin. Based on the simulated and observed river discharges, a quantitative analysis of the water resources in this basin over a long-term period of 50 years was carried out, which focuses on understanding the variability in the water resources due to climate change and human activity. In contrast to what might be believed based on common sense, it is found that the main reason for the aggravation of the drying up of the main river along the lower reach that occurred in the 1990s is climate change.

Key words Yellow River; water shortage; climate change; human activity; distributed hydrological model

INTRODUCTION

The scarcity of water resources is a major constraint for social and economic development in northern China and has led to widespread degradation of the natural ecosystem and environment. During the last 30 years the Yellow River has become a seasonal river. The drying up in the main river along the lower reaches started in 1972 and increased rapidly. In the most serious situation in 1997, the main river close to the sea dried up for more than 226 days, and the no-flow distance reached 704 km from the river mouth. Water shortages, especially the drying up of the main river along the lower reach in the Yellow River, have drawn a lot of attention from all over the world. Many studies have focused on finding the reason for this event. As a result of these studies, climate change and an increase in artificial water uses have been identified as the two main factors leading to the drying up of the river (Cheng *et al.*, 1999; Yang *et al.*, 2004). However, the details are not clear as to the effects of climate change and human activity over the long term.

On the other hand, historical climate data are generally available and provide a reference for studying the impact of climate change on water resources. However, due to the nonlinear behaviour of hydrological processes, changes in climate can not easily be converted into amounts of water resource change. Natural water resources need to be assessed by hydrological models using historical climate data as the input. Water balance models are commonly used for such purposes (Arnell, 1999). Middelkoop *et al.* (2001) compared a water balance model and a physically-based model for water resources assessment in the Rhine basin. It was found that a physically-based model did a better job than the water balance model in many cases. For the present study, an assessment of the natural river discharge was carried out by applying a distributed physically-based model for simulating the hydrological cycle over the last 50 years in the Yellow River basin. Based on the long-term hydrological simulation, the variability of water resources due to climate change and artificial uses are discussed.

METHODOLOGY

Data used in the study

The geographical information concerning the Yellow River basin used in this research is obtained from a number of global data sets. The digital elevation data of 1-km resolution is obtained from the USGS HYDRO1k data set. Land cover is obtained from the USGS Global Land Cover

Characteristics Data Base Version 2.0. This data has a spatial resolution of 1-km. For each vegetation type, a monthly leaf-area-index (LAI) is calculated from the monthly NDVI. A global data set of monthly NDVI with 8-km resolution is obtained from the DAAC of GSFC/NASA. This data set is available from 1982 onwards. The soil type is obtained from the Digital Soil Map of the World and Derived Soil Properties (FAO, 2003). It is developed at 5-minute resolution using the FAO-UNESCO soil classification. The soil properties used for the hydrological simulation include the porosity, the saturated hydraulic conductivity, and the other soil water parameters corresponding to each soil type in this map, are obtained from the Global Soil Data Task (IGBP-DIS, 2000). The water-retention relationship and unsaturated hydraulic conductivity are represented by Van Genuchten's formula, and the parameters are available in this data set.

The meteorological data from 1951 to 2000 were obtained from the China Administration of Meteorology, and is available at a daily temporal resolution at 108 gauges inside and close to the Yellow River basin (see Fig. 1). The discharge data collected before 1990 is from the "Hydrological Year Book" published by the Hydrological Bureau of the Ministry of Water Resources of China.

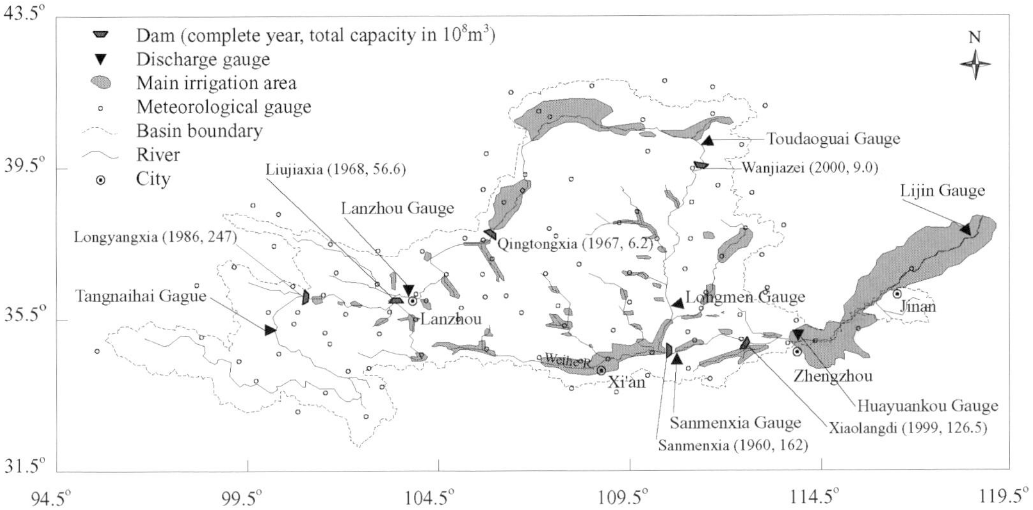

Fig. 1 The Yellow River basin.

Description of the distributed hydrological model-GBHM

The present study employs a distributed model for estimating the natural runoff in this basin over the last half century. The model uses a grid system with a 10-km spatial resolution, and runs in hourly time steps. Since there are significantly less drainage areas in the lower reach, most of the runoff is generated from the upstream of the Huayuankou gauge (see Fig. 1), and therefore the hydrological simulation only covers this area.

For subdividing the Yellow River basin, the Pfafstetter scheme (Yang *et al.*, 2003) is applied in the present study and a total of 137 sub-basins have been identified in the upstream of the Huayuankou gauge. The sub-grid parameterization used includes representations of the sub-grid variability in topography and land-cover. The topographical parameterization uses the catchment geomorphologic properties, which represent a grid by a number of hillslopes. The hillslopes located in a 10-km grid are grouped according to the land cover types. The hillslope is a fundamental computational unit for hydrological simulation. The geomorphology-based hydrological model (GBHM) (Yang *et al.*, 2002) is used for simulating the hillslope hydrology. The hydrological processes included in this model are snowmelt, canopy interception, evapotranspiration, infiltration, surface flow, subsurface flow and the exchange between the groundwater and the river (Yang *et al.*, 2002). The actual evapotranspiration is calculated from the potential evaporation by considering seasonal variation of LAI, root distribution and soil moisture availability. This is computed individually from the canopy water storage, root zone and soil surface. Infiltration and water flow in the subsurface in the vertical direction and along the hillslope are described in a quasi-two-dimensional (2-D) subsurface model. The vertical water flow in the topsoil is represented by Richards equation and solved by an implicit numerical solution scheme. In this scheme, the topsoil is divided into a near surface layer of 5 cm, a root zone and a deep zone. The root zone and deep zone are again derived into sub-layers in the present model. The first layer is expected to be saturated during the rainfall period. Therefore, the upper boundary condition is given as constant soil water content for the rainfall cases. During the non-rainfall period, evaporation from the soil surface occurs, and the upper boundary condition is given as a constant flux. The soil water distribution along the hillslope is treated as uniform. The surface runoff is from the infiltration excess and saturation excess calculated by solving Richards equation. The surface runoff flows through the hillslope into the stream by a kinematic wave. The groundwater aquifer is treated as an individual storage corresponding to each grid. The exchange between the groundwater and the river water is considered as steady flow and is calculated by Darcy's law (Yang *et al.*, 2002). The runoff generated from the grid is the lateral inflow into the river at the same flow interval. Flow routing in the river network is solved using the kinematic wave approach.

Model calibration and validation

A 5-year test run from 1981 to 1985 was carried out to calibrate the model parameters. One of the calibrated parameters is the snowmelt factor in the temperature-based snowmelt equation. Another calibrated parameter, the hydraulic conductivity of the groundwater, is calibrated by checking the baseflow in different sub-basins. Model validation was carried out from 1986 to 1990 in the upstream of the Tangnaihai discharge station, where the human activity is insignificant, and the snowmelt runoff and groundwater flow are the main sources of river discharge. Figure 2 shows a comparison between the simulated and observed daily discharges at the Tangnaihai gauge for both the calibration and validation periods. Based on the daily discharges, the ratio of the absolute error to the mean and the Nash coefficient are calculated to be 19% and 0.88, respectively, for the calibration period and to be 17% and 0.89 for the validation period, respectively. A good agreement between the simulated daily hydrograph with the observed and a consistency of the simulations in both the calibration and validation periods are achieved.

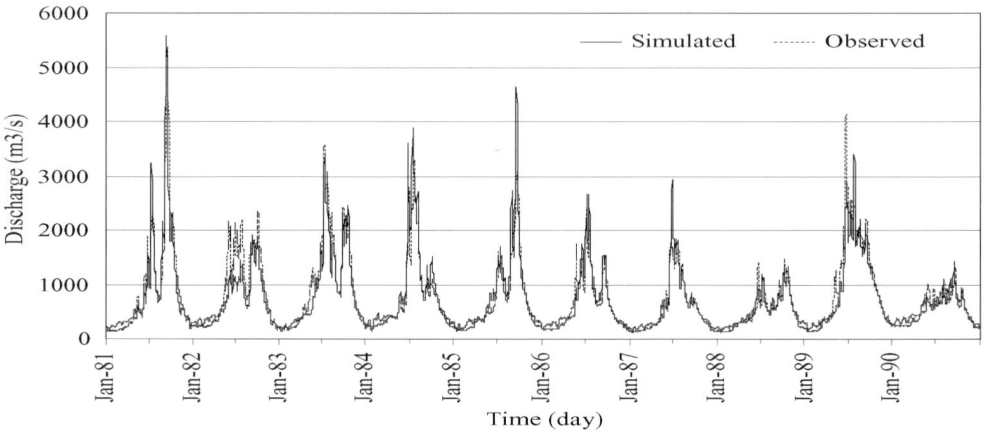

Fig. 2 Comparison of simulated and observed daily river discharge at Tangnaihai station.

ANALYSIS OF WATER RESOURCES IN THE LAST HALF CENTURY

Spatial and seasonal distribution of water resources for a long-term mean

By checking the 50-year average water balances covering the total study area, it was found that the annual precipitation, evaporation and runoff are 439 mm, 362 mm and 77 mm, respectively. The annual runoff consists of only 17.5% of the annual precipitation. High spatial variability of the hydrological characteristics can be seen in this basin. The annual precipitation ranges from 154 to 764 mm, which increases from north to south and from west to east. The annual actual evapotranspiration ranges from 137 to 589 mm (Fig. 3(a)). The Weihe basin and downstream of the Sanmenxia Dam have the highest evapotranspiration. The annual runoff ranges from 0 to 345 mm (Fig. 3(b)). The major source areas are located upstream of the Lanzhou gauge, the southern part of the Weihe River basin, and downstream of the Sanmenxia dam.

Considering the seasonal precipitation and water uses, the year may be divided into a dry season from November to June and a wet season from July to October. Table 1 shows the seasonal characteristics of the water resources in the study basin. The highly uneven distribution of the seasonal precipitation can be seen from Table 1, in which about 64% of the annual precipitation concentrates within the wet season from July to October. By the land surface hydrological processes, this seasonally-uneven distribution of precipitation is mediated in the river discharge in the upper regions but is amplified in the lower regions. For the whole of the simulated area, about 60% of the annual river discharge flows within the wet season from July to October.

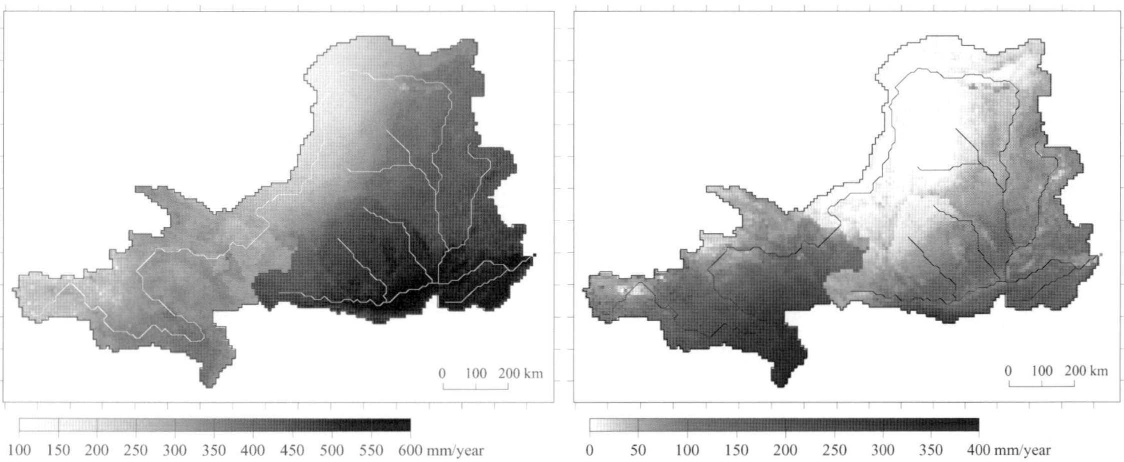

Fig. 3 Spatial distributions of 50-year means: (a) annual evapotranspiration, (b) annual runoff.

Table 1 Long-term average water balances from 1951 to 2000 for the dry season (November–June) and the rainy season (July–October).

Section *	Annual precipitation (mm)		Annual actual evaporation (mm)		Simulated annual river discharge (mm)	
	Dry	Rain	Dry	Rain	Dry	Rain
Upstream of Lanzhou	173	274	121	189	57	79
Lanzhou–Toudaoguai	83	180	113	139	6.6	5.6
Toudaoguai–Longmen	135	285	184	198	14.8	23.6
Longmen–Sanmenxia	204	344	238	230	27	53.7
Sanmenxia–Huayuankou	247	384	258	253	43.3	77
Whole study area	159	280	168	194	30	47

* The Yellow River basin is divided into six sections by the discharge gauges which correspond to the different characteristics of hydrology and water uses (see Fig. 1).

Reason for the drying-up of the main river along the lower reach over the last 30 years

Table 2 gives the historical records of river drying-up along the lower reaches over the last 30 years. A drying-up of the main river along the lower reach occurred from 1972, the situation continued during the 1980s and became worse in the 1990s. Figure 4 compares the annual available runoff and consumption with the no-flow duration along the lower reach in this basin over the last 30 years (1971–2000). It can be seen that the dry-ups occurred in the relatively dry years, especially during the whole of the 1990s, during which time the annual available runoff was less than the 30-year mean of 55 billion cubic meters. However, in contrast to common sense, the "drying-up" years do not correspond to the years of relatively high consumption of surface water. The annual variability in the consumption corresponds to the runoff availability. When less runoff is available, less surface water is consumed. The annual consumption of surface water shows a slightly decreasing trend is relation to the annual available runoff. It is concluded here that the aggravation of the drying-up situation in the 1990s is mainly caused by climate change. Table 3 summarizes the decadal averages of the water balance components for the whole simulation area. The variability of the natural runoff has a 10-year cycle. Comparing the basin averages in the two decades of the 1980s and the 1990s, the annual precipitation decreased by 38 mm and the annual mean temperature increased by 0.6°C. Consequently, the naturally available annual runoff decreased 24 billion cubic meters (36%) in the 1990s. In fact, the annual consumption of surface water in the 1990s decreased by 6.4 billion cubic meters (18%) compared to the 1980s. The ratio of the annual water consumption to the annual water available is 55% in the 1980s, but reaches 67% in the 1990s. As the result of both climate change and artificial consumption, annual runoff flowing into the sea decreases by nearly 43%. Keeping the similar trend of climate change, the Yellow River might cease to flow into the sea in future.

Table 2 Historical records of river drying up in the last 50 years.

Year	1971	1972	1973	1974	1975	1976	1977	1978	1979	1980
Days of no flow		19		20	13	8		5	21	8
Maximum distance of no flow (km)		310		316	278	166		104	278	104
Year	1981	1982	1983	1984	1985	1986	1987	1988	1989	1990
Days of no flow	36	10	5				17	5	24	
Maximum distance of no flow (km)	662	278	104				216	150	277	
Year	1991	1992	1993	1994	1995	1996	1997	1998	1999	2000
Days of no flow	16	83	60	74	122	136	226	142	42	
Maximum distance of no flow (km)	131	303	278	380	683	579	704	449	278	

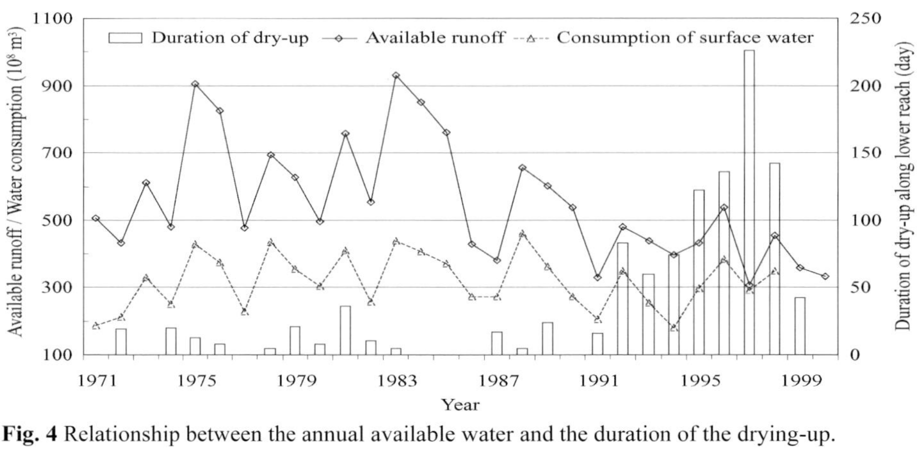

Fig. 4 Relationship between the annual available water and the duration of the drying-up.

Table 3 Decadal variations of water resources in the whole simulation area.

Component	1950s	1960s	1970s	1980s	1990s
Precipitation (mm/yr)	440.2	466.0	436.3	446.6	408.5
Annually mean temperature (°C)	6.22	6.38	6.61	6.69	7.31
Natural available water ($\times 10^8$ m^3/yr)	572.3	763.8	607.1	648.5	407.8
Artificial water consumption* ($\times 10^8$ m^3/yr)	134.0	236.4	312.4	355.1	273.4
Ratio of consumption to available (%)	23.4	30.9	51.5	54.8	67.1

* The difference between simulated and observed river flows.

For improving the no-flow condition along the lower reaches, two urgent countermeasures should be carried out: (1) protection of the water source areas, especially the upstream of the Lanzhou gauge, and (2) saving water usage along the middle and lower reaches. Protection of the upstream of the Lanzhou gauge should be emphasized here because the Chinese government is now promoting western development.

CONCLUSION

The spatial and seasonal distributions of water resources show high variability in the Yellow River basin. The major source area of the basin total annual runoff is located in the upstream of the Lanzhou gauge. About 60% of the annual precipitation is concentrated within the 4 months from July to October. The aggravation of the drying up situation in the 1990s is mainly a result of climate change because the 1990s were the driest period experienced in the last half century. A reduction in the water requirements from in the Yellow River basin is the only solution to the water shortage problem in the near future. The prediction of climate change and its effect on future water resources is also helpful for their management.

Acknowledgements This research was supported by the National Natural Science Foundation of China (no. 50721140161). The authors would like to express their appreciation for the grants received to aid in this research.

REFERENCES

Arnell, N. W. (1999) Climate change and global water resources. *Global Environ. Change* **9**, S31–S49.

Cheng, X., Li, X. & Lu, G. (1999) Characteristics of the river dry-up and variation of water resources in the Yellow River basin. *Adv. Sci. & Technol. of Water Resour.* **1**, 34–37 (in Chinese).

FAO (2003) Digital soil map of the world and derived soil properties. *Land and Water Digital Media Series Rev.* **1**.

IGBP-DIS (2000) Global Soil Data Products CD-ROM. International Geosphere-Biosphere Programme, Data and Information System, Potsdam, Germany. Available from Oak Ridge National Laboratory Distributed Active Archive Center, Oak Ridge, Tennessee, USA. http://www.daac.ornl.gov.

Middelkoop, H., Daamen, K., Gellens, D., Grabs, W., Kwadijk, J., Lang, H., Parmet, B., Schadler, B., Schulla, J. & Wilke, K.. (2001) Impact of climate change on hydrological regimes and water resources management in the Rhine basin. *Climate Change* **49**, 105–128.

Yang, D., Herath, S. & Musiake, K. (2002) Hillslope-based hydrological model using catchment area and width functions. *Hydrol. Sci. J.* **47**(1), 49–65.

Yang, D. & Musiake, K. (2003) A continental scale hydrological model using distributed approach and its application to Asia. *Hydrol. Processes* **17**, 2855–2869.

Yang, D., Li, C., Hu, H., Lei, Z., Yang, S., Kusuda, T., Koike, T. & Musiake, K. (2004) Analysis of water resources variability in the Yellow River basin during the last half century using the historical data. *Water Resour. Res.* **W06502**.

Application of the grey self-memory neural network model for annual runoff forecasting

XIAOWEI ZHANG, LINGMEI HUANG & BING SHEN

Northwest Key Lab of Water Resources and Environment Ecology, Ministry of Education, Xi'an University of Technology, Xi'an 710048, China

dbh0913@163.com

Abstract A runoff time series represents the hydrological response to a catchment, which is a nonlinear, weakly dependent, and highly complicated dynamic system. The key to improving the accuracy of runoff forecasting is to dig the information from the limited sample sufficiently. Grey system modelling reveals the law of system dynamics by processing grey information; self-memory theory emphasizes the fore-and-aft relation of system status, which urges upon evolutional rules of system itself. Then differential equations of the dynamic system could be built for the self-memory models. Combination of grey theory and self-memory theory can effectively represent the trend of the extreme values for the time series, but there is a phase lag between observed values and predicted values. The neural network has an advantage of solving nonlinearity of the system. Based on the idea of integrative forecasting, the grey self-memory neural network model was established, which can be used to predict the annual runoff series.

Key words grey system; self-memory model; neural network; annual runoff

INTRODUCTION

Runoff forecasting, being a very important problem in the hydrology and water resources system, is the basic reference for basin planning and design, water resources assessment, reservoir regulation and inter-basin water transfer.

The hydrology system is a complicated nonlinear system (Li *et al.*, 2006) and the runoff process is a non-stationary random process with an annual period (Zhang *et al.*, 2005). To improve the forecasting precision, it is necessary that the information and the variation law in the runoff time series are sufficiently investigated.

For processing the small samples by grey theory, the new series are generated from the primordial data to find out the change in law within the system without the statistical characteristic values, based on distinguishing the difference among the systemic factors and the correlation analysis. It is short of the rigorously theoretic foundation, while solving the forecasting equation (Zhang *et al.*, 2002; Li *et al.*, 2005; Shang *et al.*, 2005). Much ameliorating arithmetic has been put forward, but they are not valid for the series with the obvious fluctuation. The development of self-memory theory originates from the retrieved model. Self-memory theory is a measure on the basis of time series to process the data, and is a realization of the two methods of Determinism and Uncertainty in mathematics (Cao, 2002). The combination of grey and self-memory theory can effectively reflect the extreme trend of series (Shen, 2005), but there are the phenomena of phase lag between the forecasting values and observed data. The neural network method is a highly nonlinear dynamical system with the stronger fault-tolerant, self-learning ability and adapting. Depending on the BP neural network to process the errors from the combination model with the grey theory, and self-memory theory to further explain the information from the runoff data, one can enhance the forecasting precision.

GREY SYSTEM THEORY AND MODELLING

The research object of the grey system is small samples. It reveals the dynamic laws in the system by processing grey information (Guo *et al.*, 2003). The grey system research attaches importance to the correlation analysis so that the system information can be fully used to make the disorderly and unsystematic data translate into ordered series that fit for differential equation modelling.

Suppose there is an original series $\boldsymbol{X}^0 = [\boldsymbol{x}^0(1), \boldsymbol{x}^0(2), \ldots, \boldsymbol{x}^0(n)]$, it becomes $\boldsymbol{X}^1 = [\boldsymbol{x}^1(1), \boldsymbol{x}^1(2), \ldots, \boldsymbol{x}^1(n)]$ by accumulating time series at a time. The differential equation with the whitening form is set up as follows:

$$dx^1 / dt + ax^1 = u \tag{1}$$

in which a stands for developing grey number and u stands for endogenesis controlling grey number. $\hat{\alpha} = [\boldsymbol{a}, \boldsymbol{u}]^T$, supposing $\hat{\alpha}$ is a parameter to be determined by using linear analysis. Then the following equation is acquired according to the least two multiplication:

$$\hat{\alpha} = (\boldsymbol{B}^T \boldsymbol{B})^{-1} \boldsymbol{B}^T \boldsymbol{Y}_N \tag{2}$$

in which $\boldsymbol{Y}_N = [\boldsymbol{x}^0(2), \boldsymbol{x}^0(3), \ldots, \boldsymbol{x}^0(N)]^T$; $\boldsymbol{B}$ is data matrix as follows:

$$\boldsymbol{B} = \begin{bmatrix} -0.5[\boldsymbol{x}^1(1) + \boldsymbol{x}^1(2)] & 1 \\ -0.5[\boldsymbol{x}^1(2) + \boldsymbol{x}^1(3)] & 1 \\ \vdots & \vdots \\ -0.5[\boldsymbol{x}^1(N-1) + \boldsymbol{x}^1(N)] & 1 \end{bmatrix}$$

if we suppose $dx^1 / dt = F$ in equation (1), then

$$F = -ax^1 + u \tag{3}$$

SELF-MEMORY THEORY AND MODELLING

The memory concept is brought forward when the irreversible process is studied, i.e. the future not only has something to do with the present, but also with times past (Cao, 1995). The self-memory principle can be used to set up a self-memory model corresponding according to the dynamical system, which can be described with differential equations.

Suppose the differential equation of system evolution is:

$$\frac{\partial x}{\partial t} = F(x, \lambda, r, t) \tag{4}$$

In which, x is variable, λ is parameter, r is space and t is time. Define a memory function $\beta(r,t)$ and $|\beta(r,t)| \le 1$. The inner product is defined in Hilbert space as follows:

$$(f, g) = \int_a^b f(\varepsilon) g(\varepsilon) d\varepsilon \qquad f, g \in L^2 \tag{5}$$

Transform the differential equation of system with equation (5), suppose x and β are continuous, differentiable, integrable, make $\beta_t \equiv \beta(t)$, $\beta_0 \equiv \beta(t_0)$, $x_t \equiv x(t)$, $x_0 \equiv x(t_0)$, and the rest is by analogy with this method. A backdate order of p finite difference-integral equation, named as memory equation, is obtained by means of subsection integral and mean value theorem as follows:

$$\beta_t x_t - \beta_{-p} x_{-p} - \sum_{i=-p}^{0} (\beta_{i+1} - \beta_i) x_i^m - \int_{-p}^{t} \beta(\tau) F(x, \lambda, r, \tau) d\tau = 0 \tag{6}$$

in which, x_i^m represents the mean value of t_i and t_{i+1}, ss $x_i^m \equiv x(t_m)$ when $t_i < t_m < t_{i+1}$. Equation (6) emphasizes the relationship between fore and after time of system status, that is, self-memory of the system itself. After determining memory function, simulation or forecasting can be carried out.

$dx^1 / dt = F$, set up by grey system theory, is considered as the dynamic kernel, time series model is set up with system self-memory principle. X is replaced with x^1 in the equation of dynamic kernel. For equation (6), dispersion is carried out, integral is replaced by summation, differential becomes difference, mean value $x^m(t_i)$ is replaced by two moments, suppose $x(t-p) = x^m(t-p-1)$, $\beta(t-p-1) = 0$ and the moments are the same interval, suppose

$\Delta t = t_{i+1} - t_i = 1$, the dispersion form is obtained as follows:

$$x_t = \sum_{i=-p-1}^{-1} \alpha_i y_i + \sum_{i=-p}^{0} \theta_i F(x,i) \tag{7}$$

in which, α_l and β_l are coefficients.

$$x_i^{\ m} = 0.5(x_{i+1} + x_i) \equiv y_i \tag{8}$$

$$F(x,i) = -ax_i + u \tag{9}$$

According to the historical data, memory coefficients α_l and β_l are obtained with the least square method.

$\boldsymbol{X}_t = [\boldsymbol{x}_{t1}, \boldsymbol{x}_{t2}, \cdots \boldsymbol{x}_{tL}]^T$, $\boldsymbol{W} = [\alpha_{-p-1}, \alpha_{-p}, \cdots, \alpha_{-1}]^T$, $\Theta = [\theta_{-p}, \theta_{-p+1}, \cdots, \theta_0]^T$

$\boldsymbol{Y} = \begin{bmatrix} y_{-p,1} & y_{-p+1,1} & \cdots & y_{0,1} \\ y_{-p,2} & y_{-p+1,2} & \cdots & y_{0,2} \\ \vdots & \vdots & \vdots & \vdots \\ y_{-p,L} & y_{-p+1,L} & \cdots & y_{0,L} \end{bmatrix}$, Matrix $\boldsymbol{F}_{L\times(P+1)}$, made up of $\boldsymbol{F}(\boldsymbol{x},\boldsymbol{i})$, can express similarly following the pattern of $\boldsymbol{Y}_{L\times(p+1)}$. Therefore equation (7) can be expressed as matrix form as follows:

$$\boldsymbol{X}_t = \boldsymbol{YW} + \boldsymbol{F}\Theta \tag{10}$$

if suppose $\boldsymbol{M} = [\boldsymbol{Y}, \boldsymbol{F}]$, $\Omega = [\boldsymbol{W}, \Theta]^T$, equation (11) can be obtained:

$$\boldsymbol{X}_t = \boldsymbol{M}\Omega \tag{11}$$

After the coefficient matrix Ω is calculated with the least square method (equation (12)), simulation or forecasting can be conducted.

$$\Omega = (\boldsymbol{M}^T\boldsymbol{M})^{-1}\boldsymbol{M}^T\boldsymbol{X}_t \tag{12}$$

$\hat{x}^1(t)$ represents the fitting or predicting values by accumulating at a time; its reducing value $\hat{x}_0(t)$ can be obtained by equation (13):

$$\hat{x}^0(t) = \hat{x}^1(t) - \hat{x}^1(t-1) \tag{13}$$

in which t = 1, 2, …, N and $\hat{x}^0(0) = 0$.

BP NEURAL NETWORK

An artificial neural net is a flexible mathematical model, which can simulate the large-scale, continuous and complicated system with high nonlinearity. The BP network falls into the feedforward type. Because BP arithmetic is easy to operate, has the merit of little computational complexity and fine parallelism, it is one of the most adopted, the widest to be applied, and the most developed arithmetic in neural networks at present. Its essence is to solve the minimum of error function. The structure of a tri-layer BP network model is set up in Fig. 1.

It has been theoretically proved (Xu, 2002) that a tri-layer BP network model could realize the discretional continuous mapping (Kitahara *et al.*, 1992). Because learning of BP network is carried out, based on the error reverse spread arithmetic of gradient descent (Xu *et al.*, 2002), network training speed is often slow, and it is easy to arrive at local minima during network training. Therefore, some improved algorithm with high speed could be used to solve the problem.

In the practical application, the design of BP neural networks is conducted in virtue of the neural network tool box in the Matlab software bag. In order to overcome the defect of general BP learning algorithm, the "trainbfg" function of Quasi-Newton optimization numerical algorithm in Matlab toolbox is used. This function is suitable to fitting (Xu *et al.*, 2002), and its convergence rate is fast.

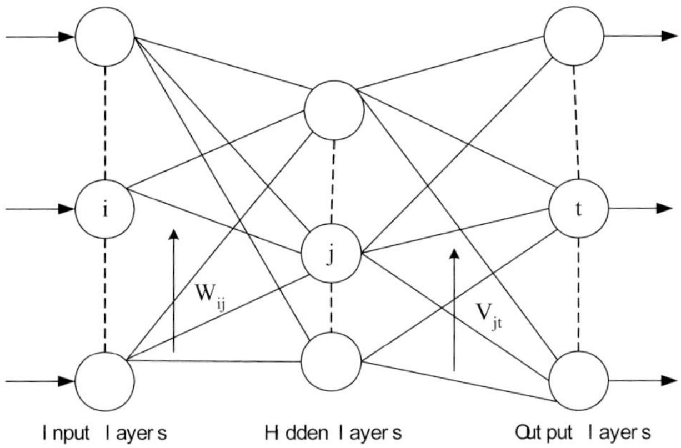

Fig. 1 Structure of a tri-layer BP neural network.

APPLICATION

Weihe River is the biggest tributary of the Yellow River, which is the longest river in Guanzhong Plain. Linjia village hydrometric station is the first station which the Weihe River flows into in Guanzhong. Runoff in the middle reaches of Weihe River is analysed and predicted to be significant for water resources utilization and economic development in Guanzhong Plain.

According to above modelling theories, the grey self-memory neural network model is set up. The observed data from 1956 to 1995 is used for parameter calibration, and from 1996 to 2000 for the model test. For the sake of the comparison, the grey self-memory model is also established.

According to grey theory, the differential equation is set up as follows:

$$F = -0.0081x^1 + 0.6158 \tag{14}$$

A self-memory equation is set up whose backdate order is 3. Its self-memory coefficient is calculated with the least square method for the discretization form of the self-memory equation. At last, the forecasting equation is achieved as follows:

$$x_t = \sum_{i=0}^{3} \alpha_i y_i + \sum_{i=0}^{3} \theta_i F(x,i) \tag{15}$$

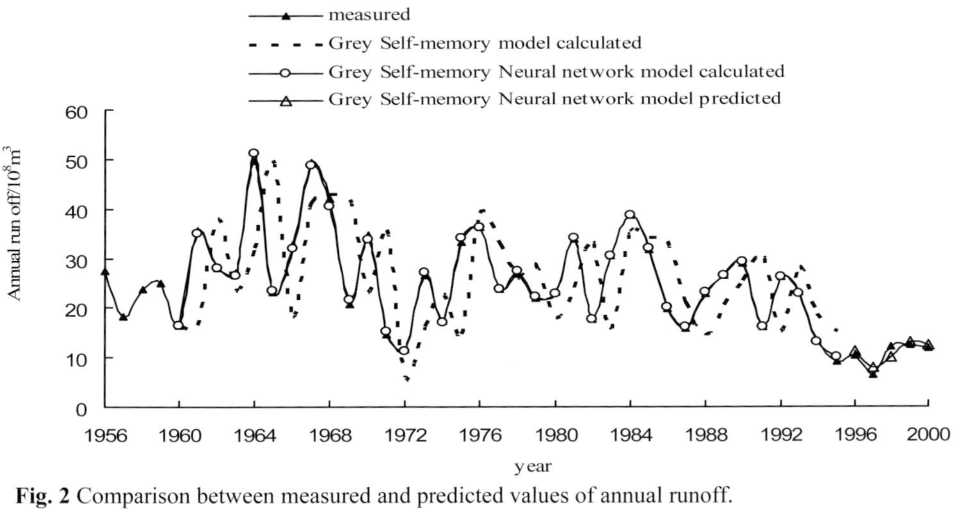

Fig. 2 Comparison between measured and predicted values of annual runoff.

Self-memory coefficients are as follows:

$$\Omega = [\alpha_0, \alpha_1, \alpha_2, \alpha_3, \theta_0, \theta_1, \theta_2, \theta_3]^T \quad (16)$$

$$= [-1.83, 3.17, -2.53, 2.18, 0.25, 0.17, 0.34, 0.09]^T .$$

Error series of runoff fitting is trained and fitted by tri-layer BP neural network. Annual runoff is predicted with grey self-memory neural network model. Model fitting and predicting results are shown in Fig. 2.

From Fig. 1, it can be shown that the grey self-memory model can reveal the long-term trend and fluctuation change by virtue of the grey model and self-memory theory so that it can effectively reflect the extreme trend of runoff series, but there is the phenomenon of phase lag between forecasting values and observed data. Based on the idea of integrative forecasting, the grey self-memory neural network model is set up. The error series from the grey self-memory model is nonlinear; it can be estimated with neural network model to accord with the changing characteristics of runoff. So the model can be used to predict the annual runoff series whose degree of accuracy is much higher than that of the grey self-memory model.

CONCLUSIONS

The runoff process is a complicated, nonlinear, dynamical system; runoff forecasting is still at the probing stage. Improving the accuracy of runoff forecasting depends on making full use of the information of runoff series and extracting what can reflect the change in law. Based on the combination predicting idea, grey theory, self-memory and neural network methods are combined to set up a grey self-memory neural network model. From the sample, it is available to predict the annual runoff.

REFERENCES

Cao, H. (1995) Self-memorization equation and self-memory model. *Meteorological Monthly* **21**(1), 9–13.

Cao, H. (2002) *The Self-Memory Theory About Dynamic System – Forecasting and Computational Application*. Geological Publishing House, Beijing, China.

Guo, Q., Guo, X. & Li, G. (2003) *The Theory and Method about the System Modelling*. University of National Defence Science and Technology Press, Changsha, China.

Li, R., Shen, B. & Zhang, J. (2006) Self-memory hydrologic forecasting model based on phase-space reconstitution. *J. Hydraul. Engng* **37**(5), 583–587.

Li, Z., Xu, J. & Zou, X. (2005) Application of grey system theory in agricultural water demand forecasting. *China Rural Water & Hydropower* **2**(11), 24–26.

Shang, J., Wang, B. & Cui, H. (2005) Application of the grey system to the analysis and forecasting of concrete cracks. *J. Xi'an Univ. Architecture & Technology (Natural Science Edition)* **37**(1), 100–104.

Shen, B., Li, R. & Huang, L. (2005) Predicting annual runoff with grey self-memory model. *J. Northwest A & F Univ. (Natural Science Edition)* **33**(4), 132–134.

Xu, J. (2002) *Mathematical Methods in Contemporary Geography*. High Education Press, Beijing, China.

Xu, D. & Wu, Z. (2002) *System Analysis and Design Based on the MATLAB 6.x Neural Network*. Xidian University Press, Xi'an, China.

Zhang, D., Zhang, S. & Shi, K. (2002) Theoretical defect of grey forecasting formula and its improvement. *Systems Engineering-Theory & Practice* **8**, 140–142.

Zhang, M., Liao, S. & Gu, Z. (2005) A mixed stochastic model for runoff process and its application. *J. Hydraul. Engng* **24**(3), 1–5.

Study on countermeasures for water resources shortage and changes of ecological environment in Shiyang River basin

YONGMING ZHANG[1,2], JUN JIANG[1], BING SHEN[1] & QINGLING SHEN[2]

1 *Northwest Key Lab. of Water Resources and Environment Ecology, Xi'an University of technology, Xi'an 710048, China*
yongmingzhang123@163.com

2 *Shiyang River Basin Administration Branch, Gansu Bureau of Water Resources, Lanzhou 730000, Gansu, China*

Abstract The Shiyang River basin is a special inland river basin in China, in which the population density is the biggest, per capita GDP is higher, and the water resources average per person is the least. This paper shows that the basin water resources have already been seriously overloaded, and the basin comprehensive governing programme ought to be more urgent, by analysing the present water supply volume, groundwater equilibrium, the present water consumption level, the water use efficiency, and the present supply and demand balance of the Shiyang River basin. Using countermeasures to reserve the source in the southern basin, developing the oasis in the middle part, controlling the sandstorms in the northern part by protecting the headwater, and reforming the middle river, it will support the sustainable development of society, and promote the harmonious relationship between man and nature.

Key words Shiyang River basin; GDP; water resources; sustainable development

INTRODUCTION

The backward economy of the northwest endorheic drainage basins and the development of urbanization (Fang *et al.*, 2004) following the ecological environment security problem (Cheng *et al.*, 2002) was threatened by the shortage of water resources, but its sustainable development received people's attention (Shao *et al.*, 2005). The Shiyang River basin was a special inland river basin in China, of which the population density was the biggest, per capita GDP was higher, and the average per person water resources possession was the least.

The ecology question caused by "lack of water" was mainly concentrated downriver of Minqin oasis, which was highly regarded by the central leader authorities. Many scholars also studied the question from different angles in the Shiyang River basin in recent years: Feng (1963) researched into the problem of water system evolvement of Minqin oasis; Chen (1982) studied water and soil resources and their utilization in the Shiyang River basin; Xiao (1988) researched Minqin land degeneration; Yang (2002) studied human mechanism research of Minqin Oasis change in the lower reaches of the Shiyang River; Tang (1994) used the multi-stratum fuzzy comprehensive evaluation of strategies for exploitation and utilization of water resources in the Shiyang Watershed; Gao (1995) analysed the multiple transformation characteristics of water resources and proposed the mathematical model about transformation relations of water resources; Liu (2001) took the Minqin Basin of Shiyang River as an example to conduct research on water resources development and oasis evolution; Zhong (2002) conducted the research into the water resources and environment protection of downstream Shiyang River in Minqin downstream; Chang (2005) had done research on the natural and artificial factors and their transfer on sandy desertification of lower reaches of the Shiyang River basin.

The basin is situated at the important location of the Silk Road and Europe-Asia continental bridge; the development of agriculture and animal husbandry history is old. It has been a rich, well populated place since ancient times. The downriver Minqin oasis extends to the crevice between the Tengger and Badain Jaran desert. It has prevented the two deserts closing up and has become a natural barrier over Wuwei, Lanzhou and even the north of China. The eco-environment protection and construction in Shiyang River basin not only have something to do with the basin of living space and economy development, but also have a bearing on the northwest and north of China. Therefore, it is extremely significant to study the question of water resources in the Shiyang River basin.

RESEARCH TECHNIQUES

Areas survey

Shiyang River, emanating from the Qilian mountains, is one of three big river systems in Hexi inland river basin. It begins at Wu Shaoling in the east, stops at Dahuangshan in the west, and links up Tengger and Badain Jaran desert in the north. The entire basin is constituted of eight tributaries upstream, including the Dajing River, Gulang River, Huangyang River, Zamu River, Jinta River, Xiying River, Dongda River, Xida River from east to west, and main current of Shiyang River downriver (Fig. 1). The total area of this basin is 41 600 km^2. The total population was 2 230 000 and its urbanization level was 27.1% in 2000. GDP was 9 470 000 000 Yuan. Average GDP per person was 4243 Yuan. Gross industrial output value was 10 020 000 000 Yuan. Total cultivated area was 368 550 ha^2, irrigated area was 309 800 ha^2, total agricultural output value was 4 290 000 000 Yuan, gross foodstuff output was 997 000 t (2004).

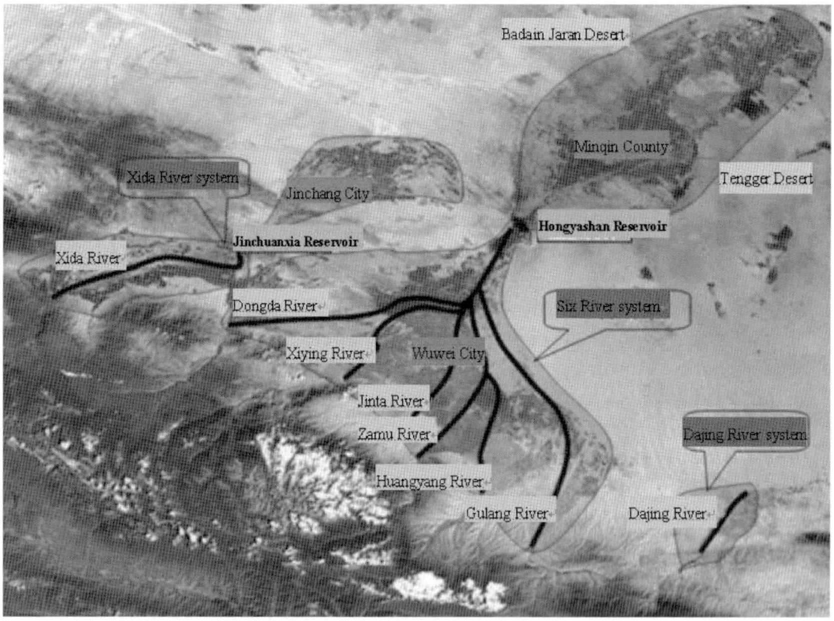

Fig. 1 Shiyang River basin map.

Table 1 Contrast between water supply in Shiyang River Basin in 1980, 1995 and 2000.

Year	Project	Surface water supply				Groundwater	Other	Total
		Storage reservoir	Water diversion	Water-lifting	Sub-total	Water supply	Water supply	
1980	Water volume (10^8 m^3)	10.32	5.62	–	15.94	9.8	–	25.74
	Ratio(%)	40	21.8	–	62	38	–	100
1995	Water volume (10^8 m^3)	9.2	4.32	0.24	13.76	12.1	0.007	25.87
	Ratio(%)	35.6	16.7	0.9	53.2	46.78	0.002	100
2000	Water volume (10^8 m^3)	9.95	3.94	0.01	13.9	14.47	0.17	28.54
	Ratio(%)	34.86	13.81	0.05	48.7	50.7	0.58	100

Assessment of current situation development and utilization of water resources

Present situation of water supply There were many water supply projects in this basin, the water supply capacity of which was 3 100 000 000 m^3. The types of projects mainly contain storage reservoir works, water diversion works, and water-lifting works. The actual water supply was 2 854 000 000 m^3 in 2000, increased by 10% and 9.4% compared to 1980 and 1995, respectively, caused by the exploitation of groundwater (Table 1).

Analysis on groundwater balance After calculating the groundwater balance in this plain in 2000, the total groundwater recharge was 1 193 000 000 m^3. If deducting 251 000 000 m^3 regression water quantities from wells, the recharge of groundwater will become 942 000 000 m^3. Total output water is 1 667 000 000 m^3, if subtracting the regression of irrigation from spring water; the actually groundwater consumption become 1 416 000 000 m^3, so the overdraft quantity of groundwater is 474 000 000 m^3. The groundwater resources efficiency of Shiyang River basin is 121%. It means the groundwater has been seriously overexploited.

Analysis on the present situation of water consumption level Total water consumption was 2 135 000 000 m^3 in the entire basin in 2000. Water consumption was 46 000 000 m^3 in the mountainous area. Production, life, and ecological water consumption was 1 718 000 000 m^3. The natural ecological and evaporation of nullity water consumption was 368 000 000 m^3. Actual water resources consumption was 769.6 m^3 per person and 1975.0 m^3 per acre. Water consumption was 742.1 m^3 per person and 1911.8 m^3 per acre in Liangzhou area. Water consumption was 1653.8 m^3 per person and 2800.6 m^3 per acre in Minqin. It is 2.2 times and 1.5 times more than the middle reaches areas, respectively. Its water resources consumption level of per person and per acre was the highest endorheic drainage (2002).

Analysis of the water use efficiency Water volume per 10 000 Yuan GDP was approximately 3014 m^3 in the Shiyang River basin, which was four times more than the national average level. Stere water produced 0.41 kg/m^3 grain, which was lower than the national average level and ranged from 0.6 kg/m^3 to 1.0 kg/m^3. Water volume per 10 000 Yuan GDP reached as high as 8487 m^3 in Minqin downriver, which was 12 times more than the national average level. Stere water only produced 0.23 kg/m^3 grain. From these data, it explained that Shiyang River basin water resources allocation was mainly on low output of industry and Stere water use efficiency was somewhat low.

Development and use factor of water resources

According to the water supply gauge in 2000, the development and use factor of water resources (the ratio of water supply and total water resources), which was 20.0% in China, 53.3% in northwest China, 52.5% in inland river basin, 92% in the Hexi Corridor, and 162% in Shiyang River. It is shown that the development and use factor of water resources in Shiyang River basin was the highest because of the water resources recycling and groundwater over-exploiting without completely considering water for environment.

Table 2 Major index of synthesis assessment.

Evaluation index	Unit	Numerical value
Irrigation rate of infield	%	84.1
Average per person cultivated area in countryside	(ha^2 per person)	0.23
Average per person irrigated area in countryside	(ha^2 per person)	0.19
Average per person water resources	(m^3 per person)	789
Average per person annual water consumption	(m^3 per person)	1278
Storing and regulating rate of reservoir project	%	21.85
Development and use factor of surface water	%	89
Development and use factor of groundwater	%	121
Use factor of total water resources	%	162
Mining potential index of groundwater resources	%	52.1

Comprehensive evaluation

Based on the research achievements on exploitation and use of water resources, unifying the characteristics of Hexi inland river basin and the Shiyang River water resources exploitation and use, 10 targets has been put forward to take comprehensive evaluation on exploitation and use of the present water resources in Shiyang River basin (Table 2).

PROMINENT PERFORMANCE AND ORIGIN ANALYSIS OF WATER RESOURCES

Prominent performance of water resources

Scarcity of water resources, contradiction of supply and demand Total water resource of Shiyang River basin is only 1 661 000 000 m^3 and available water is 1 762 000 000 m^3. Water resource is 789 m^3 per person and 52 m^3 per acre, which is lower than 1150 m^3 per person and 62 m^3 per acre in Gansun province. So it is typical of water resource shortage areas. Water requirements status is 3 186 000 000 m^3, available water is 2 661 000 000 m^3 and total water consumption is 2 135 000 000 m^3. The water deficit is 525 000 000 m^3 and the lack of water degree is as high as 31%. The supply and demand contradiction is incisive. With the development of the social economy, the contradiction of for water resources supply and demand will be greater.

Sudden dropping of groundwater table, sharp degrading of ecological environment Along with the enhancement of water resources development and use degree in upper and middle reaches, the water entering into the downriver of Minqin oasis was reduced year by year. The present oasis area of Hongyashan reservoir in Minqin was 1312.66 km^2, which was 288.94 km^2 less than that in the 1950s. The groundwater was over-exploited in the middle and lower reaches of the basin. In recent years, 9000 ha^2 narrow-leaved oleaster forests got spike top and downfall, natural vegetation such as 23 330 ha^2 Bai Ci, the Chinese tamarisk, etc., was dead or half dead condition in Minqin. Sandbreak forest and farmland protection forest in the north of the basin have reduced by nearly 200 000 ha^2. The desert pushed 8–10 m to the oasis every year. The average desertification area reached 15 000 ha^2 every year. Sandstorm and dust storm harm is aggravated as a result of shrinkage and desertification of Minqin oasis, which not only seriously threatened people's production and living in Shiyang River basin, but also underestimated the influence on the ecological environment of the entire province and even the northwest area of north China.

Degeneration of the Qilian mountainous area ecology, the weakness of water source self-control function Forest vegetation and the alpine steppes of the Qilian mountains is a water source self-control area in the Shiyang River basin. Prominent contradiction among the forest worker, forest herd and forest mine is created as a result of the population increase and of overdevelopment and overgrazing. The lower Qilian mountainous bush forest line has moved up 40 m since the 1950s, 30% bush forests appear as grassland desertlization and desertification. The forest land and percentage of vegetation coverage is reduced. The natural grassland degenerates. The water source self-control ability continues to decline.

Irrationality of economic structure, low efficiency of the water used The industry construction of the Shiyang River basin took agriculture as the primary structure, the increased value of the primary industry occupies the proportion of GDP; for Wuwei city it was 35.2% and for Minqin County was 62.9%. It is far higher than the entire province, which was 19.63%. The industrialization coefficient of Wuwei city and Minqin County were lower than the entire province average level. The water used in the farmland and forest-grass occupied almost 90.4%, and the industry, urban domestic and countryside person livestock only accounted for 9.6%. The agriculture water used was obviously high, and the primary industry was on the high side; secondary and tertiary industries were on the low side.

Dysfunction of the management system, weakness of water resources unification management In recent years, the authorities of country and province enlarged the dynamics to dangerous reservoir and water-saving transformation irrigation area, but the river basin program authorized by the provincial government in 1990 had not been comprehensively carried out because the work of the basin unified management had not followed. The region water allocation plan was not in accord with the plan implementation, which caused the water entering downriver of Minqin to become less and less, and the oasis ecological problems to become more and more serious.

Origin analyses

Drought climate and shortage of water resources in local Shiyang River basin is located deep in the centre of the mainland. Its climate is continental temperate dry. The water resources average per person measured 789 m^3, which was only 1/3 of the nation and far lower than the average per person 1000 m^3 for the inadequacy standard of international water resources. The Chinese Mou water resources measured 318 m^3, which was only 1/5 of the nation. Water resources quantity of average per person and infield of unit area possess was seriously insufficient, so it was the main root for scarcely water used and deterioration of the eco-environment in the Shiyang River basin.

Consideration of water resources condition was insufficient in social area economy development, the social economy water using had occupied the eco-environment water The basin has reclaimed wasteland and planted crops to develop foodstuff production base on a large scale since the 1970s, because of the insufficient overall planning consideration of water resources condition. It pricked up the eco-environment worsening. Water used contradictory among upper, lower reaches and each national economy department was extremely incisive. The basin average volume of eight main branches' runoff at mountain outlet was 1 456 000 000 m^3. The water volume entering into Minqin represented 31.1–31.6% of branches at mountain outlet before the 1970s. The proportion between the water volume entering to Minqin and various branches at the mountain outlet reduced year by year after the 1970s, because water volume consumption in the middle reaches increased yearly. The water volume actually entering to lower reaches did not achieve 100 000 000 m^3 in the late 1990s, which was only 8.3% of branches at mountain outlet. Drying up of the downstream riverway and the reduction of surface water volume, caused by increase of the middle reaches' water consumption, results in over-exploitation of the groundwater and the continuous decline of groundwater table. Some vegetation which depended on the groundwater gradually started to degenerate or wither away. That was why the eco-environment in the lower reaches had deteriorated.

Reclamation was disordered and ecological environment was seriously destroyed downriver of Minqin There is no self-produced surface water downriver of Minqin; the water resources is almost all foreign water from the middle reaches and its eco-environment is being

Table 3 Contrast between average yearly volume of upper branches' runoff at mountain outlet and inflow of Hongyashan Reservoir.

Year	Average yearly volume of upper branches' runoff at mountain outlet (billion m^3)	Runoff enter to downriver of Minqin (billion m^3)	Ratio of lower branches' runoff at mountain outlet (%)	Ratio of water using between lower and middle reaches
1955–1959	17.524	5.527	31.5	1:0.5
1960–1964	14.376	4.471	31.1	1:0.4
1965–1969	14.034	4.437	31.6	1:0.5
1970–1974	13.776	3.737	27.1	1:0.4
1975–1979	14.357	2.712	18.9	1:0.2
1980–1984	14.938	2.359	15.8	1:0.2
1985–1989	15.529	2.215	14.3	1:0.2
1990–1995	13.716	1.65	12	1:0.1
1996–2000	12.87	1.07	8.3	1:0.1
Average	14.56	3.131	21.5	1:0.3

worsened day by day because the discharge in the middle reaches is reduced year by year. However, the Minqin irrigated infield has developed in a disorderly way in recent years and the irrigated area scale has enlarged. Statistics have indicated that there were 40 000 ha^2 in the irrigated area in Minqin in the middle 1980s. The irrigated area had once achieved about 80 000–867 000 ha^2, driven by the economics of "heat melon seeds" actuation in the 1990s. Along with falling of the melon seed quotation, massive farmlands were abandoned, so the irrigated area still maintains about 68 000 ha^2 in Minqin County at present. Farmlands have been abandoned again caused by predatory exploitation of groundwater, the continuing decline of groundwater table, and the deterioration of groundwater quality. Some of the people are unable to survive in Minqin lake region at present. They are forced to leave their native home and become "ecological refugees".

The pressure of the still-growing very large population The population yearly rate was 17.5‰ in this basin for the last 50 years. The population mainly concentrates in the oasis area. The oasis's bearing population had more than 300 people per square kilometre. As a result of water resources condition limitation and land resource development potential limitation, the area suitable for human living was 10% of the total land area (2004), developing the suitable agricultural wasteland also become a much bigger pressure on water resources supply. Therefore, the increase of population is an important factor, which affects the deterioration of region eco-environment.

COUNTERMEASURES ON BASIN COMPREHENSIVE CONTROL

Basic clues

The basic clues of comprehensive administration is: insist on taking the ecological protection and the improvement as the basis, taking the water resources of the scientific management, the reasonable allocation, the high-efficiency utilization and the effective protection as the core, consolidating the engineering measures and non-engineering measures, giving attention to two or more effects of ecology benefit and economic benefit, making overall plans, and taking all factors into consideration on upper, middle and lower reaches, regulating waterways together, colluding development, corresponding to the life, production and ecological waterwell, carrying on the comprehensive administration to keep the ecosystem worsening tendency of Shiyang River basin within limits, and urging the harmonious coexistence between man and nature.

COUNTERMEASURES

Advancing water saving, establishing water saving society

We should carry out the industry and life water saving technology vigorously; expanding the region of town centralism water supply. To develop the industrial enterprise water balance test positively, to strictly restrict new high water consumption and pollution project, to popularize new water saving technology, new craft and new equipments, and to reduce unit water demand. Push agriculture convention water saving technology comprehensively; unceasingly improving its quality and standard. So we need to diligently carry out sprinkler irrigation, drip irrigation and ooze irrigation as the main content new agriculture water saving technology, construct water saving society.

Adjusting economic structure reasonably, optimizing disposition of water resources

Taking the market as a guide, take the redistribution of water resources as the core, speed up secondary and tertiary industrial development, and promote the industrial structure reasonable adjustment. Unify the readjustment of the economic structure in the basin, develop small towns vigorously, and adjust population distribution, through annex the village and spot, move town distribution in resettlement. This causes water conservation in the forest region in the south and serious desertification in the north, the local population disagrees with oasis to be suitably centralized to the middle of the corridor. Optimize agriculture industrial structure, increase forest-grass planting area, and adjust the dual of grain and economic crop to the three-plant dimension structure of grain, economic crop and grass. Use optimal allocation to region water resources to promote industrial structure adjustment gradually and use industrial structure adjustment to promote the concatenation system of optimal allocation to region water resources.

Developing the ecology construction vigorously, improving the ecological environment diligently

Take the soil and water conservation and comprehensive administration as the key point. Persist in making, controlling, developing simultaneously, and combine with sealing and forestation. Administrate mountain water, forest, field and road comprehensively. Develop closing hillsides (sand) to facilitate afforestation (grass), return cropland to forestry, push natural forest protection effort, sands controlling and afforestation and the construction of farm-shelter forest, expand woods and grass vegetation, improve ecological environment vigorously.

Actualizing water resources unification management and attemper

Perfectly authoritative, highly effective, harmonious basin management system and the operating mechanism should be established. Take the rationally water resources allocation as the core, practice centralized management and unified allocation. First, constitute rules and regulations of Shiyang River basin, delimit the hydrant permission management system permission legally and reasonably, implement the system, and unify total quantity control and quota management by norms. Second, Shiyang River year water volume real-time scheduling plan should be established and supervised, actualized responsibly by the basin management structure. Third, close the motor-pumped well designedly, and gradually reduce the quantity of groundwater exploitation. Fourth, establish the advanced hydrological measurement and reporting and water resource information management and attemper system, according to the request of unification management Shiyang River basin water volume and water quality dispatch.

REFERENCES

Chang, Z. & Han, F. (2005) Natural and artificial factors and their transfer on sandy desertification of lower reaches of Shiyang River basin. *Arid Land Geography* **28**(2), 150–155.

Cheng, J. & Xu, Z. (2002) Study on assessment of the ecological security in the continental watershed in north west China – A case study at the Middle Reaches of Heihe River Watershed – Zhang Ye Prefecture. *Arid Land Geography* **25**(3), 219–224.

Cheng, L., Qu, Y. & Chen H. (1982) Resources of water and soil and its utilization in the Shiyang River drainage basin. *J. Desert Res.* **4**, 6–13.

Fang, C., Huang, J. & Bu W. (2004). The oretical study on urbanization process and ecological effect with the restriction of water resource in arid area of north west China. *Arid Land Geography* **27**(1), 1–7.

Feng, S. (1963) Water system evolvement of Minqin oasis. *Acta Geography Sinica* **3**, 241–249.

Gansu Water Resources Plan and Design Institute, Shiyang River Basin Management Bureau (2002) *Report on Comprehensive Plan on Shiyang River Basin Development.* Lanzhou, China.

Gao, F. & Zhang, Y. (1995) The model and the cross-correlationship of water resources for the inland Shi-Yang River basin – Hexi corridor. *J. Hydraul. Engng* **11**, 76–83.

Liu, H., Zhong, H. & Gu, Y. (2001) Water resources development and oasis evolution in inland river basin of arid zone of northwest China – a case study: Minqin Basin of Shiyang River. *Adv. Water Sci.* **9**(3), 378–384.

Shao, B. & Chen, X. (2005) Research on the actuality of the coordinated development of economy and ecological environment in northwest China. *Arid Land Geography* **28**(1), 136–141.

Shiyang River Basin Authority of Gansu Water Conservancy (2004) *Catchment Area Planning of Shiyang River Basin in Gansu Province* .Lanzhou, China.

Tang, B. (1994) The Nulti-Stratum fuzzy comprehensive evaluation of strategies for exploitation utilization of water resources in the Shiyang watershed. *Agric. Res. in Arid Areas* **12**(3), 85–94.

Xiao, H. (1988) A study on land degeneration in Minqin oasis in the low reaches of the Shiyang River . *J. Desert Res.* **2**, 33–38.

Yang, Y. & Li, J. (2002) The Human mechanism research of Minqin oasis change in the lower reaches of the Shiyang River. *Geogr. Res.* **4**, 449–458.

Zhong, H. & Liu, H. (2002) On water resources and environment protection of downstream Shiyang river in Minqin downstream. *Northwest Water Resources & Water Engng* **13**(1), 10–13.

Determination of root-zone water storage in a desert woodland using a two-layer moisture balance model

YONGHUA ZHU[1], LILIANG REN[1], HAISHEN LÜ[1] & T. H. SKAGGS[2]

1 *State Key Laboratory of Hydrology-Water Resources and Hydraulic Engineering, Hohai University, Nanjing 210098, China*
yonghua321@yahoo.com.cn

2 *US Salinity Laboratory, 450 W. Big Springs Rd., Riverside, California 92507, USA*

Abstract Root-zone water storage (RWS) is a fundamental component of the soil–plant–atmosphere continuum. In the lower reaches of arid river basins in inland China, low water recharge and hence low RWS have been associated with a series of ecological and environmental problems. Developing an improved understanding of RWS, as well as methods for estimating RWS, will permit improved water management and help achieve sustainable development in these arid basins. In this study, the temporal variation of RWS under the natural undershrub *Sophora alopecuroides* was examined at an experimental site located in the Ejina basin of northwest China. RWS was modelled using a two-layer soil moisture balance model comprising a shallow surface layer (0–10 cm) and a root-zone layer. Model simulations were found to be in very good agreement with experimental data. Possible future applications include regional RWS assessments in northwestern China in which the model surface layer is correlated with remote-sensing measurements of surface moisture content.

Key words root-zone water storage; two-layer soil moisture balance model

INTRODUCTION

In many agricultural and hydrological applications, RWS is an important variable because of its impact on plant growth and on the water balance at the land surface. Defined as the amount of water per unit area in the root-zone, RWS is commonly determined based on field measurements of soil water content. However, such field measurements can be time consuming and expensive to obtain. Furthermore, when considering regional scale RWS, these data are point measurements that are difficult to interpolate or extrapolate to the larger scale because of the spatial variability of soil properties and the scarcity of data in time and space.

Microwave remote sensing of soil moisture over large areas is a possible alternative to field measurements. However, microwave remote sensing usually only provides data for surface soil layers (0–10 cm). A method for estimating RWS based on remotely-sensed measurements of the surface soil moisture content would be extremely useful for the study of regional scale soil moisture. A number of authors previously investigated this estimation problem using a two-layer model, where the first layer is a shallow surface layer and the second is the root zone (e.g. Deardorff, 1977; Wigneron *et al.*, 1999; Montaldo & Albertson, 2003). For the most part, these studies have considered only grasslands (Montaldo & Albertson, 2003) or cropped areas (Ragab, 1995; Wigneron *et al.*, 1999) where moisture enters the soil predominately from the surface (due to precipitation or irrigation) and where root zones are shallow (~50 cm).

In extremely arid regions of China where precipitation is very low, plant growth is correspondingly small and is dependent on phreatic evaporation, which transfers water from shallow water tables up to the root zone. The two-layer moisture models noted previously did not consider phreatic evaporation, nor were they tested for conditions of deep-rooted plants. The objective of this paper is to test a two-layer model for conditions of phreatic evaporation and deep plant rooting. We estimate the temporal evolution of RWS under a perennial riverside undershrub *Sophora alopecuroides* (*S. alopecuroides*) in the Ejina basin of northwest China. The model is found to be in good agreement with experimental data. The success of the two-layer model under these conditions suggests the model may be useful in future applications involving remote-sensed soil moisture data in western China and elsewhere.

EXPERIMENTAL LOCATION AND METHODS

An investigation of the temporal variability of soil moisture was conducted at a Chinese Academy of Sciences eco-hydrological research site in the extremely arid woodlands (40°10′–42°30′N, 99°25′–102°00′E, 940.5 m above sea level). The Ejina basin is near the northeastern edge of the Badanjilin desert and in the lower reaches of the Heihe River basin, the second largest inland river basin in arid northwest China (Fig.1). The experimental site is 4 km southwest of Ejina City. Mean annual precipitation is 38.24 mm, mean annual free water surface evaporation is 3631.9 mm, and average temperature is 8°C. The undershrub *S. alopecuroides* in the experimental woodland has not been significantly affected by human activities. The experimental site is 10 000 m^2 and is 200 m from the river. The groundwater depth ranges from 1.5 to 3 m.

Fig. 1 The scheme of the study region.

The experimental area was divided into 10 sections. Within each section, a site with soil representative of the whole section was chosen for measuring soil moisture. Soil moisture measurements continued for three years (2003–2005), beginning each year on 1 May and ending on 31 October. Soil cores to a depth of 140 cm were obtained with an augur every 5 days. The cores were sectioned into eight layers: 0–10 cm, 10–20 cm, 20–40 cm, 40–60 cm, 60–80 cm, 80–100 cm, 100–120 cm, and 120–140 cm. From within each layer, three replicate subsamples were taken near the midpoint of each layer. The subsamples were weighed, dried at 105°C, and re-weighed to determine soil moisture.

Starting on 23 September 2004 and ending 31 September 2004, measurements were made of effective root length. One shrub was selected from each of the 10 experimental sections. A soil column centred beneath the shrub and measuring 60 cm in diameter and 140 cm deep was excavated and sectioned into eight layers: 0.0–10.0 cm, 10.0–20.0 cm, 20.0–31.5 cm, 31.5–61.5 cm, 61.5–94.5 cm, 94.5–127.5 cm, 127.5–135.0 cm, and 135.0–140.0 cm. The soil layers were sieved and root diameters and lengths were measured. All roots with diameter less than 0.4 cm were classified as effective roots, presumed active in absorbing moisture from the soil (Zhu, 2002).

The soil at the experimental site is mainly fine sand, so the soil was treated as a homogeneous fine sand with a field capacity soil moisture content of 31.7% and a wilting point water capacity of

0.73% (Feng & Chen, 1999). Daily weather data were obtained from the Ejina County Weather Station.

SIMULATION METHOD

Two-layer soil moisture balance model

The two-layer soil moisture balance model is given by:

$$\begin{aligned} \frac{\partial \theta_s}{\partial t} &= \frac{P + E_{g_s} - E_s}{d_s} - C(\theta_s - \theta_{rz}) \\ \frac{\partial \theta_{rz}}{\partial t} &= \frac{P + E_{g_{rz}} - R - E_{rz}}{d_{rz}} \end{aligned} \qquad (1)$$

where θ_s is the volumetric water content of the surface layer (0–10 cm) (cm^3 cm^{-3}), θ_{rz} is the volumetric water content in root-zone (cm^3 cm^{-3}), P is the precipitation rate (mm day^{-1}), E_g is the rate of phreatic evaporation reaching to the surface layer (mm day^{-1}), E_s is the surface evaporation (mm day^{-1}), d_s is the depth of the surface layer (mm), C is the pseudo-diffusivity coefficient (day^{-1}) that is a function of the surface and root-zone soil moisture content and the thickness of the two layers, $E_{g_{rz}}$ is the phreatic evaporation reaching to root-zone (mm day^{-1}), E_{rz} is the root-zone evaporation (mm day^{-1}), R is deep drainage (mm day^{-1}), and d_{rz} is the root-zone depth (mm).

Determination of input parameters

Phreatic evaporation The phreatic evaporation rate was determined using the following formulas (Mao *et al.*, 1999):

$$E_{g_{rz}} = E_0 \left[1 - \frac{h - d_r}{h_0} \right]^m \qquad (2)$$

$$E_{g_s} = E_0 \left[1 - \frac{h - 0.01}{h_0} \right]^m \qquad (3)$$

where d_r is the depth of the effective root-zone (m), E_0 is the daily average potential evaporation during the period Δt (mm day^{-1}), h is the average depth of the water table during Δt (m), h_0 is the water table depth below which phreatic evaporation ceases (m), and m is an empirical parameter that is a function of soil characteristics, climate, etc. and which takes on different values during the growth periods of *S. alopecuroides*. In this paper, m was determined for each month of its growth period by fitting equations (2) and (3) to data measured by Liu *et al.* (2005). The fitted parameter values are given in Table 1.

Table 1 Index m for different months during the growth period of *S. alopecuroides*.

Month	May	June	Jul.	Aug.	Sep.	Oct.
m	6.86	5.85	4.30	6.30	6.52	9.00

For the *S. alopecuroides* woodland we studied, d_r was 1.4 m and h_0 was 4.72 m, the latter estimate being based on the work of Feng (1998).

Actual evapotranspiration Actual evapotranspiration (*ET*) was calculated by multiplying potential evapotranspiration (*PET*) by a reduction factor that depends on the moisture content of the soil. For these purposes, the soil profile was divided into the same 8 layers indicated above for the soil moisture measurements. Transpiration and uptake were assumed to be at potential levels when the soil was at field capacity, and decline linearly with decreasing available. Letting *i* index the day of measurement (from 1 May to 31 October) and *j* the soil layer ($1 \leq j \leq 11$), available water was calculated as (Ragab, 1995):

$$SI(i,j) = \frac{\theta(i,j) - \theta_{\text{wilting point}}}{\theta_{\text{field capacity}} - \theta_{\text{wilting point}}} \tag{4}$$

The actual evapotranspiration, *ET*, was then calculated as:

$$ET(i,j) = SI(i,j)PET(i) \tag{5}$$

Potential evapotranspiration, *PET*, was calculated from weather data using the method of Stockle *et al.* (1992).

Initial values of moisture content According to Camillo & Schmugge (1983) and Ragab (1995), under conditions of a fully grown root system and sufficiently dry soil, root-zone soil moisture storage W_{rz} (cm) can be obtained by multiplying the volumetric moisture content of the surface layer θ_s by a constant of proportionality $I(z_s)$. As shown in equation (6), the constant of proportionality is calculated from the root density distribution:

$$W_{rz} = \theta_s I(z_s)$$
$$I(z_s) = \int_{\substack{\text{root}\\\text{zone}}} \left[\frac{R(z_s)}{R(z)}\right]^{\frac{1}{bn}} dz \tag{6}$$

Here, $R(z_s)$ is the relative root density (m^{-2}) in the surface layer, $R(z)$ is the relative root density (m^{-2}) in the root-zone, *b* is a soil texture parameter, and *n* is a parameter that depends on the functional form of $R(z)$.

The growth period of *S. alopecuroides* in the Ejina basin is from May to October, with the dry period beginning in September. Thus from September to October, the soil moisture distribution is controlled mainly by phreatic evaporation and the sorption of moisture by roots, and equation (6) can be used to estimate the profile moisture content from measurements of the surface layer if $I(z_s)$ can be determined. Following Camillo & Schmugge (1983), the soil texture parameter *b* at the experimental site is given as 4.05 and, we calculated $n = 2 + \frac{2}{b} = 2.49$. The root measurement data (length of root per volume of soil) resulted in the following root distribution density function (cm^{-2}) for *S. alopecuroides* (average distribution for the 10 sections):

$$R(z) = \begin{cases} 0.0157 & 0.0 \leq z \leq 10.0 \text{ cm} \\ 0.0101 & 10.0 < z \leq 20.0 \text{ cm} \\ 0.0398 & 20.0 < z \leq 31.5 \text{ cm} \\ 0.0898 & 31.5 < z \leq 61.5 \text{ cm} \\ 0.1630 & 61.5 < z \leq 94.5 \text{ cm} \\ 0.0393 & 94.5 < z \leq 127.5 \text{ cm} \\ 0.4561 & 127.5 < z \leq 135.0 \text{ cm} \\ 0.9279 & 135.0 < z \leq 140.0 \text{ cm} \\ 0.0000 & z > 140.0 \text{ cm} \end{cases}$$

With this root distribution, the proportionality coefficient in equation (6) for the 0–10 cm surface layer $I(z_s)$ was:

$$I(z_s) = 1211.91 \tag{7}$$

Pseudo-diffusivity coefficient *C* Ragab (1995) discussed several possible values for the pseudo-diffusivity coefficient *C*. We took *C* to be equal to 0.0492, a value roughly in the middle of those discussed by Ragab (1995).

Simulation procedures

The initial day of the simulation was 1 September 2005. Although the initial value of the surface layer moisture content could in principle be obtained by remote sensing, our object was to first verify that the two-layer model was appropriate for situations with phreatic evaporation, and we initialized the surface layer with the measured gravimetric water content. The initial value of the root-zone moisture content was then determined by equations (6) and (7). Finally, equation (1) was solved using divided differences and a daily time step to compute the time variation of root-zone soil moisture. Because 1 September is in the middle of the growth period of *S. alopecuroides*, the simulation was done both forward and backward from this date so that RWS would be computed for the entire growth period.

RESULTS AND DISCUSSION

A comparison of simulated $W_{rz_{simulated}}$ and observed $W_{rz_{iobserved}}$ RWS is shown in Fig. 2. The data are the averaged values from the 10 experimental sites. The root-mean-square error (r.m.s.) and the relative mean error (r.m.e.) are calculated by the following formulas, respectively:

$$r.m.s. = \left\{ \frac{1}{N} \sum_{i=1}^{N} \left[W_{rz\ i\ observed} - W_{rz\ i\ simulated} \right]^2 \right\}^{\frac{1}{2}} \qquad r.m.e. = \frac{1}{N} \sum_{i=1}^{N} \left[\frac{W_{rz_{i\ observed}} - W_{rz_{i\ simulated}}}{W_{rz_{i\ observed}}} \right]$$

The calculated r.m.s. and r.m.e. for the root-zone were 1.15 cm and 1.25%, respectively. These goodness-of-fit values, as well as visual inspection of Fig. 2, indicate that prediction errors are low and that the simulated values agree very well with the data. Figure 2 also shows that the simulated RWS has larger fluctuations in the early part of the growing season than in the later. The reason is that it in the early part of the growing season, *S. alopecuroides* grows fast and, with a daily time step, phreatic evaporation may lag behind evapotranspiration, leading to oscillatory behaviour. Later in the season, the plant grows more slowly and the system is less dynamic and closer to an equilibrium condition.

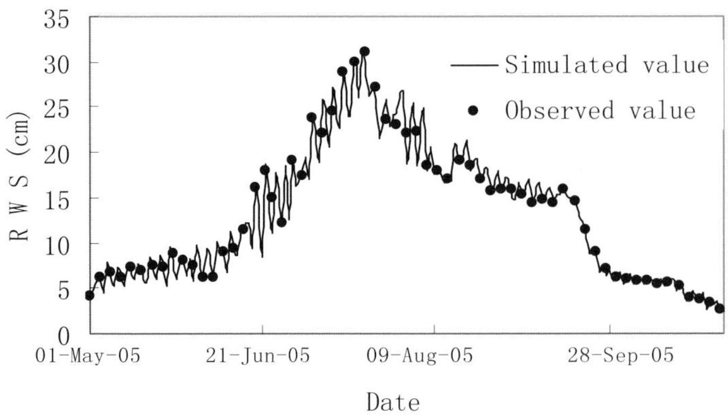

Fig. 2 The comparison of simulated and observed RWS for *S. alopecuroides*.

CONCLUSIONS

This study investigated temporal changes in RWS in a natural desert woodland with undershrub *S. alopecuroides*. Located in Ejina Basin, China, the experimental woodland is extremely arid and plants rely on phreatic evaporation as a primary water source. A two-layer model of RWS was developed and tested using surface soil moisture measurements to initialize the simulated root zone storage. The simulated RWS was in good agreement with experimental data. The observation errors were low (the r.m.s and r.m.e were 1.15 cm and 1.25%, respectively), indicating that the two-layer model is reliable for landscapes in which phreatic evaporation is a significant component of the water balance. Because the two-layer model can be initialized with measurements of surface soil moisture, the model may be used in conjunction with remotely-sensed soil moisture data.

This study considered only temporal changes in RWS at a single (averaged) location during the growth period. Future studies should focus on the spatial component of RWS and the incorporation of the two-layer model into a distributed hydrological model. Additionally, this study considered only the undershrub *S. alopecuroides* growing in an arid desert region. Future studies should focus on additional plants and climates.

Acknowledgement This research is supported by National Key Basic Research Program of China (2006CB400502). The project "Study on Coupled Processes of Atmosphere-Land-Hydrology" (IRT0717) has been supported by the Program for Changjiang Scholars and Innovative Research Team in University (PCSIRT).

REFERENCES

Camillo, P. & Schmugge, T. J. (1983) Estimating soil moisture storage in the root zone from surface measurements. *Soil Sci.* **135**(4), 245–264.

Deardorff, J. W. (1977) A parameterization of ground-surface moisture content for use in atmospheric prediction models. *J. Appl. Meteorol.* **16**, 1182–1185.

Feng, Q. (1998) The study of water balance, the optimized ecological groundwater level and ecological water consumption in Ejina oasis. Post-doctorate Thesis, Cold and Arid Regions Environmental and Engineering Research Institute, CAS, Lanzhou, China.

Feng, Q. & Chen, G. (1999) Moisture distribution and movement in sandy lands of China. *Acta Pedologica Sinica* **36**(5), 225–236.

Liu, T., Chi, D., Yu, W., Wang, D. & Chen, S. (2005) Experimental research on phreatic evaporation rules in Liaoning Province. *J. Shenyang Agricultural University* **36**(1), 76–79.

Mao, X. M., Lei Z. D., Shang, S. H. & Yang, S. X. (1999) Method of equivalent phreatic evaporation by lowering evaporation surface for estimation of the phreatic evaporation from farm land based on that from bare soil. *Guan Gai Pai Shui* **18**(2), 26–29.

Montaldo, N. & Albertson, J. D. (2003) Multi-scale assimilation of surface soil moisture data for robust root zone moisture predictions. *Adv. Water Resour.* **26**, 33–44.

Ragab, R. (1995) Towards a continuous operational system to estimate the root-zone soil moisture from intermittent remotely sensed surface moisture. *J. Hydrol.* **173**, 1–25.

Wigneron, J. P., Calvet, J. C., Olioso, A., Chanzy, A. & Bertuzzi, P. (1999) Estimating the root-zone surface soil moisture and SVAT modeling. *Phys. Chem. Earth* (B) **24**(7), 837–843.

Zhu, Y. H. (2002) Water transfer laws in GSPAC in arid regions, north-western China. PhD Thesis, Cold and Arid Regions Environmental and Engineering Research Institute, CAS, Lanzhou, China.

Impact of climate change on the streamflow in the headwater catchment of the Yellow River basin

FANGFANG ZHAO, ZONGXUE XU & JUNXIONG HUANG

Key Laboratory of Water and Sediment Sciences, Ministry of Education, College of Water Sciences, Beijing Normal University, Beijing 100875, China

zongxuexu@vip.sina.com

Abstract A coupled system integrating GCMs output, a downscaling model and a distributed hydrological model are proposed in this study. First, the Statistical DownScaling (SDS) model for the simulation of historical climate (1961–1990) and three future scenarios (2020s, 2050s and 2080s) are calibrated and validated. Then the distributed hydrological model, Soil and Water Analysis Tool (SWAT), is applied to simulate the streamflow for headwater catchment of the Yellow River basin. Finally, the coupled system is applied to investigate the hydrological response to climate change in the study area. The results show: (a) the SDS model is successful in reproducing the main features of the observed hydrometeorology from the baseline climate simulation, when it is used to the HadCM3 GCM output; (b) the monthly streamflow simulated by SWAT corresponds well with the measured ones, and the model can satisfactorily capture the seasonal tendency; (c) the hydrological processes in the study area are very sensitive to climate changes in the future.

Key words climate change; downscaling; scenario; streamflow; SWAT model; Yellow River

INTRODUCTION

With the scientific understanding of global warming, the concerns are growing about the impact of climate change on the water cycle among the public, governmental and non-governmental authorities in the world. Streamflow, especially extreme floods and droughts, are several of the hydrological processes easily affected by global warming (Bürger & Chen, 2005). The issue of the impact of climate change on runoff in ungauged and poorly gauged basins is a great challenge. The current approach for this problem is the coupling of climate model with hydrological models. It is noted that these kinds of studies cannot be regarded as predictions, but should be regarded as a scenario analyses (Piling & Jones, 1999). Downscaling methods are also usually used because of the coarse spatial resolution of GCMs. Presently, the method of GCM-downscaling, especially statistical downscaling, is well developed as an appropriate tool for regional climate impact studies (Wilby & Wigley, 1997).

In this study, the statistical downscaling (SDS) technique to drive local climatic parameters on a daily time scale for input in the distributed hydrological model SWAT (Soil and Water Assessment Tool) is used. The investigation is focused on the impact of climate change on streamflow characteristics and related processes in the headwater catchment of the Yellow River basin. The objectives of this study include: (a) to generate three climate scenarios for future periods (2020s, 2050s and 2080s); (b) to calibrate and validate the hydrological component of SWAT over a 15-year period (1986–2000) by using historical climate data, and comparing the simulated output with observed streamflow measured at Tangnaihai station; and (c) to evaluate the impact of climate change on monthly and annual streamflow in response to future climate scenarios with the SWAT model.

METHODOLOGY

Statistical downscaling

Statistical downscaling (SDS) is based on the fact that the regional climate may be regarded as a physical process by two factors: the large scale climatic state and regional/local physiographic features (IPCC, 2001). From this viewpoint, regional or local climate information is derived by first determining a statistical model which relates large-scale climate variables (or "predictors") to regional and local variables (or "predictands"). Then the predictors from a GCM simulation are fed into this statistical model to estimate the corresponding local and regional climate characteristics.

In this paper, a software package, Statistical DownScaling Model (SDSM), is used. It can directly employ GCM output in the scenario construction processes, so this model is widely used in the hydrological and agricultural research communities (Wilby *et al.*, 2002, 2004). The key procedure to use the SDSM are as follows: (a) selection of downscaling predictor variables; (b) model calibration; (c) synthesis of observed data; (d) generation of future climate scenarios; (e) result analysis including diagnostic testing, statistical analyses and graphing model output, etc.

Hydrological model

Most of the watershed models, such as HSPF, HEC-HMS, and CREAMS, have a lot of limitations, including scale problems, continuous-time problems, simulation problems, number of sub-watershed problems, etc. (Saleh *et al.*, 2000). The SWAT model was developed to overcome these limitations. It incorporates features of several ARS models and is a direct outgrowth of the SWRRB model. It was developed to predict the impact of land management practices on water, sediment and agricultural chemical yields in large complex watersheds with varying soils, land use and management conditions over long periods of time. In SWAT, a watershed is divided into multiple sub-watersheds, which are then further subdivided into a large number of homogeneous hydrological response units (HRUs) representing unique combinations of soil and land use. The hydrological model is based on the water balance equation in the soil profile where the processes simulated include precipitation, infiltration, surface runoff, evapotranspiration, lateral flow and percolation (Bouraoui *et al.*, 2005; Neitsch *et al.*, 2005). The SWAT is used worldwide and has been validated for a lot of watersheds for its strong merits (Arnold *et al.*, 1999; Saleh *et al.*, 2000; Santhi *et al.*, 2001) and also successfully applied in the Yellow River basin (Li *et al.*, 2003; Zhang *et al.*, 2003).

STUDY CATCHMENT AND DATA DESCRIPTION

The headwater catchment of the Yellow River basin (HYRB), a poorly gauged basin, is upstream of Tangnaihai station, with a drainage area of 122 000 km^2 (see Fig. 1). It is the main part of the Qinghai-Tibetan plateau at the elevation ranging from approx. 3000–4500 m, formed by great rising of Himalayas movement. The climate belongs to the Qinghai-Tibetan plateau climate system. The annual mean runoff is 203.9 × 10^8 m^3, but the runoff has obviously decreased during past decades.

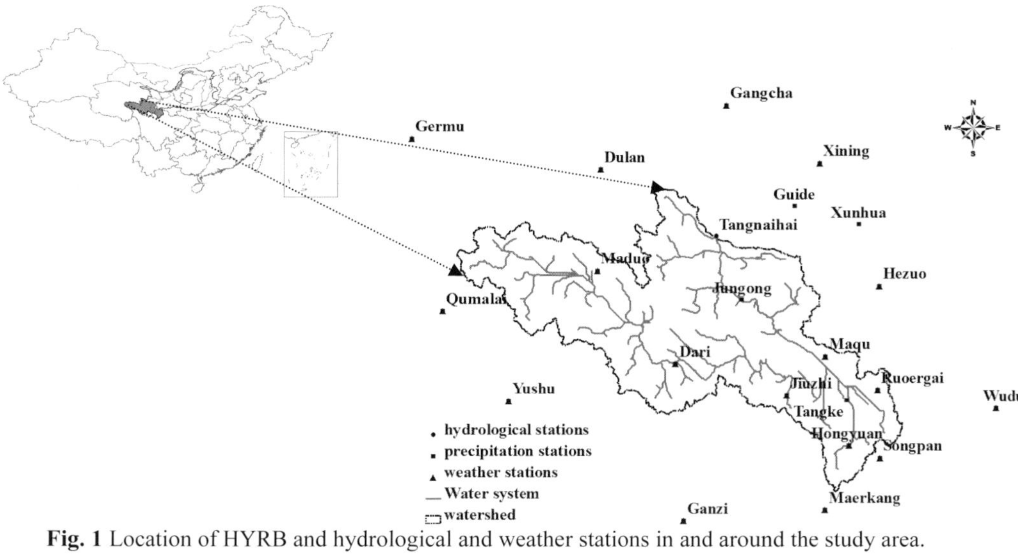

Fig. 1 Location of HYRB and hydrological and weather stations in and around the study area.

Three data sets were compiled for the purpose of climate scenario generation and hydrological model application: (a) observed station data; (b) NCEP reanalysis data; and (c) HadCM3 GCM simulations of current and future climatic conditions. Daily precipitation (PRCP) and maximum and minimum temperatures (TMAX and TMIN) in and around HYRB and obtained from the Meteorological Data Center, China Meteorological Administration. The atmospheric predictor variables used to calibrate the SDS model were extracted or calculated from the daily grid point data of the NCEP reanalysis (Table 1). All variables were regridded to conform to the HadCM3 GCM grid and were standardized by their respective 30-year averages and standard deviations (Wilby *et al.*, 2002). The first 15 years data (1961–1975) were used for model calibration, the remaining 15 (1976–1990) for independent model validation. Two time series of predictor variables, the baseline (1961–1990) and the future climate conditions (2020s, 2050s and 2080s), were extracted from the HadCM3 GCM grid box nearest to the study area.

Table 1 Candidate predictor variables.

	PRCP	TMAX	TMIN
Predictor variables	Specific humidity	Mean sea level pressure	Mean sea level pressure
	Lag-1 mean sea level pressure	Zonal velocity component	500 hPa geopotential height
	500hpa relative humidity	500 hPa geopotential height	850 hPa geopotential height
	Vorticity	850 hPa geopotential height	850 hPa vorticity
		850 hPa zonal velocity component	2 metre temperature
		2 metre temperature	850 hPa zonal velocity component

Table 2 Statistics for the simulated daily PRCP, TMAX and TMIN.

Statistics	Month mean wet-day amounts (mm/d)	Monthly wet-day frequencies (%)	Monthly maximum dry-spell length (d)	Monthly maximum wet-spell length (d)	Monthly variation of wet-day amounts (mm^2)	Daily TMAX (°C)	Daily TMIN (°C)
Observed	3.22	41.22	76.83	32.26	11.52	5.92	–7.97
Downscaled	3.19	28.32	84.10	22.64	12.51	6.24	–8.44
Bias	–0.03	–12.90	7.27	–9.62	0.99	0.32	–0.47
R^2	0.97	0.96	0.92	0.98	0.95	1.00	1.00
Ens	0.90	0.70	0.79	0.75	0.94	0.99	1.00

MODEL APPLICATION AND RESULT ANALYSIS

Future climate scenarios generation

The compiled data sets were used to develop three future climate scenarios to investigate the impact of climate change on streamflow in the PUB region, HYRB in the future. The downscaling model based on multiple linear regression equations is developed during the model calibration process. Combining the observed data with the predictor variables, the empirical relationships between meso-scale atmospheric variables and sub-grid scale surface climate variables (daily PRCP, TMAX and TMIN) are established. To evaluate the dependability of the downscaling model, a validation process should be made. In order to quantitatively describe the above results, model bias, correlation coefficient (R^2) and Nash-Suttcliffe coefficient (*Ens*) are applied to evaluate the model performance. Table 2 lists the statistics to evaluate the model for the downscaled daily PRCP, TMAX and TMIN. The results show that the SDS model can capture the main characteristics of the daily PRCP, TMAX and TMIN.

The future climate scenarios are generated by downscaling the HadCM3 GCM output with the relationships between predictand and predictors. Figure 2 shows the climate change for the PRCP in the future periods (2020s, 2050s and 2080s). It shows an obviously decreasing tendency in December and February for the future PRCP. At the same time, a distinct increasing trend is represented by June and September, other months show little change. For the annual mean change, the PRCP will increase by 3.47%, 6.42% and 8.67% in the future periods compared with the baseline. In conclusion, the PRCP will show an increasing trend with a distinct season cycle in the future, but the increasing trend is not significant. Figure 3 lists the monthly mean daily TMAX and TMIN and their changes, respectively. The downscaled data for the daily TMAX and TMIN in the future are shown in Fig. 3(a) and Fig. 3(b). The downscaled daily TMAX will increase quickly. Almost parallel increasing tendency resulted in the future climate scenarios as shown in Fig. 3(c), and the daily TMAX will increase 1.34°C, 2.60°C and 3.90°C in the future, respectively. Different seasons and months show different features, and it is most obvious for the TMAX in spring and autumn. Compared with scenarios of daily TMAX, the scenarios of daily TMIN in the future are not very obvious, it will increase by 0.87°C, 1.49°C and 2.27°C, respectively, and the distinct seasons are summer and autumn. In a word, it is inevitable that the temperature will increase for the study area in the future.

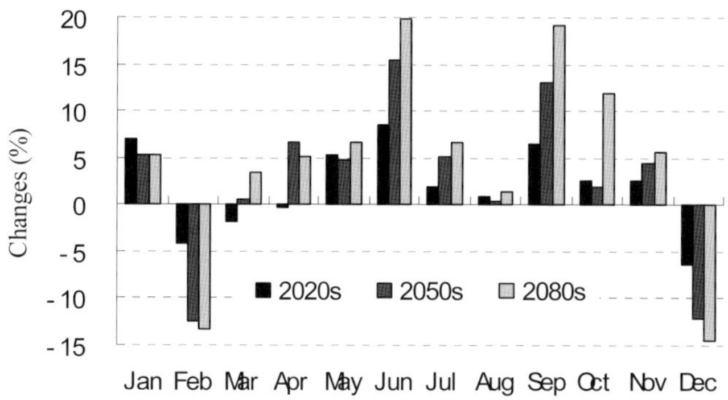

Fig. 2 Changes (%) in monthly mean wet-day amounts in headwater of the Yellow River basin for three typical periods.

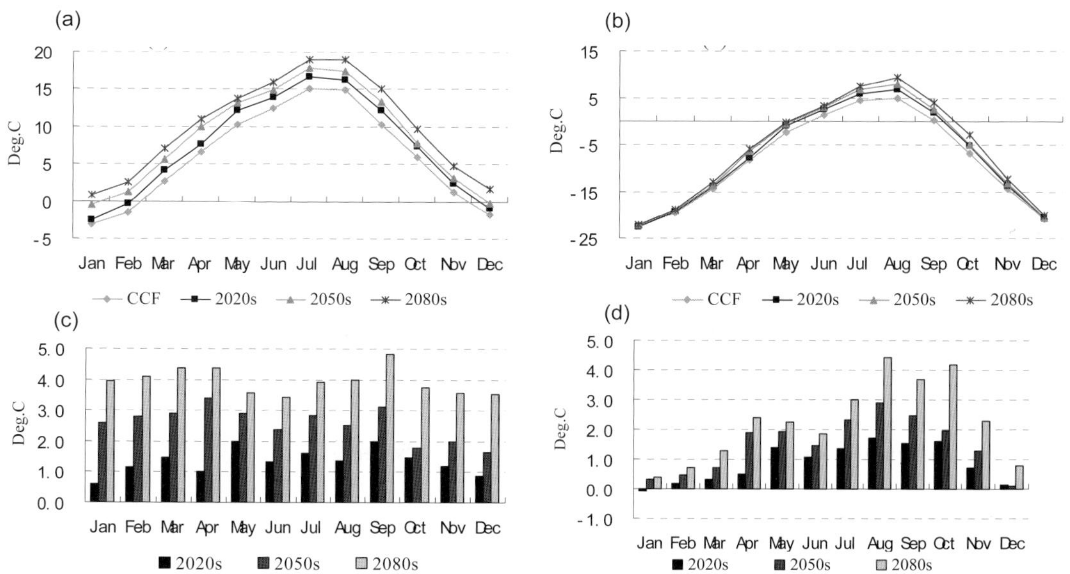

Fig. 3 Downscaled data and scenarios of TMAX and TMIN in headwater catchment of the Yellow River basin: (a) downscaled TMAX; (b) downscaled TMIN; (c) TMAX scenarios; (d) TMIN scenarios.

Streamflow simulation with measured climate variables

Based on the integrity of obtained data, considering wet, average and dry years, the data from 1986–2000 at Tangnaihai hydrological station is selected, and 1986–1995 for calibration, 1996–2000 for validation. In this study, combined with the auto-sensitivity of SWAT model, nine parameters were chosen (CN2, SURLAG, TIMP, CH_K2, ESCO, ALPHA_BF, SMTMP, SMFMN and SOL_AWC). Then the model was calibrated and validated for streamflow using the measured data at Tangnaihai station in the HYRB.

A time-series plot of the measured and simulated monthly streamflow (Fig. 4) shows that the magnitude and trend in the simulated monthly flows closely followed the measured data well, although some peak-flow months were over predicted (e.g. 1986 and 1995) and some were under predicted (e.g. 1989 and 1993). Because the snowmelt parameters (TIMP, SMTMP and SMFMN) were calibrated, the model simulated the streamflow in dry periods well. The statistical evaluation yielded an R^2 value of 0.80, an *Ens* value of 0.73, and an annual mean error (*Re*) of 9.54%, as shown in Table 3, indicating a strong correlation between the measured and simulated flows.

The validation was conducted using the streamflow data for the period from 1996 to 2000. In the validation process, the model was run with parameters obtained during the calibration process without any change. Figure 5 shows the time series plot of monthly measured and simulated streamflows, and indicates an acceptable correspondence of simulated streamflows with the measured values. The R^2, *Ens* and *Re* values between the measured and simulated streamflows are 0.77, 0.59 and 20.12%, respectively, as shown in Table 3.

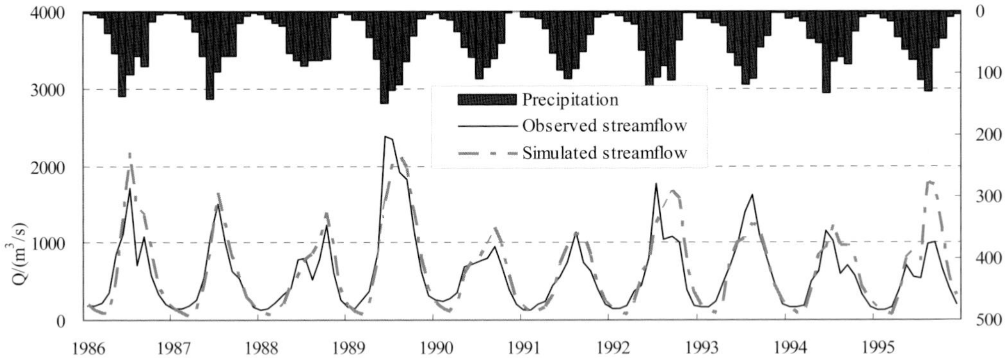

Fig. 4 Comparison between the measured and simulated monthly mean streamflow during the calibration period.

Table 3 Evaluation of the simulation results for monthly mean streamflow.

Periods	Monthly mean flow (m^3/s)		*Re*	R^2	E_{ns}
	Observed	Simulated			
Calibration	605.36	666.32	9.54%	0.80	0.73
Validation	543.40	655.48	20.12%	0.77	0.59

Long-term streamflow simulation under different climate scenarios

The impact of climate change on the hydrological cycle was quantified by driving the calibrated SWAT model with climate scenarios generated by SDSM corresponding to the contemporary and future conditions. A coupled atmospheric-hydrological model was developed, with the regionalized scenarios for future periods (2020s, 2050s and 2080s) served as input to the SWAT model. The future streamflow conditions in HYRB clearly depend on both temperature and

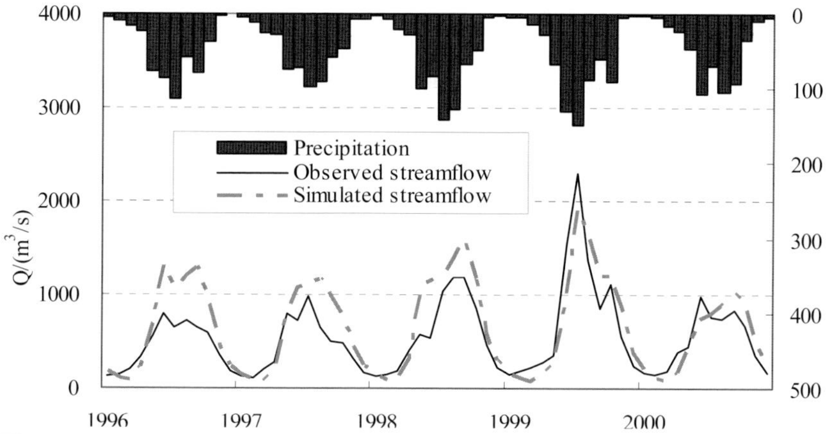

Fig. 5 Comparison between the measured and simulated monthly mean streamflow during the validation period.

precipitation. Figure 6 shows the simulated monthly mean streamflow for the baseline and future periods. The results can explain the seasonal characteristics for the hydrological response to future climate change and indicate a general decrease of flow over several months. Annual average streamflow will decrease by 88.61 m^3/s (24.15%), 116.64 m^3/s (31.79%) and 151.62 m^3/s (41.33%) due to future climate change, with the largest decrease occurring in summer and autumn. This disproportionate change, i.e. 24.15%, 31.79% and 41.33% decrease in average annual streamflow *vs* 3.47%, 6.42% and 8.67% increase in average annual precipitation, can be attributed to more increased temperature rates. Apparently, in summer, increased temperature leads to an increase in evapotranspiration and thus results in a general tendency towards reduced streamflow (Menzel *et al.*, 2002). The monthly simulation shows that future streamflow are different in each month, and the streamflow decreases for three benchmark periods during May to December, especially for the 2080s. According to the estimation of streamflow, it is concluded that the streamflow will decrease greatly in flood seasons, which is very disadvantageous for the development of industry and agriculture in the Yellow River basin. In conclusion, the streamflow in HYRB is very sensitive to future climate change. The streamflow in the study area will greatly decrease in the future. Similar results were also obtained by Hao *et al.* (2006). Therefore, it is inevitable that the streamflow in HYRB will obviously decrease, and water scarcity will be more serious in the future.

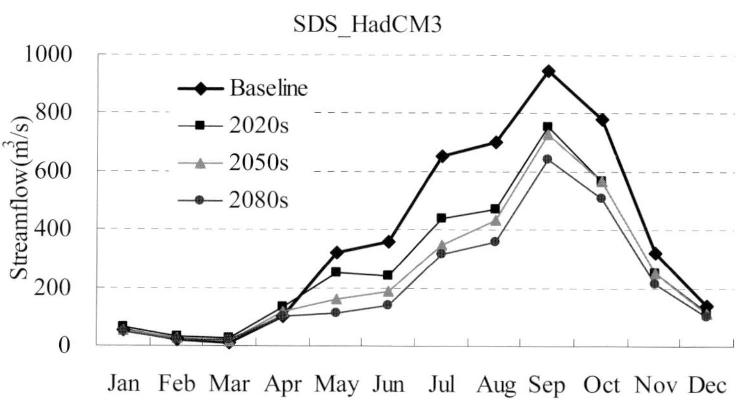

Fig. 6 Comparison among simulated streamflow for SDS scenarios in different periods.

As the important runoff generation region of the Yellow River basin, the HYRB plays a pivotal function. Therefore, it is very important to analyse the spatial distribution of the runoff yield under future climate scenarios. In this study, the spatial distribution of HYRB runoff yield is predicted by using the SWAT for the scenarios in the baseline and three future benchmark periods, as shown in Fig. 7. From the baseline distribution, it is known that the large runoff yield region is mainly located in the southeastern part of the study area, especially the southern part from Maqu station. The runoff yield is small in the northeastern and northwestern part of the study area, and it is smallest in the downstream of HYRB. The comparison between the baseline and future runoff scenarios clearly reveals that the predicted runoff yields will decrease significantly across most of the area of the HYRB in response to the precipitation and temperature changes simulated in the statistical downscaling model. These results underline that the impact of climate changes within HYRB may be widespread and would not be limited to only local areas.

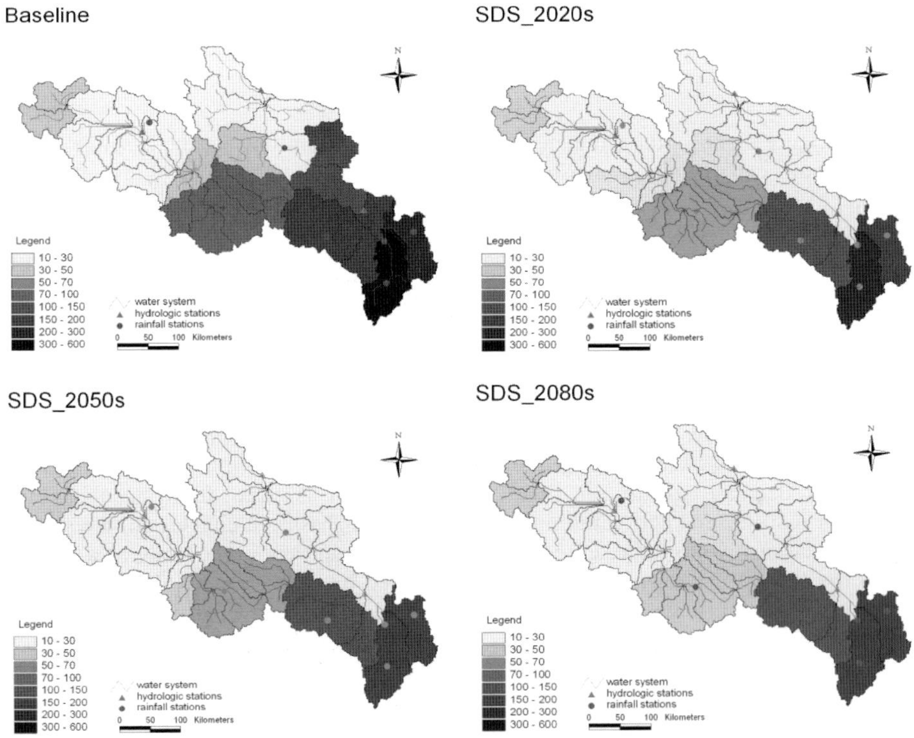

Fig. 7 Spatial distribution of predicted runoff yields in HYRB for future scenarios (mm).

DISCUSSION AND CONCLUSIONS

The study in this paper shows that the statistical downscaling model was successful in reproducing the main features of the observed hydrometeorology from the baseline climate simulation, when it was used with HadCM3 GCM output. It shows that the local scaling of the simulated large-scale precipitation and temperature is quite successful. There is a small increasing trend for the precipitation and an obvious increasing trend for the maximum/minimum temperature in the future.

The combination of different methods, the coupling of the HadCM3 GCM output to SDS, the application of downscaled climate data to the SWAT model and the subsequent simulation of streamflow has been demonstrated to be a useful tool for climate change studies. The investigation of the climate change scenarios and its impact on climate and streamflow in the headwater catchment of the Yellow River basin shows a distinct sensitivity of hydrological response to future climate changes. The results show that the streamflow in the study area will decrease greatly in the future.

A single-direction coupling model was developed in this study, but the feedback process from the atmosphere to the land surface needs further study. The spatial variability of the hydrological process is far greater than that of the atmospheric process. A lot of uncertainties exist in the downscaling process. The uncertainties in downscaling result from: (a) the basic equation on which the downscaling models are based, and (b) the data used (Mohammad *et al.*, 2006). Climate models have some uncertainties when they are used to predict future climate changes. The coupling model of hydrological processes and atmosphere should be developed. In addition, climate change might cause secondary effects such as the changes in vegetation composition; future investigation on the impact of climate change should be extended by considering land cover, the modification of hydrological processes and its feedback to climate. The coupling model and the impact of anthropogenic activity and climate change on hydrological processes should be fully considered.

Acknowledgements This study is jointly financed by the project "Distributed Soil Erosion Model and Variability of Streamflow and Sedimentation in the Wei River Basin" (50579003) supported by the National Natural Science Foundation of China (NSFC) and "Jingshi scholar" Leading Professor Program, Beijing Normal University. The authors gratefully acknowledge the data provision and technical support provided by the developer of the SDS model and the SWAT model. The authors also gratefully thank the editors whose constructive comments and suggestions greatly improved the manuscript.

REFERENCES

Arnold, J. G., Srinivasan, R., Muttiah, R. S. & Allen, P. M. (1999) Continental scale simulation of the hydrologic balance. *J. Am. Water Resour. Assoc.* **35**(5), 1037–1051.

Bouraoui, F., Benabdallah, S., Jrad, A. & Bidoglio, G. (2005) Application of the SWAT model on the Medjerda River basin (Tunisia). *Phys. Chem. Earth.* **30**, 497–507.

Bürger, G. & Chen, Y. (2005) Regression-based downscaling of spatial variability for hydrologic applications. *J. Hydrol.* **311**, 299–317.

Hao, Z. C., Wang, J. H., Li, L., Wang, Z. H. & Wang, L. (2006) Impact of climate change on runoff in source region of Yellow River. *J. Glaciol. & Geocryology* **28**(1), 1–7(in Chinese).

Li, D. F., Tian, Y., Liu, C. M. & Hao, F. H. (2004) Impact of land-cover and climate changes on runoff of the source regions of the Yellow River. *J. Geograph. Sci.* **14**(3), 330–338.

Menzel, L. & Bürger, G. (2002) Climate change scenarios and runoff response in the Mulde catchment (Southern Elbe, Germany). *J. Hydrol.* **267**, 53–64.

Mohammad, S. K., Paulin, C. & Yonas, D. (2006) Uncertainty analysis of statistical downscaling methods. *J. Hydrol.* **319**, 357–382.

Neitsch, S. L., Arnold, J. G., Kiniry, J. R. & Williams, J. R. (2005) *Soil and Water Assessment Tool Theoretical Documentation, Version 2005*. Blackland Research Center, Texas Agricultural Experiment Station, Temple, Texas, USA.

Piling, C. & Jones, J. A. A. (1999) High resolution climate change scenarios: implications for British runoff. *Hydrol. Processes* **13**, 2877–2895.

Saleh, A., Arnold, J. G., Gassman, P. W., Hauck, L. M., Rosenthal, W. D., Williams, J. R. & Mcfarland, A. M. S. (2000) Application of SWAT for the Upper North Bosque River Watershed. *Trans. ASAE* **43**(5), 1077–1087.

Santhi, C., Arnold, J. G., Williams, J. R., Dugas, W. A., Srinivasan, R. & Hauck, L. M. (2001) Validation of the SWAT model on a large river basin with point and nonpoint sources. *J. Am. Water Resour. Assoc.* **37**(5), 1169–1188.

Wilby, R. L. & Wigley, T. M. L. (1997) Downscaling General Circulation Model output: a review of methods and limitations. *Prog. Phys. Geogr.* **21**, 530–548.

Wilby, R. L. & Dawson, C. W. (2004) User Manual: Using SDSM Version 3.1 – a decision support tool for the assessment of regional climate change impacts. http://www-staff.lboro.ac.uk/~cocwd/SDSM/ManualSDSM.pdf.

Wilby, R. L., Dawson, C. W. & Barrow, E. M. (2002) SDSM – A decision support tool for the assessment of regional climate change impacts. *Environ. Modell. Software* **17**, 147–159.

Zhang, X. S., Hao, F. H., Cheng, H. G. & Li, D. F. (2003) Application of SWAT model in the upstream watershed of the Luohe River. *Chinese Geograph. Sci.* **13**(4), 334–339.

Key word index